42

Springer Series in Chemical Physics

Edited by Fritz Peter Schäfer

Springer Series in Chemical Physics

Editors: Vitalii I. Goldanskii   Fritz P. Schäfer   J. Peter Toennies

Volume 40 **High Resolution Spectroscopy of Transient Molecules**
Editor: E. Hirota

Volume 41 **High Resolution Spectral Atlas of Nitrogen Dioxide 559–597 nm**
By K. Uehara and H. Sasada

Volume 42 **Antennas and Reaction Centers for Photosynthetic Bacteria**
Structure, Interactions, and Dynamics
Editor: M. E. Michel-Beyerle

Volumes 1–39 are listed on the back inside cover

Antennas and Reaction Centers of Photosynthetic Bacteria

Structure, Interactions, and Dynamics

Proceedings of an International Workshop
Feldafing, Bavaria, F.R.G., March 23–25, 1985

Editor: M. E. Michel-Beyerle

With 168 Figures

Springer-Verlag
Berlin Heidelberg New York Tokyo

Professor Dr. Maria Elisabeth Michel-Beyerle
Sonderforschungsbereich 143, Technische Universität München, Lichtenbergstraße 4,
D-8046 Garching, Fed. Rep. of Germany

Series Editors

Professor Dr. Fritz Peter Schäfer
Max-Planck-Institut für
Biophysikalische Chemie
D-3400 Göttingen-Nikolausberg
Fed. Rep. of Germany

Professor Vitalii I. Goldanskii
Institute of Chemical Physics
Academy of Sciences
Vorobyevskoye Chaussee 2-b
Moscow V-334, USSR

Professor Dr. J. Peter Toennies
Max-Planck-Institut für Strömungsforschung
Böttingerstraße 6–8
D-3400 Göttingen
Fed. Rep. of Germany

ISBN 3-540-16154-6 Springer-Verlag Berlin Heidelberg New York Tokyo
ISBN 0-387-16154-6 Springer-Verlag New York Heidelberg Berlin Tokyo

Offset printing: Beltz Offsetdruck, 6944 Hemsbach/Bergstr. Bookbinding: J. Schäffer OHG, 6718 Grünstadt
2153/3150-543210

Preface

The workshop on "Antennas and Reaction Centers of Photosynthetic Bacteria" was held at Feldafing, Bavaria (F. R. G.), March 23–25, 1985. This workshop focussed on primary processes with emphasis on structure, interactions and dynamics. It assessed structural, spectroscopic and dynamic data which have accumulated recently, providing an overview of the mechanism of the acquisition, storage and useful disposal of energy in bacterial photosynthesis.

This volume is a record of the invited papers presented at the workshop. The material was organized into five sections:

I. *Antennas:* Structure and Energy Transfer
II. *Reaction Centers:* Structure and Interactions
III. *Electron Transfer:* Theory and Model Systems
IV. *Reaction Centers:* Structure and Dynamics
V. Model Systems on Function of Antennas and Reaction Centers

I would like to express my gratitude to all the participants in the workshop for their contributions, and to the authors for the timely preparation of their manuscripts. I am indebted to the members of the organizing committee, Professors Sighart F. Fischer and Hugo Scheer for their most valuable assistance and advice. The workshop would not have been so successful without the help of my secretary, Frau Petra Kahlfuß, and my coworkers in its organization. I thank Frau Kahlfuß particularly also for her assistance in the preparation of these proceedings.

The workshop was organized under the auspices of the Technical University of Munich, the Max-Planck-Society and the University of Munich. It was supported by the "Deutsche Forschungsgemeinschaft" within the frame of Sonderforschungsbereich 143. In this context we would like to thank especially Dr. Klaus Fleischmann for his efficient cooperation in helping to make this conference a success. Contributions of the Siemens AG and Wacker-Chemie allowed our younger student coworkers to participate in this international meeting. We express our sincere thanks to these organizations for their generous support.

Garching, October 1985 *M. E. Michel-Beyerle*

Contents

Part I	**Antennas: Structure and Energy Transfer**

Structure of Antenna Polypeptides. By H. Zuber 2

The Crystal and Molecular Structure of C-Phycocyanin
By R. Huber 15

C-Phycocyanin from *Mastigocladus laminosus.* Isolation and Properties of Subunits and Small Aggregates. By W. John, R. Fischer, S. Siebzehnrübl, and H. Scheer (With 9 Figures) 17

Picosecond Time-Resolved, Polarized Fluorescence Decay of Phycobilisomes and Constituent Biliproteins Isolated from *Mastigocladus laminosus*
By S. Schneider, P. Geiselhart, T. Mindl, F. Dörr, W. John, R. Fischer, and H. Scheer (With 4 Figures) 26

Fluorescence Behaviour of Crystallized C-Phycocyanin (Trimer) from *Mastigocladus laminosus*
By S. Schneider, P. Geiselhart, C. Scharnagl, T. Schirmer, W. Bode, W. Sidler, and H. Zuber (With 4 Figures) 36

Energy-Transfer Kinetics in Phycobilisomes
By A.R. Holzwarth (With 2 Figures) 45

Exciton State and Energy Transfer in Bacterial Membranes: The Role of Pigment-Protein Cyclic Unit Structures
By R.M. Pearlstein and H. Zuber 53

Carotenoid-Bacteriochlorophyll Interactions
By R.J. Cogdell (With 1 Figure) 62

Bacteriochlorophyll *a*- and *c*-Protein Complexes from Chlorosomes of Green Sulfur Bacteria Compared with Bacteriochlorophyll *c* Aggregates in CH_2Cl_2–Hexane. By J.M. Olson, P.D. Gerola, G.H. van Brakel, R.F. Meiburg, and H. Vasmel (With 8 Figures) ... 67

Reverse-Phase High-Performance Liquid Chromatography of Antenna Pigment- and Chlorosomal Proteins of *Chloroflexus aurantiacus.* By R. Feick (With 2 Figures) 74

Fluorescence-Detected Magnetic Resonance of the Antenna Bacteriochlorophyll Triplet States of Purple Photosynthetic Bacteria. By A. Angerhofer, J.U. von Schütz, and H.C. Wolf (With 1 Figure) 78

High-Resolution ^{1}H NMR of Light-Harvesting Chlorophyll-Proteins. By C. Dijkema, G.F.W. Searle, and T.J. Schaafsma 81

Crystallization and Linear Dichroism Measurements of the B800–850 Antenna Pigment-Protein Complex from *Rhodopseudomonas sphaeroides* 2.4.1
By J.P. Allen, R. Theiler, and G. Feher (With 2 Figures) 82

Crystallization of the B800–850-complex from *Rhodopseudomonas acidophila* Strain 7750
By R.J. Cogdell, K. Woolley, R.C. Mackenzie, J.G. Lindsay, H. Michel, J. Dobler, and W. Zinth (With 6 Figures) 85

Linear Dichroism (LD) and Absorption Spectra of Crystals of B800–850 Light-Harvesting Complexes of *Rhodopseudomonas capsulata*. By W. Mäntele, K. Steck, T. Wacker, W. Welte, B. Levoir, and J. Breton (With 5 Figures) 88

Part II **Reaction Centers: Structure and Interactions**

The Crystal Structure of the Photosynthetic Reaction Center from *Rhodopseudomonas viridis*
By J. Deisenhofer and H. Michel (With 2 Figures) 94

Single Crystals from Reaction Centers of *Rhodopseudomonas viridis* Studied by Polarized Light. By W. Zinth, M. Sander, J. Dobler, W. Kaiser, and H. Michel (With 3 Figures) 97

On the Analysis of Optical Spectra of *Rhodopseudomonas viridis* Reaction Centers
By E.W. Knapp and S.F. Fischer (With 3 Figures) 103

Orientation of the Chromophores in the Reaction Center of *Rhodopseudomonas viridis*. Comparison of Low-Temperature Linear Dichroism Spectra with a Model Derived from X-Ray Crystallography. By J. Breton (With 4 Figures) 109

Calculations of Spectroscopic Properties of Bacterial Reaction Centers. By W.W. Parson, A. Scherz, and A. Warshel (With 5 Figures) 122

On the Temperature-Dependence of the Long Wavelength Fluorescence and Absorption of *Rhodopseudomonas viridis* Reaction Centers. By P.O.J. Scherer, S.F. Fischer, J.K.H. Hörber, M.E. Michel-Beyerle, and H. Michel (With 3 Figures) 131

Local Environments of Pigments in Reaction Centers of Photosynthetic Bacteria from Resonance Raman Data
By M. Lutz and B. Robert (With 4 Figures) 138

The Spin-Polarization Pattern of the $\Delta m = 1$ Triplet EPR Spectrum of *Rps. viridis* Reaction Centers
By F.G.H. van Wijk, P. Gast, and T.J. Schaafsma 146

Triplet State Investigation of Charge Separation and Symmetry in Single Crystals of *R. viridis* Reaction Centers
By J.R. Norris, D.E. Budil, H.L. Crespi, M.K. Bowman, P. Gast, C.P. Lin, C.H. Chang, and M. Schiffer 147

Triplet-minus-Singlet Absorbance Difference Spectroscopy of Photosynthetic Reaction Centers by Absorbance-Detected Magnetic Resonance. By A.J. Hoff (With 11 Figures) 150

ENDOR Studies of the Primary Donor in Bacterial Reaction Centers. By W. Lubitz, F. Lendzian, M. Plato, K. Möbius, and E. Tränkle (With 6 Figures) 164

ENDOR of Semiquinones in RCs from *Rhodopseudomonas sphaeroides*. By G. Feher, R.A. Isaacson, M.Y. Okamura, and W. Lubitz (With 10 Figures) 174

Photoinduced Charge Separation in Bacterial Reaction Centers Investigated by Triplets and Radical Pairs
By J.R. Norris, D.E. Budil, S.V. Kolaczkowski, J.H. Tang, and M.K. Bowman (With 5 Figures) 190

Spin Dipolar Interactions of Radical Pairs in Photosynthetic Reaction Centers. By A. Ogrodnik, W. Lersch, M.E. Michel-Beyerle, J. Deisenhofer, and H. Michel (With 4 Figures) 198

Protein/Lipid Interaction of Reaction Center and Antenna Proteins. By J. Riegler, W.M. Heckl, J. Peschke, M. Lösche, and H. Möhwald (With 6 Figures) 207

The Architecture of Photosystem II in Plant Photosynthesis. Which Peptide Subunits Carry the Reaction Center of PS II?
By A. Trebst and B. Depka (With 3 Figures) 216

Part III Electron-Transfer: Theory and Model Systems

Application of Electron-Transfer Theory to Several Systems of Biological Interest. By R.A. Marcus and N. Sutin 226

Effects of Distance, Energy and Molecular Structure on Long-Distance Electron-Transfer Between Molecules
By J.R. Miller (With 4 Figures) 234

Ultrafast Electron Transfer in Biomimetic Models of Photosynthetic Reaction Centers. By M.R. Wasielewski, M.P. Niemczyk, W.A. Svec, and E.B. Pewitt (With 5 Figures) 242

Electron Transfer Through Aromatic Spacers in Bridged Electron-Donor-Acceptor Molecules. By H. Heitele and M.E. Michel-Beyerle 250

Electron Transfer in Rigidly Linked Donor-Acceptor Systems By S.F. Fischer, I. Nussbaum, and P.O.J. Scherer (With 3 Figures) . 256

Electron Conduction Along Aliphatic Chains By R. Bittl, H. Treutlein, and K. Schulten (With 4 Figures) 264

Part IV **Reaction Centers: Structure and Dynamics**

Kinetics and Mechanisms of Initial Electron-Transfer Reactions in *Rhodopseudomonas sphaeroides* Reaction Centers By W.W. Parson, N.W.T. Woodbury, M. Becker, C. Kirmaier, and D. Holten (With 3 Figures) 278

Femtosecond Studies of the Reaction Center of *Rhodopseudomonas viridis*: The Very First Dynamics of the Electron-Transfer Processes. By W. Zinth, M.C. Nuss, M.A. Franz, W. Kaiser, and H. Michel (With 5 Figures) 286

Analysis of Time-resolved Fluorescence of *Rhodopseudomonas viridis* Reaction Centers By J.K.H. Hörber, W. Göbel, A. Ogrodnik, M.E. Michel-Beyerle, and E.W. Knapp (With 3 Figures) 292

The Characterization of the Q_A Binding Site of the Reaction Center of *Rhodopseudomonas sphaeroides*. By M.R. Gunner, B.S. Braun, J.M. Bruce, and P.L. Dutton (With 2 Figures) 298

Part V **Model Systems on Structure of Antennas and Reaction Centers**

Structure and Energetics in Reaction Centers and Semi-synthetic Chlorophyll Protein Complexes. By S.G. Boxer (With 4 Figures) .. 306

Small Oligomers of Bacteriochlorophylls as *in vitro* Models for the Primary Electron Donors and Light-Harvesting Pigments in Purple Photosynthetic Bacteria By A. Scherz, V. Rosenbach, and S. Malkin (With 7 Figures) 314

Experimental, Structural and Theoretical Models of Bacteriochlorophylls a, d and g. By J. Fajer, K.M. Barkigia, E. Fujita, D.A. Goff, L.K. Hanson, J.D. Head, T. Horning, K.M. Smith, and M.C. Zerner (With 6 Figures) 324

ENDOR Characterization of Hydrogen-Bonding to Immobilized Quinone Anion Radicals. By P.J. O'Malley, T.K. Chandrashekar, and G.T. Babcock (With 3 Figures) 339

Concluding Remarks. Some Aspects of Energy Transfer in Antennas and Electron Transfer in Reaction Centers of Photosynthetic Bacteria. By J. Jortner and M.E. Michel-Beyerle (With 6 Figures) 345

Index of Contributors 367

Part I

Antennas: Structure and Energy Transfer

Structure of Antenna Polypeptides

H. Zuber

Institut für Molekularbiologie und Biophysik,
Eidgenössische Technische Hochschule Zürich, CH-8093 Zürich, Switzerland

1. Introduction

Photosynthesis begins with the uptake of light energy by a number of pigment molecules specifically bound to polypeptides within the antenna complexes [1]. The pigments become excited and transfer excitation energy to the special pair of the reaction center [2,3]. In the antenna complex, which is a multimolecular transport system for delivering excited-state energy, the polypeptides have a most important function [4]. They are responsible for:

1.) The specific binding of the pigment molecules (their position, orientation and separation).
2.) The specific environment of the pigments (spectral properties).
3.) The structural organization of the pigment-protein-complexes and antenna, (the arrangement of the polypeptides within these complexes).

Comparing the structures of antenna polypeptides from various organisms, it should be possible to understand and differentiate the multiplicity of antenna and energy-transfer systems and of their phylogenetic relationships. On this basis, it may also be possible to deduce a general structural principle of energy-transfer. With this object in mind, we have compared the primary structures of antenna polypeptides from cyanobacteria and photosynthetic bacteria. These data allowed statements on the possible structural organization of the antenna complexes and on the energy transfer.

From an evolutionary point of view we can assume that the reaction center by itself was the first light-harvesting molecule. Very early it was coupled with the electron transfer system of fermentation (dark reactions) [5]. To increase the light-harvesting capacity (more and various types of pigments for heterogenetic energy-transfer) the antenna system evolved, creating a multiplicity of antenna in photosynthetic organisms. This antenna multiplicity is correlated with the multiplicity of photosynthetic organisms which live in different ecological niches. Therefore ecological and intracellular factors (constraints) should also determine the structure and function of antenna systems. Of primary ecological influence are the light conditions: the spectral range which determines the type of pigments, and the light intensity which influences the size of the antenna. Of secondary influence are metabolic factors (symbiotic or competitive factors) which determine the position of the organism in the ecosystem: The ecological niches are different for oxygenic or anoxygenic, aerobic or anaerobic life, for primary or secondary primary producers [6].

The intracellular factors influencing the antenna structure are: 1) the overall cell - structure and organization and 2) the structure of the pigment and polypeptide phylogenetic precursors and their evolution. For functional reasons the reaction center and the electron transfer system must be located within the membrane. Different types of photosynthetic membranes are found in the various photosynthetic organisms with increasing complexity from primitive to higher organisms: Simple cytoplasmic membranes (green bacteria) to intracytoplasmic membranes (purple bacteria) to simple thylakoids (Cyanobacteria) and to complex thylakoid lamellae and grana (plants). The antenna are either mainly intramembrane (photosynthetic bacteria) or partly intramembrane (plants) or extramembrane (phycobilisome: cyanobacteria, red algae, chlorosome: green bacteria). The structure of the antenna polypeptides is adapted to the various intra- or extramembrane environments. The important tetrapyrrole precursor molecule is protoporphyrin IX. It was modified during evolution by reduction of ring IV and formation of ring V (chlorophylls) or tetrapyrrole ring opening (phycobilins). Further evolutionary changes of the Chl, BChl or bilins are strongly correlated to the evolution of the antenna polypeptide structures to develop specific interaction sites with these polypeptides (basis for the structure of the antenna complexes).

Each antenna system of the various photosynthetic organisms is composed of several (1-5) pigment-protein complexes with different absorption maxima, the basis for heterogeneous energy-transfer (from short-wave to long-wave absorbing species). 3 types of complexes can be differentiated with respect to their position in the heterogeneous energy-transfer to the reaction center: 1) Those close to the reaction center within the membrane (Chla: 678, BChla: B- 870/890, BChlb: B- 1015, BChlg: B-788). 2) More distant complexes within the membrane (Chlb: 652, BChla: B- 800-850/820). 3) Complexes at the membrane surface (BChlc, d, e: B- 715-760, phycobilins: 545-670). The size of the antennae is between 24-1000 KD/per reaction center.

The large extramembrane antennae (phycobilisomes, chlorosomes) form large heterogeneous energy-transfer systems ($\Delta\lambda$: 125-150 nm). Heterogeneous energy-transfer is smaller in intramembrane antennae: ($\Delta\lambda$: 10-65 nm). The red shift of the tetrapyrrole pigments in the polypeptide environment is larger in intramembrane than in extramembrane antennae (lipid environments, charges?).

The size of the antenna polypeptides of these complexes is between 5-7 KD in photosynthetic bacteria and between 18-25 KD in cyanobacteria, algae and plants. As the basic structural and functional unit they bind from 2-7 (13) pigment molecules/per polypeptide chain. These polypeptide units aggregate to build up the pigment-protein-complex and finally the whole antenna. Why are the antenna composed of small polypeptides? Probably for regulatory reasons. To regulate the size of the antenna and energy-transfer within the antenna (dynamic system).

2. Antenna Polypeptides of Purple Photosynthetic Bacteria

The anoxygenic purple photosynthetic bacteria are classified into 3 groups on the basis of their infrared absorption maxima, which are representative of 3 types of antenna complexes [1].

Group I (e.g. Rs. rubrum, Rp. viridis) has one complex: B-870/ 890 or B-1015.

Group II (e.g. Rp.sphaeroides, Rp.capsulata) possesses two complexes: B-870, B-800-850.

Group III (Chromatinum vinosum, Rp.acidophila) has 3 complexes: B-870, B-800-820 and B-800-850.

All these complexes are part of a specific intracytoplasmic membrane system. As we found also by sequence studies, these antenna complexes contain two types of small polypeptides α- and β-polypeptides. They are the basic structural units of these complexes, and bind a characterisitc number of BChl molecules (non-covalently): α-polypeptide: 1 BChl, β-polypeptide: 1-2 BChl.

We determined the primary structures of the α- and β-polypeptides of the B-870, B-1015 and B-800-850 complexes of a number of purple bacteria (Rs.rubrum, Rp.viridis, Rp.sphaeroides, Rp.capsulata, Rp.gelatinosa) [7-12]. They are composed of 50 - 60 amino acid residues, and probably originate from a common ancestor molecule: the sequence homology of α and β of the same complex is 7 - 13% (evolution of α- and β-chains), the homology between α or β of the different complexes is 13 - 18% (evolution of different complexes)[4].

If one aligns the α- and β-polypeptides for sequence homology, a typical domain structure can be demonstrated [7,8,9]:

a) a polar, charged N-terminal domain (α : 12-19 residues, β : 20-22 residues)
b) a hydrophobic, central domain (α : 21 residues, β : 23 residues).
c) a polar, charged C-terminal domain (α : 19-25 residues, β : 3-10 residues).

The most homologous (conserved) amino acid residue is His found in the hydrophobic domain of the α- and β-chains. It is most probably the main binding site for BChla(b) molecules (via Mg-atom)[7,8,9]. A second His residue is present in all β-chains near the N-terminal domain. In agreement with the hypothetical BChla binding sites 2-3 active site regions can be demonstrated on the basis of higher sequence homologies (BChl, carotenoid). Of interest is also the distribution of aromatic amino acid residues (π-π-interactions), charged and polar amino acid residues which could be interaction sites between polypeptides and pigments, or polypeptides and polypeptides [8,9] (polar residues in the hydrophobic domain, see discussion).

The domain structure of the antenna polypeptides indicates a specific organization of the polypeptides within the membrane [4,8,9]. It is reasonable to assume that the hydrophobic domain is located within the membrane: The N- and C-terminal domains should be found at the polar head region or at the membrane surface. Assuming a thickness of the photosynthetic membrane between 40-60 Å (30 - 40 Å hydrocarbon tail region) the hydrophobic domain of the polypeptides would fit only in the α-helical form and only with a transmembrane orientation (helix length 31-34 Å). The domain structure is supported by the hydropathy index and the α-helix structure by IR-data [13]. The transmembrane orientation of the light-harvesting polypeptides was supported in part by digestion experiments of intact vesicles with the cytoplasmatic site outside [14]. Only the N-terminal parts accessible at the cytoplasmic site were split off, demonstrating also the identical orientation of the α- and β-chains (N-terminus: Cytoplasm, C-terminus: periplasm).

On the basis of the transmembrane model, we can postulate that the BChla(b) molecules of the hydrophobic domain of the α - and β-polypeptides are probably excitation-coupled (distance 10-15 Å): Either within the α-β-polypeptide pair or between two pairs (depending on the orientation)[4]. Of particular importance is the amino acid residue environment: polar, charged, or aromatic amino acid residues, lipid environment.

The molecular organization of the α- and β -polypeptides is responsible for the specific arrangement of the BChla(b) pair within the BChl-protein-complexes[4]. This molecular organization of the polypeptides is based on their aggregation properties in a lipid environment. The two transmembrane α-helices aggregate by polypeptide-polypeptide interactions[4]: 1) at the N-terminal half of the hydrophobic domain by polar (Ser, Thr) and hydrophobic residues. 2) Via the N- or C-terminal domains near or at the membrane surface. With respect to the hydrophobic domain contacts the α-β-polypeptide pair (α-helices) is in a tightly packed and twisted arrangement, which is the basic structural unit of the complex. This helix-pair can further aggregate to form large arrays of antenna polypeptides. The amino acid side chain interactions of tightly packed helices probably results in a cyclic arrangement of polypeptides (e.g. in cyclic hexamers of the B-800-850 complex (12 polypeptides) or the large cyclic array of the B-870 (B-1015) complex (24 polypeptides without the γ-polypeptide) surrounding the reaction center, including in the case of the B-1015 complex (e.g. Rp.viridis) a third γ-polypeptide (6 additional polypeptides)).

To form the whole intramembrane light harvesting antenna, the cyclic arrays of the BChl-protein complexes could aggregate further via interactions of the N- or C-terminal domains (B-870, B-1015 and B-800-850 complex)[4]. By the arrangement of B-870 plus B-800-850, a concentric system (cycles of complexes) for heterogeneous energy-transfer is set up, combining the inner (B-870) and outer (B-800 - 850) complexes and focusing the energy migration to the reaction center located in the middle of the B-870 complex. The distances between the cyclic complexes are larger: 40 Å. In this cyclic arrangement the BChl are fixed rather strongly to the polypeptide matrix, probably partly covered by the C-terminal of the polypeptide domain and partly exposed to lipid, yielding a special environment for spectral properties (red shift etc.).

3. Extramembrane Antenna Polypeptides of Green Photosynthetic Bacteria

With respect to the antenna system from anoxygenic green photosynthetic bacteria, we investigated (in collaboration with R.C. Fuller, Amherst, the extramembrane antenna polypeptides from the gliding green bacterium Chloroflexus aurantiacus [15]. In green bacteria the antenna system in the form of the BChla B-808-865 complex is part of the cytoplasmic membrane. In addition, at the cytoplasmic site there are additional (accessory) extramembrane antenna, the chlorosomes composed of the BChlc B-740 complex [15,16]. These extramembrane chlorosomes of Chloroflexus aurantiacus contain the 5.6 KD BChlc binding polypeptide, in addition to a 18 KD and 11 KD polypeptide. We determined the primary structure of the 5.6 KD polypeptide [17]. The polypeptide shows no domain structure, but secondary structure predictions on the basis of the primary structure data indicated preferentially α-helix formation between Trp 5 and Ile 48 (38 amino acid residues, α-helix length 57 Å). Up to 8 BChlc molecules were found to bind to the 5.6 KD polypeptide chain [16]. Most interestingly (and supporting the α-helical structure) 7 of 8 Gln or Asn residues found in the α-helix are located at one side, whereas the residual polar residues (Ser, Thr, His) are located on the other side [17].

The number of Gln and Asn residues (7-8), which corresponds to the number of BChlc molecules, and their polar distribution indicates a specific BChlc binding to these two residues. On the basis of the distances between the Gln and Asn residues in a Nicholson molecular model, it can be postulated that 7 BChlc are bound via the central Mg-atom (as electron-acceptor) to Gln or Asn (electron donor, nucleophile). Thus, in this case there is no binding to His residues. In addition, the arrangement of the BChl molecules indicates specific interaction between the pigments: BChlc acts both as electron-donor (C=O group ring V) and electron acceptor OH-group (hydroxyethyl-group) in ring I (hydrogen-bonded nucleophilic interaction). Interestingly only BChl c, d, e (present in organisms with chlorosomes) contain the hydroxyethyl-group, which probably is the basis for such a BChl arrangement. Also an interesting correlation exists with the basic work of FAJER [18] on bacteriochlorophyll aggregates in model systems (oligomeric arrangement). The aromatic amino acid residues (Phe 21, Trp 25, 27) in the center of the α-helix are probably also of functional importance (π-π-interactions).

The sideness of the α-helix of the BChlc binding polypeptide of Chloroflexus aurantiacus with respect to Ser or Thr residues indicates possible polypeptide-polypeptide interactions via these residues to form high aggregates of the BChlc antenna polypeptide. Crosslinking experiments show the possible existence of dimers [16]. Higher polypeptide aggregates or chlorosome subunits (BChlc-protein complex) or so-called globular units of long rod-shaped elements were shown in chlorosomes by electron microscopic studies [19]. These elements are 60 Å in length, which would correspond nicely to the length of the α-helix of the BChlc antenna polypeptide. The diameter of the rod-shaped elements is about 52 Å which would allow the arrangement of 6 dimers (12 polypeptides with 84 BChlc). It was estimated by OLSON [20] that in each Chlorobium chlorosome about 10'000 BChlc are bound per reaction center. With the dimensions of the Chloroflexus chlorosome of 1060($\pm$240) x 320($\pm$10) x 120($\pm$20)Å 130 globular units of the rod-shaped elements are arranged within the chlorosome. On the basis of the BChlc complex proposed, this means that 130 times 84 or about 10'000 BChlc molecules are bound in the chlorosome, corresponding to the Olson data.

The arrangement and aggregation of the helices and rod shaped elements takes place in an area which is protected from the bulk water by a membrane. On the surface of the α-helix the BChlc molecules could be organized in an oligomeric arrangement having their transition dipoles oriented about parallel (average angle 40°) to the α-helix and therefore to the long chlorosome axis, as was postulated by FULLER and BLANKENSHIP [21] on the basis of spectroscopic studies. An interesting feature is the cyclic arrangement of the BChlc molecules or pairs (excitons ?). From the terminal BChlc molecules the energy can migrate in the direction of the membrane surface to the base plate, antenna and reaction center BChla.

4. Phycobiliproteins: Extramembrane Antenna Polypeptides of Cyanobacteria

Phycobiliproteins are extramembrane antenna complexes of oxygenic cyanobacteria, red algae and cryptomonads. In contrast to the examples we have already discussed, the phycobiliproteins are water-soluble, globular pigment-protein complexes folded in an aqueous environment with covalently bound open chain tetrapyrroles of the bilins. Three main types of these antenna complexes absorbing at different wave lengths, the basis for heterogeneous energy-transfer to the reaction center, are [22,23]: 1.) Phycoerythrin or phycoerythrocyanin absorbing at 560 nm. 2.) Phycocyanin absorbing at 620 nm and 3.) allophycocyanin absorbing at 650-680 nm. All phycobiliproteins are composed of α- and β-polypeptides (mol.weight 16-20 KD). This polypeptide pair is the basic structural/functional unit of the antenna system. Each polypeptide carries a characteristic number of bilin-pigments. (α-polypeptide: 1-2, β-polypeptide: 2-5). The phycobiliprotein-polypeptide pair ("monomer") has the specific property to aggregate further to form trimers, hexamers and stacks of hexamers. This aggregation tendency is the basis for the in vivo phycobiliprotein aggregate of the phycobilisomes located at the surface of the thylakoid membrane of cyanobacteria and red algae [24].

In the following we will concentrate on the phycobilisomes and phycobiliproteins from the cyanobacterium Mastigocladus laminosus. In this organism electron microscopy reveals phycobilisomes with a hemidiscoidal structure of about 40x70 nm (mol weight: ~5 mill.,containing some 300 pigment molecules, WEHRMEYER and co-workers[24]). For heterogeneous energy transfer from outside to inside, hexamer-discs (12 polypeptides) of phycoerythrocyanin, C-phycocyanin and trimers of allophycocyanin in the center region have a specific arrangement in the phycobilisome. This inward energy-transfer system is modulated by linker polypeptides bound to the hexamers or trimers [25,26]. In allophycocyanin energy is trapped by specific polypeptides (allophycocyanin B, 80-100 KD anchor-polypeptide).

Further data on the molecular structure of this extramembrane energy-transfer system were obtained by primary and three-dimensional structure analysis. We determined the primary structures of the α- and β-chains of allophycocyanin [28], C-phycocyanin [27] and phycoerythrocyanin [29] of Mastigocladus laminosus. All chains show homologous amino acid sequences when compared pairwise (21 - 67%). They are

functionally and phylogenetically (common precursor) related. A phylogenetic tree of all polypeptide chains can be constructed (protein differentiation).[23]
The main binding site for the tetrapyrrole pigments in all polypeptide chains of C-phycocyanin, phycoerythrocyanin and allophycocyanin is the cysteine in position 84. It is the most essential pigment for energy-transfer. In this active site region the sequence homology is high (hydrophobic and charged residues). An additional pigment binding site (position 155) has been found in the β-chains of the C-phycocyanin and phycoerythrocyanin, the outer phycobiliproteins of the phycobilisome, to increase the energy uptake capacity. It is located in a 10 amino acid insertion region. In other phycobiliproteins located on the phycobilisome surface, e.g. in phycoerythrin from Fremyella diplosiphon (living in dark areas), even more pigments are bound (α-polypeptide: 2, β-polypeptide: 3)[30]. Dr. Sidler in our laboratory could prepare suitable crystals of C-phycocyanin for X-ray analysis. The three-dimensional structure was determined by SCHIRMER, BODE and HUBER [31,32]. In this analysis the folding of the α- and β-chains in the monomer (α-helices), the position of the pigment 84 and 150, the structure of the trimer-aggregate and the position and orientation of the pigments in the C-phycocyanin-trimer could be derived.
Electron microscopic and spectroscopic data and the primary and three-dimensional structure analysis indicate that during aggregation of the α- and β-polypeptide chains to form trimers, hexamers and the phycobilisomes, an efficient energy-transfer system is built up, based on a bringing together of the pigments 84 (α, β) and the formation of cyclic trimers and hexamers and therefore cyclic pigment arrangements. The hexamer-cycles of different phycobiliproteins to form stacks of hexamers (trimers) yield ultimately the heterogeneous energy-transfer system of the phycobilisomes.

The antenna complexes of the phycobiliproteins can aggregate not only to form various types of phycobilisomes at the thylakoid surface, but they are found as different aggregates within the cytoplasmic membrane (Gloebacter violaceus) and also intrathylakoidal in the case of cryptomonads. Cryptomonad extramembrane antenna show a different energy-transfer system [33]: 1.) They do not contain allophycocyanin and only either phycocyanin 645 (Chroomonas) or phycoerythrin 545 (Rhodomonas). 2) They form small aggregates of 45 - 57 KD of α- and β-chains with the arrangement $[(\alpha_1\beta)\ (\alpha_2\beta)]$ or $[(\alpha_1\beta_1)\ (\alpha_2\beta_2)]$. We determined part of the primary structures of phycocyanin 645 and phycoerythrin 545 and compared these with those of allophycocyanin, C-phycocyanin and phycoerythrocyanin of Mastigocladus laminosus [34]. Cyanobacterial and cryptomonad phycobiliproteins are phylogenetically related: The sequence homology of the β-chains are homologous between 30-50%, that of the α-chains is only about 20%. However, at the N-terminus of the α-chain of phycocyanin 645 or phycoerythrin 545 60 amino acids residues are missing (close to the pigment cystein 84) compared to allophycocyanin, C-phycocyanin, phycoerythrocyanin. This causes probably the different aggregation properties (no hexamer formation, but possible tetramers). Interestingly, these different aggregation behaviors seem to correlate to differences in the energy-transfer system. Phycocyanin 645 contains 2 phycocyanobilins (662 nm) and one cryptoviolin (590 nm) covalently bound to the β-chain and one green (unknown) PXB-bilin (697 nm) in the α-chain. In the antenna complex $[(\alpha_1\beta)(\alpha_2\beta)]$ heterogeneous energy-transfer can take place within the same phycobiliprotein (not between several biliproteins as in the phycobilisome). Energy should migrate from pigment 590 nm to pigment pair 662 nm (exciton?) and to pigment 697 nm, which is an energy trap like allophycocyanin.

To sum up, the question arises if it is possible to derive a common structural and functional principle from such a multiplicity of antenna systems (there may well be other as yet unknown examples). Here I would like to postulate the following:

The antenna polypeptides carrying 1 - 10 pigments are the basic structural and functional units to organize antenna complexes for energy-migration between the specifically bound and arranged antenna pigments [4]. They aggregate specifically:

First, to couple 2 pigments more strongly (possible excitons).

Secondly, to produce a cyclic pigment arrangement to store energy (unknown large exciton system?)

Thirdly, to link antenna cycles with different absorption maxima (energy levels) to the reaction centers.

5. References

1. J.P. Thornber, R.J. Cogdell, B.K. Pierson and R.E.B. Seftor: J.Cellular Biochemistry 23, 159 (1983)
2. R.M. Pearlstein: Photochem.Photobiol. 35, 835 (1982)
3. A.Yu. Borisov: "Energy-Migration Mechanismus in Antenna Chlorophylls" in "The Photosynthetic Bacteria" (Eds.R.K. Clayton, W.R. Sistrom) Plenum Press, New York, London 1978, p.323
4. H. Zuber: "Structural Organization of Tetrapyrrole Pigments in Light-Harvesting Pigment-Protein Complexes" in "Optical Properties and Structure of Tetrapyrroles (Eds. G. Blauer, H. Sund) W. de Gruyter, Berlin, New York 1985, p.425
5. H. Gest:"Evolutionary Roots of Anoxygenic Photosynthetic Energy Conversion" in "The Phototrophic Bacteria: Anaerobic Life in the Light" (Ed. J.G. Ormerod), Blackwell Scientific Publications, Oxford, 1983, p.215
6. N. Pfennig: "General Physiology and Ecology of Photosynthetic Bacteria" in "The Photosynthetic Bacteria" (Eds. R.K. Clayton, W.R. Sistrom) Plenum Press, New York, London, 1978, p.3
7. R.A. Brunisholz, P.A. Cuendet, R. Theiler and H. Zuber: FEBS Lett. 129, 150 (1981)
8. R.A. Brunisholz, F. Suter and H. Zuber: Hoppe Seyler's Z. Physiol. Chem. 365, 675 (1984)
9. R. Theiler, F. Suter, V. Wiemken and H. Zuber: Hoppe Seyler's Z. Physiol. Chem. 365, 703 (1984)
10. M.H. Tadros, F. Suter, G. Drews and H. Zuber: Eur.J.Biochem. 129, 533 (1983)
11. M.H. Tadros, F. Suter, M.H. Seydewitz, J. Witt, H. Zuber and G. Drews: Eur. J. Biochem. 138, 209 (1984)
12. R.A. Brunisholz, F. Jay, F. Suter and H. Zuber: Biol. Chem. Hoppe Seyler 366, 87 (1985)
13. R. Theiler and H. Zuber: Hoppe Seyler's Z. Physiol.Chem. 365, 721 (1984)
14. R.A. Brunisholz, V. Wiemken, F. Suter, R. Bachofen and H. Zuber: Hoppe Seyler's Z.Physiol.Chem. 365, 689 (1984)
15. B.K. Pierson and R.W. Castenholz: Arch.Microbiol. 100, 5 (1974)
16. R.G. Feick and R.C. Fuller: Biochem. 23, 3693 (1984)
17. T. Wechsler, F. Suter, R.C. Fuller and H. Zuber: FEBS Lett. 181, 173 (1985)
18. K.M. Smith, L.A. Kehres and J. Fajer: J.Am.Chem.Soc. 105, 1387 (1983)

19. A. Staehelin, J.R. Golecki, R.C. Fuller and G. Drews: Arch.Microbiol. 119, 269 (1978)
20. J.M. Olson, R.C. Prince and D.C. Brune: "Reaction Center Complexes in Green Bacteria" in "Chlorophyll Proteins, Reaction Centers and Photosynthetic Membranes" (Eds. J. Olson and G. Hind) Proceedings of the Brookhaven Symposia in Biology Nr.29, 1976, p.244
21. J.A. Betti, R.E. Blankenship, L.V. Natarajan, L.C. Dickinson and R.C. Fuller: Biochim.Biophys.Acta 680, 194 (1982)
22. A.N. Glazer: Biochim.Biophys.Acta 768, 29 (1984)
23. H. Zuber: "Structure and Function of the Light-Harvesting Phycobiliproteins from the Cyanobacterium Mastigocladus laminosus" in "Photosynthetic Procaryotes: Cell Differentiation and Function" (Eds. G.C. Papageorgiou and L. Packer), Elsevier Publishing Co. 1983, p.23
24. M. Nies and W. Wehrmeyer: Arch.Microbiol.129, 374 (1981)
25. A.N. Glazer: Ann.Rev.Microbiol. 36, 173 (1982)
26. P. Füglistaller, R. Rümbeli, F. Suter and H. Zuber: Hoppe Seyler's Z.Physiol.Chem. 365, 1085 (1984)
27. G. Frank, W. Sidler, H. Widmer and H. Zuber: Hoppe Seyler's Z.Physiol.Chem. 359, 1491 (1978)
28. W. Sidler, J. Gysi, E. Jsker and H. Zuber: Hoppe Seyler's Z.Physiol.Chem. 362, 611 (1981)
29. P. Füglistaller, F. Suter and H. Zuber: Hoppe Seyler's Z.Physiol.Chem. 364, 691 (1983)
30. W. Sidler, B. Kumpf, G. Frank, F. Suter, W. Rüdiger and H. Zuber: Photochem.Photobiol. 39, 195 (1984)
31. T. Schirmer, W. Bode, R. Huber, W. Sidler and H. Zuber: J.Mol.Biol. in press
32. T. Schirmer, W. Bode, R. Huber, W. Sidler and H. Zuber: "The Crystal Structure and its Implications for Function and Evolution" in "Optical Properties and Structure of Tetrapyrroles" (Eds. G. Blauer, M. Sund) W. de Gruyter, Berlin, New York, 1985, p.445
33. R. Mc Call and D. Guard-Friar: J.Biol.Chem. 258, 14327 (1983)
34. W. Sidler, B. Kumpf, F. Suter, W. Morisset, W. Wehrmeyer and H. Zuber: Biol.Chem.Hoppe Seyler 366, 233 (1985)

The Crystal and Molecular Structure of C-Phycocyanin

R. Huber

Max-Planck-Institut für Biochemie, D-8033 Martinsried, F. R. G.

Cyanobacteria possess light-harvesting organelles attached to the outer surface of the photosynthetic membrane, which conduct the energy of light to the membrane-associated reaction centers, in which charge-separation across the membrane occurs (1). These organelles, the phycobilisomes, form rod-like substructures composed of different, but chemically and structurally related components. A main component is C-phycocyanin. Amino acid sequences of phycocyanins of different bacteria have been determined and compared with each other and with sequences of other phycobilisome components, phycoerythrin and allophycocyanin (2, 3). These show extensive homology indicating common three-dimensional foldings.

The basic building unit of phycocyanin is a (αβ) dimer. Each subunit is composed of about 160 amino acid residues. α and β chains are homologous in amino acid sequence. The α subunit has one, the β subunit two chromophores covalently attached. The chromophores are open chain tetrapyrroles.

The crystal structure of phycocyanin of the cyanobacterium Mastiglocladus laminosus has been determined at high resolution (4, 5, 6). In these crystals three (αβ) units are arranged in a hollow disk with inner and outer diameters of about 40 and 110 Å and a thickness of about 30 Å. The folding of the polypeptide chain of the subunits is mainly α-helical and similar in α- and β-subunits. There are eight helical segments in each polypeptide chain, six of which are folded into a globular part, and an N-terminal U-shaped pair of helices. The arrangement of the helices in the globular part resembles the fold of the globins. The N-terminal helices of the α-subunit associate with the globular domain of α and vice versa to establish a tightly linked (αβ) pair. (αβ) units associate less tightly to the trimeric disk structure.

The chromophores are linked to equivalent cysteine residues of helices E in the α- and β-subunits (α1 and β1). The second chromophore (β2) in β is bound to a cysteine residue in the GH bend. The three chromophores are embedded in different protein environments. α1 and β2 are located on the periphery of the disk, while β1 points into the central cavity. α1 is close to β1.

Electron micrographs show that a rod of the phycobilisome is built from two to three dimers of discs, which are 60 A thick. Crystals of the Mastigocladus laminosus C-phycocyanin are assembled from $(\alpha\beta)_3$ discs and cannot provide information about the mode of dimer formation. Crystals of C-phycocyanin of the cyanobacterium Agmenellum quadruplicat um (8), however, are built from $2^*(\alpha\beta)_3$ dimeric discs (9). The crystal structure analysis of this protein at high resolution (9) shows that discs dimerize face to face and exhibit 32 point group symmetry. The buried interface involves exclusively the α-subunit. In the $2*(\alpha\beta)_3$ dimeric disc the β1 chromophores come close in space along the rod. It is likely that they preferentially conduct the light energy along the antenna to the photosynthetic membrane.

References

(1) J. Deisenhofer, H. Michel, R. Huber: Trend Biochem. Sci. 10, 243 (1985)

(2) A.N. Glazer: Annu.Rev.Biochem. 52, 125 (1983)

(3) P. Flueglistaller, F. Sutter, H. Zuber: Hoppe Seyler`s Z. Physiol. Chem. 364, 691 (1983)

(4) T. Schirmer, W. Bode, R. Huber, W. Sidler, H. Zuber, in: "Optical Properties and Structure of tetrapyrroles", ed. by G. Blauer and H. Sund, Walter de Gruyter, Berlin (1985)

(5) T. Schirmer, W. Bode, R. Huber, W. Sidler, and H. Zuber: J.Mol.Biol. 184, (1985) in press

(6) T. Schirmer, Thesis, Technische Universität München (1985)

(7) N. Nies and W. Wehrmeyer: Planta 150, 330 (1980)

(8) M.L. Hackert, C. Abad-Zapatero, S.E. Stevens and J.L. Fox: J.Mol.Biol. 111, 365 (1977)

(9) R. Huber, T. Schirmer, M. Schneider, M.L. Hackert (1985) in preparation

C-Phycocyanin from *Mastigocladus laminosus*. Isolation and Properties of Subunits and Small Aggregates

W. John, R. Fischer, S. Siebzehnrübl, and H. Scheer

Institut für Botanik der Universität München, Menzinger Str. 67
D-8000 München 19, F.R.G.

1. Introduction

Photosynthetic organisms cover most of their energy needs with sunlight. They have consequently developed a variety of adaptation mechanisms to compete efficiently for it. In higher plants, a dominant mechanism is the growth towards the light. Aquatic and microorganisms adapt commonly by chromatic adaptation of the photosynthetic antenna. The chlorophylls a and b are rather inefficient in collecting green light, and several additional pigment systems have evolved to fill this hole in the action spectrum.

The phycobiliproteins comprise one such group of antenna pigments. They are used in cyanobacteria, red alga and cryptophytes (1). In the former two, they are highly aggregated together with colorless linker polypeptides (2), in the phycobilisomes (3). These are microscopic structures situated at the outer surface of the thylakoid membrane, which transfer their excitation energy efficiently (quantum yield ≥95%) to the chlorophyllous reaction centers within the membrane. In cooperation with the group of S. SCHNEIDER (Garching), we have recently begun to study the fluorescence properties of small aggregates and subunits of phycocyanins (4). The aim of this work is an understanding of the energy transfer in these pigments in relation to the size and structure of the assembly. This report is concerned with the stability and the photochemical reactivity of C- phycocyanin (**PC**) and its subunits from M. laminosus. Time- resolved polarized fluorescence data are presented in the accompanying report of the Garching group.

2. Materials and Methods

M. laminosus was grown in suspension culture in CASTENHOLZ medium (5). C- **PC** was isolated as reported earlier (4). The subunits were isolated by preparative electrofocusing on Agarose gels (Pharmacia, München) in 8M urea, and renatured without delay on a desalting column (Biogel P6, Biorad).

Absorption spectra were recorded on a model 320 (Perkin Elmer, Ueberlingen) spectrophotometer. The cell holder was thermostated, and temperatures measured in the cuvettes with a Pt 100 resistor. Absorption difference spectra were measured with a ZWS II dual- wavelength photometer (Sigma, Berlin) in split- beam mode. The data were digitized and stored with model II+ computer (Apple, München) using a self- made program, which was developed for obtaining high wavelength accuracy.

Fluorescence spectra were recorded on a DMR 22 (Zeiss, Oberkochen) spectrofluorimeter. The cell holders were thermostated, and the temperatures recorded in the cuvettes with a Pt 100 resistor. The spectra are uncorrected for the spectral response function of the apparatus. Circular dichroism spectra were recorded on a model Mark V dichrograph (Jobin- Yvon ISA, München) equipped with a silex computer (Leanord, Lille) and a modified software. Sedimentation coefficients were obtained in a model E ultracentrifuge (Beckman, München) at 20°C with the scanner wavelength set at 620nm.

Chemicals and solvents were reagent grade unless otherwise stated. Sodium dodecyl sulfate- polyacrylamide electrophoresis was done by the method of LAEMMLI (6).

3. Results

3.1 Optical Spectra

All studies were done in potassium phosphate buffer (50 or 100mM) at pH7.5. Under these conditions, **PC** is isolated as a trimer (heterohexamer, ($\alpha\beta$), sedimentation coefficient ≈4.8S). It can be dissociated reversibly into the monomer (heterodimer, $(\alpha\beta)_3$ ≈ 2.8S) by incubation with potassium thiocyanate. The absorption spectrum of the monomer is slightly blue- shifted as compared to the trimer, but otherwise very similar (fig. 1).

The spectra shown in fig. 1 correspond to preparations, which are free of colorless linker- peptides. If the latter are present, the spectra of monomers (obtained by the same dissociation method) remain unchanged. The trimer spectra are more strongly red- shifted, however, and a series of slightly different species absorbing between 620 and 638nm can be isolated. Their sedimentation coefficients are only slightly higher (≈5-6S) than that of the trimer. This is due to the presence of varying amounts of several colorless linker- peptides. The exact stoichiometry is still under investigation. An increased chromophore-chromophore coupling in these red- shifted complexes is suggested from their circular dichroism (CD) spectra. Simultaneously with the red- shift in the absorption maximum, there develops a pronounced red- shifted feature in the CD- spectrum, which increases with an increasing red-shift (fig. 2). It has been rationalized as an S- shaped exciton couplet (positive at longer wavelengths), which is superimposed on the absorption- type CD- spectrum of the monomer.

Exciton couplings are absent in linker-peptide-free preparations of **PC** from M. laminosus, but have been discussed earlier in **PC** from S. platensis (7) and in particular in allophycocyanins (8). The induction of exciton coupling would require a changed geometry of the chromophores, which can be considered on the basis of the x-ray structure of trimeric **PC** from the same organism (9; see also SCHIRMER et al., this monograph). The closest chromophore contacts in the linker-peptide-free trimer are between the α - chromophore of the monomeric unit and the β1- chromophore of the adjacent one. The center- to - center distance (22A) is just too large for exciton interactions. Changes by only a few Angström may, however, be sufficient to allow for strong couplings, and it is possible that such changes are induced by the linker-peptides. Unless there is a gross structural change involved, the red- shifted species are then most likely due to a changed geometry in the contact region of the monomeric units.

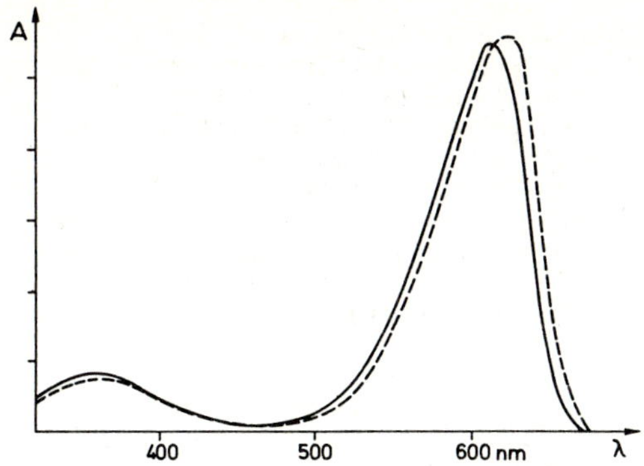

Fig. 1: Absorption spectra of monomeric ($\alpha\beta$, —) and trimeric ($(\alpha\beta)_3$;----) **PC** from M. laminosus (50mM phosphate buffer, pH 7.5)

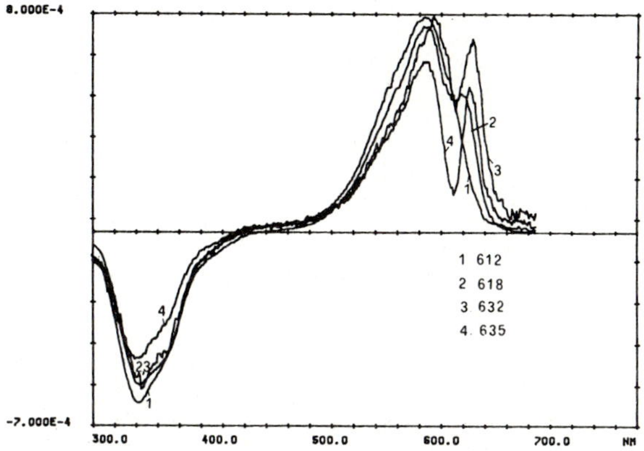

Fig. 2: Circular dichroism spectra of trimeric phycocyanin containing linker peptides: aggregates with increasing wavelength of the red absorption band. λmax see inserts.

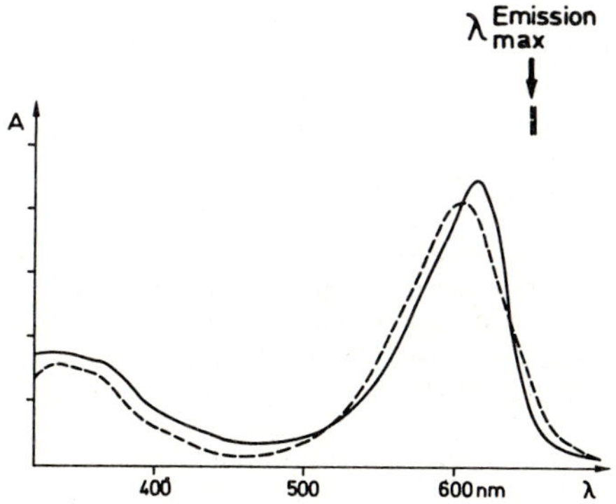

Fig. 3: Absorption spectra of the isolated subunits of **PC** from M. laminosus (α: —; β: ---). The emission maxima are indicated by the labels in the upper right.

The absorption spectra of the isolated subunits (fig.3) show considerable differences both with respect to the position and the shape of the long- wavelength band. The α - subunit (λ max= 616nm) has a narrow band similar in shape to that of integral **PC**, whereas the β-subunit has a blue- shifted (λmax= 605nm) and broadened band. The weighed sum of the two absorption spectra is very similar to that of monomeric **PC** (not shown). In spite of this blue- shifted absorption, the fluorescence maxima are almost at the same position (fig.3). The α-subunit carries one chromophore, the β- subunit two chromophores (β1, β2) at different binding sites of the apoprotein (10), the broadening should then reflect a slight difference in the excitation energies of the two chromophores. This is indeed supported by the circular dichroism spectra (fig.4). The anisotropy of the β-subunit is only $\approx$60% as compared to the α-subunit. The positive extremum is considerably blue-shifted with respect to the absorption maximum, and there is a low-intensity red-wing in the CD-spectrum. It is suggested that the β-subunit carries one optically active (λmax $\sim$595nm) and one much less active (planar ?) chromophore (λmax $\approx$615nm).

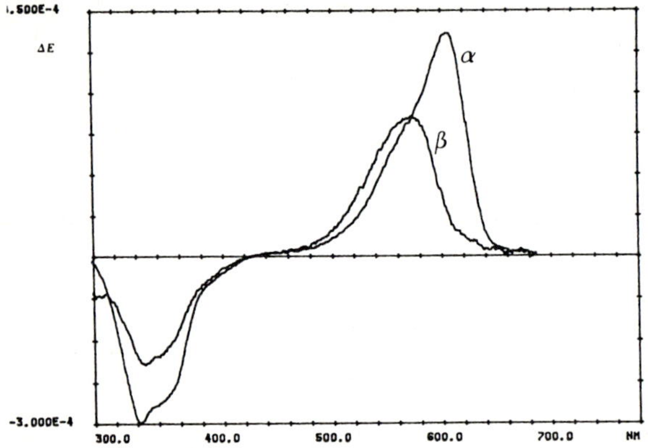

Fig. 4: CD of the isolated subunits shown in Fig. 2 (α:—;β:—). The spectra are normalized to equal absorptions in the long-wavelength maximum.

This interpretation is supported by similar CD data and additional fluorescence polarisation results of MIMURO et al. (this volume). An assignment of the individual chromophores on the β-subunit to the such defined absorptions is not yet possible. Based on the amino-acid sequence and the x-ray structure (9,10), β1 is similar to α. The absorption maxima would then relate β1 to the chromophore absorbing at $\approx$615nm. Since these chromophores come closest in the trimer, the exciton split at the red wing of the main CD-band in linker-associated trimers (vide supra) would support this assignment. However, the CD-data suggest a rather different conformation for the β- chromophore (λmax $\approx$615nm) and the α- chromophore, and biochemical labeling will be necessary to decide this point conclusively.

3.2 Stabilities

The chromophores of **PC** are held rigidly in an energetically unfavorable, extended conformation (1,9). This conformation is due to non-covalent

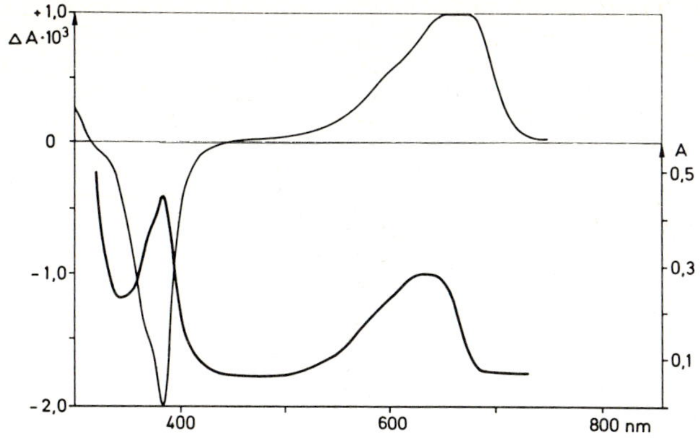

Fig. 5: Absorption (bottom) and CD- spectra (top) of a biliprotein from Pieris brassica (13). The pigment contains biliverdin IXγ bound non- covalently to an apoprotein of MW ≈25kDa.

protein-chromophore interactions, which are reversible-lost upon unfolding of the protein, e.g. with urea (11). This type of interaction is probably characteristic for all phycobiliproteins and crucial to their function as antenna pigments, because they increase the oscillator strength of the visible band by almost one order of magnitude, and decrease the radiationless decay of excited states by more than three orders of magnitude (1). The specificity of these interactions can be seen by comparison with other biliproteins, e.g. the biliverdin IXα-serum albumin complex (12), or the biliverdin IXγ-protein from Pieris brassica (13, see fig. 5). In both the latter pigments is the cyclic- helical conformation of the chromophores preserved, which is characteristic for the respective chromophores in solution (14,15). They have, accordingly, low fluorescence yields and weak absorptions in the visible range. Here, the major influence of the apoprotein is a preferential binding of one of the two helical enantiomers, which leads to intense induced CD-signals.

Due to the pronounced chromophore-protein interactions, the chromophores of **PC** can be used as very sensitive probes to monitor the state of the protein (1,11). Small aggregation-induced changes have been discussed in the first part. Much larger changes occur, if the protein is partically or fully unfolded, e.g. with urea or heat. After a previous study with **PC** from S. platensis, this method has now been used to investigate the stabilities of the **PC** and its subunits from M. laminosus.

The normalized absorption changes upon urea- denaturation are shown in fig. 6 (see legend for details). The gross changes are similar for the integral pigment (trimer) and its subunits. Small differences in the final (= 8M urea) absorptions are indicative of slightly different conformations of native chromophores, because the chromophores have the same molecular structure and are expected to have identical spectra in the denatured states. The most notable difference is the increased stability of the integral pigment, as compared to both subunits. At 4M urea, the absorption of the integral pigment is decreased by less than 4%, whereas the decrease of the subunits is 12% (α) and 15% (β). The urea concentrations necessary to induce a 50% decrease of the long-wavelength absorptions are 6.1, 5.6, 5.4M, respectively. In contrast to the results with **PC** from S. platensis, there are no indications for a stepwise unfolding evident from the absorption spectra.

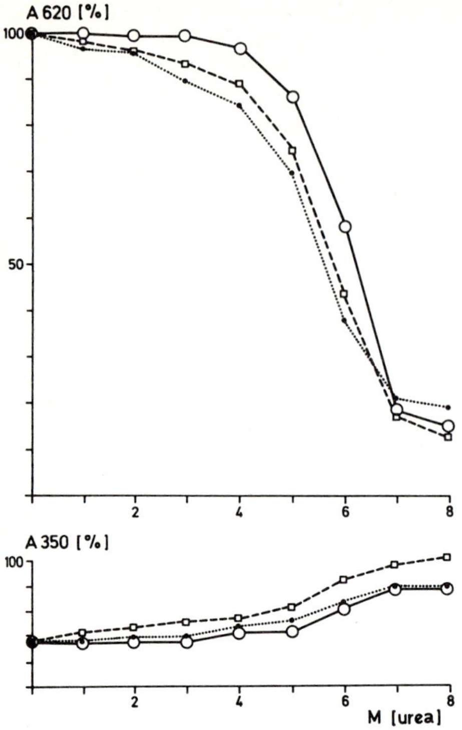

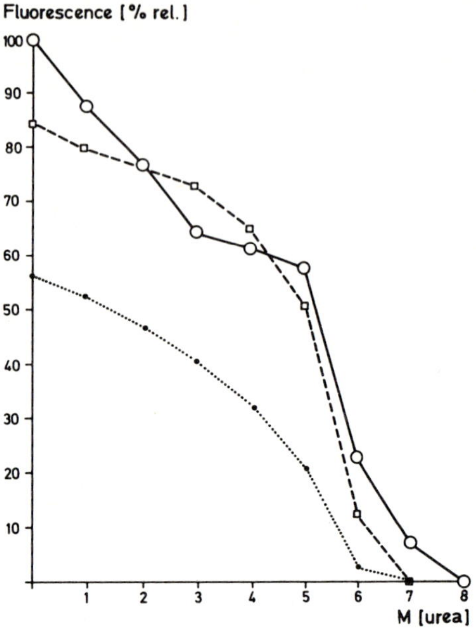

Fig. 6: Denaturation of trimeric **PC** (——) and its subunits (α:---; β:...) from M. laminosus with urea. Relative absorptions at the long- wavelength (top) and near-UV maxima (bottom), normalized to the absorptions in the absence of urea. All spectra were recorded from stock samples treated separately before the measurement with the appropriate amount of urea, either as 8M solution or as solid. All solutions in 100mM phosphate buffer, pH 7.5, T= 25°C.

Fig. 7: Denaturation of **PC** from M. laminosus and its subunits with urea. Relative fluorescence intensities corrected for absorption. λmax, exc.= 600nm;λmax, em. ≈652nm, decreasing with increasing urea concentration. Labels and details as in fig. 5.

The relative fluorescence yields during the same titration experiment are shown in fig. 7. There is already a large difference between the three species in their native states. The data for the integral pigment indicate a stepwise unfolding, with the first step being possibly related to the disaggregation of the trimer into monomers. The β- subunit is least stable (50% decrease at 3.5M urea), whereas α and the integral pigment have a 50% decrease at 5.3M urea. Although the curves of the last two species cross- over twice in the intermediate region, the sum of the subunits is always less than that of the integral pigment.

The differences of the thermal denaturation are even more pronounced among the three species (fig. 8). The "melting points" (50% decrease of the long-

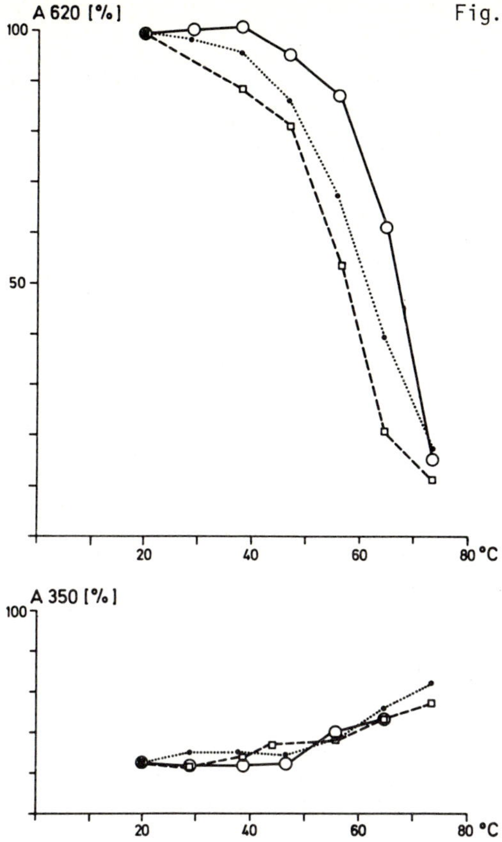

Fig. 8: Heat denaturation of **PC**. All spectra were recorded 5min after the sample had reached the temperature indicated. All measurements were done with fresh samples. Other details as in Fig. 5.

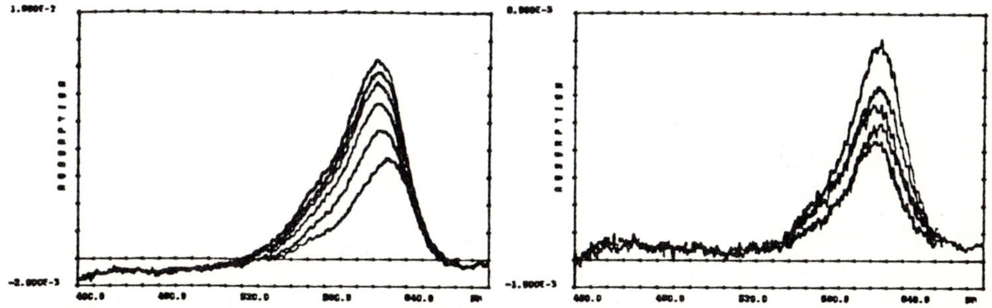

Fig. 9: Difference spectra for the photoreaction of **PC** in the presence of urea (5M). Top: Irradiation with 606nm light (interference filter). Positive bands correspond to bleaching. Bottom: Irradiation of the sample with 525nm light after saturing pre-irradiation at 606nm. Positive peaks correspond to increased absorptions.

wavelength absorption) is at 67° (**PC**), 61° (β) and 58° (α). The increase of thermal stability by almost 10° is significant with regard to the thermophilicity of M. laminosus, which grows well at temperatures as high as 55°C. At this temperature, the absorption of the integral pigment has decreased by only 10%. The "melting point" for **PC** from the mesophilic S. platensis is 52°C (11), i.e. 15°C below that of the thermophilic pigment.

The subunits differ also by their photochemical activities. **PC** is a light- harvesting pigment and its major deexcitation process in the isolated state is fluorescence. Another biliprotein, e.g. phytochrom, serves as "reaction center" pigment for photomorphogenesis in higher plants. Here, a structurally closely related chromophore is photochemically highly active. Several groups have reported recently that **PC** can undergo photochemical reactions, too, if it is partly denatured by a variety of reagents (see ref. 1 for leading references). This process is of potential significance also from a physiological point of view, because the reactions are similar to those of so- called phycochromes, which are thought be sensory photoreceptors in blue- green alga. We have studied the photochemistry of **PC** and its subunits in the presence of increasing amounts of urea. The typical response of the sample is shown in fig. 7. Upon irradiation at 606nm, there is a bleaching of the long-wavelength maximum, which saturates rapidly. This bleaching is partially reversed upon irradiation at 525nm, but about the five-fold time is required for the back reaction.

In integral **PC**, this photochemical reaction is negligible in the absence of urea. It increases up to 4.5M of the denaturant, and then decreases again. The behavior can be rationalized by a model, in which the native chromophore is so tightly bound to the polypeptide chain, that it is unable to react photochemically. A partial denaturation of the polypeptide loosens these interactions sufficiently to allow for photochemical reactions, whereas a complete unfolding opens the channel for efficient radiationless deactivation and prohibits photochemistry as is typical for free biliverdins. The β- subunit shows a similar, but much more pronounced response, and the maximum reaction is again in the same region. The α- subunit is, by contrast, almost inactive and shows an irreversible bleaching at urea concentrations $\gtrsim$5M. This indicates again a rather specific environment of the different **PC** chromophores, with the photochemical activity being mostly located on the β-subunit.

The **PC** system is well suited for a detailed study of specific non-covalent chromophore protein interactions, which are crucial for the funtion of many chromoproteins. In combination with sensitive and selective spectroscopic methods, and the recent advances in structure analysis, a detailed picture of these interactions is expected in the near future.

Acknowledgements: This work was supported by the Deutsche Forschungsgemeinschaft.

References

1. H. Scheer: In F. K. Fong (ed.) "Light Reaction Path of Photosynthesis", p. 7 - 45, Springer-Verlag (1982)

2. A.N. Glazer: Ann. Rev. Biochem. 1983. 52, 125 (1983)

3. E. Gantt: Intern. Rev. Cytol. 66, 45 (1980)

4. P. Hefferle, W. John, H. Scheer, S. Schneider: Photochem. Photobiol. 39, 221 (1984)

5. W. Castenholz: Schweizer Z. Hydrol. 35, 538 (1970)

6. U.K. Laemmli: Nature 227, 680 (1970)

7. H. Lehner, H. Scheer: Z. Naturforsch. 38c, 353 (1983)

8. R. MacColl, K. Csatorday, D.S. Berns: Arch. Biochem. 208, 42 (1981)

9. T. Schirmer, W. Bode, R. Huber, W. Sidler, H. Zuber: J. Mol. Biol. 184, 257 (1985)

10. G. Frank, W. Sidler, H. Widmer, H. Zuber: Hoppe- Seyler s Z. Physiol. Chem. 359, 1491 (1978)

11. H. Scheer, W. Kufer: Z. Naturforsch. 32c, 513 (1977)

12. G. Wagniere, G. Blauer: J. Am. Chem. Soc. 98, 7806 (1976)

13. H. Kayser: Z. Naturforsch. 39c, 938 (1984)

14. H. Falk, G. Höllbacher: Monatsh. Chem. 109, 1429 (1978)

15. S.E. Braslavsky, A.R. Holzwarth, K. Schaffner: Angew. Chem. 95, 670 (1983), ibid. Int. Ed. 22, 656 (1983)

Picosecond Time-Resolved, Polarized Fluorescence Decay of Phycobilisomes and Constituent Biliproteins Isolated from *Mastigocladus laminosus*

S. Schneider, P. Geiselhart, T. Mindl, and F. Dörr
Institut für Physikalische und Theoretische Chemie,
Technische Universität, D-8046 Garching, F.R.G.

W. John, R. Fischer, and H. Scheer
Botanisches Institut der Universität, D-8000 München 19, F.R.G.

1. Introduction

Phycobiliproteins are photosynthetic light-harvesting pigments in blue-green and red algae. They consist of 2-3 polypeptide subunits, each bearing up to 4 covalently bound linear tetrapyrrolic chromophores [1]. In vivo, the biliproteins are organized into complex structures [2], the phycobilisomes, which are attached to the outer thylakoid surface [3,4]. Energy deposited in the phycobilisomes by light absorption is transferred mainly to photosystem II, which is located in the thylakoid membrane. To achieve efficient energy-transfer, the average excitation energy of the different chromoproteins in the phycobilisomes decreases with decreasing distance to the reaction center. It is generally assumed that the energy-transfer is based upon dipole-dipole-interaction (Förster mechanism). In order to elucidate the importance of the special arrangement of the biliproteins in the phycobilisomes, i.e. the fact that the building blocs of the antenna rods are hexamers, numerous fluorescence studies have been performed on functionally intact phycobilisomes,as well as on their constituent aggregates [4-12]. Apparent differences in the results were blamed on different origin (organism) or preparation procedures, different measuring conditions (low or high light flux with the possibility of nonlinear processes), different excitation and observation wavelengths, etc. In this contribution, we summarize our measurements of the fluorescence decay of phycobilisomes from Mastigocladus laminosus and their isolated constituent biliproteins. Special emphasis is thereby given on the additional information, which can be derived from the time-resolved fluorescence polarization. In contrast to earlier measurements [9-12], the excitation wavelength could be chosen down to 545 nm; furthermore, the phycobilisomes of the present mutant were free of phycoerythrocyanin; therefore a possible interference with the decay of this pigment was excluded.

2. Measurements and data analysis

The fluorescence decay curves were measured using a synchronously pumped mode-locked dye laser in conjunction with a repetitively working streak camera (for details see e.g. [11]). The apparent time resolution of this system is approx. 25 ps without deconvolution procedure; it allows measurements with low excitation intensities (10^{13} photons per pulse and cm^2).

From the fluorescence decay curves measured with the analyzer parallel, ($I_p(t)$), and orthogonal, ($I_s(t)$), to the polarization of the exciting beam the expressions $I(t)=I_p(t)+2*I_s(t)$ and $D(t)=I_p(t)-I_s(t)$ are calculated. I(t) measures the decay of the excited state population (electronic lifetime) as does a measurement, where the analyzer is set under the magic angle to the excitation polarization. In case of several emitters, their contributions are additive. The difference function D(t) corresponds, in the case of only one kind of emitter, to the product of the excited state population I(t) and the correlation function C(t) of the absorption and emission dipoles (see e.g. [13]). In contrast to the generally discussed anisotropy function R(t)=D(t)/I(t), which for only one emitter represents the correlation function C(t), the difference function D(t) is additive and can therefore be evaluated even if more than one emitting species is present. Furthermore, there are no restrictions with respect to the time resolution of the detection system; quotient formation of convoluted functions D(t) and I(t), resp., would yield erroneous "time-resolved" anisotropy functions R(t) (for more details see e.g. [11] and [14]).

The best fits for both functions (I and D) are determined under the assumption of a bi- or tri-exponential response function by means of a Marquardt algorithm. Depending on the S/N ratio of the recorded fluorescence decay curves and their relative magnitude, the fit parameters derived may be subject to considerable error. Deduction of information on photophysical parameters (lifetimes, fluorescence quantum yields) of phycobilisomes or even monomers of biliproteins is impeded by the fact that the total fluorescence is composed of several contributions, namely:

(i) "leakage" fluorescence from directly excited, energy-transferring chromophors
(ii) fluorescence from directly excited, terminal chromophors
(iii) fluorescence from chromophors excited indirectly by energy-transfer, which may or may not transfer their excitation energy to other pigments.

Fluorescence of type (i) and (ii) will rise instantaneously and carry information on the polarization of the exciting light. Fluorescence of the third kind (iii) shows a risetime (negative amplitude in the analysis) and will in general be fairly depolarized, due to the randomizing energy-transfer process. In the simple case of a two-component system, the three fluorescence contributions must be superimposed with weighting factors, depending both on the ratio of extinction coefficients at the exciting wavelength and the spectral window of the detection system (fluorescence collection efficiency). One important consequence of this fact is that amplitudes gained from a multi-exponential analysis are without significance unless the weighting factors can be determined sufficiently precisely. E.g., it may happen that in the isotropic fluorescence the rising part of contribution (ii) cancels exactly the fast decaying contribution (i) with the effect that the decay appears essentially as single exponential with the long lifetime of the acceptor chromophor. In the difference function, on the other hand, the decay of the short-lived donor species is easily seen;

its amplitude should be about 0.4 of that of the isotropic contribution of the short-lived donor species, if a random orientational distribution of donor molecules can be assumed. With respect to the amplitude of the short-lived component in a multi-exponential fit, the fraction can be much smaller or larger depending on the extent of compensation discussed above.

In case of rotational relaxation the correlation function C(t) represents a multi-exponential; if the molecule is spherical, C(t) corresponds to a single exponential. Since no better approximations are known, we assume that the correlation function describing the energy-transfer can be approximated by a single exponentially decaying function. Under this circumstances the differences in decay time of the corresponding components of the functions I(t) and D(t), resp. can be associated to depolarizing effects like energy-transfer and orientational relaxation. The latter effect should, however, be without importance in case of protein-bound chromophors (a more detailed discussion will be given elsewhere).

3. Sample Preparation

Mastigocladus laminosus was grown at 52°C in slowly stirred 12 l fermentation flasks in Castenholz medium. Phycobilisomes were isolated by standard procedures [9,10]. The French-pressed cells were first centrifuged at 10,000 g, and the supernatant layered on a sucrose density gradient and spun over night at 140,000 g. Intact phycobilisomes banded in the lower third of the gradient. They had an absorption maximum at 635 nm with a strong shoulder around 620 nm and an emission maximum at 670 nm. Phycoerythrocyanin was isolated from green-light-grown cells following the protocol described for the isolation of phycocyanin [12]. It elutes from DEAE-cellulose column as the first band. The aggregation state was determined by ultracentrifugation from the sedimentation coefficients. The data were analyzed as described earlier [12].

4. Results and Discussion

The isotropic fluorescence decay curves displayed in figure 1a are recorded with 600 nm excitation but different spectral observation windows (cut-off filters). They clearly demonstrate the effect of partial compensation mentioned in chapter 2 onto the amplitude of the short-lived component (see table 1). With a S/N ratio typical for synchroscan streak camera measurements, a tri-exponential fit seems adequate from the point of view of residuals only for trace A; the other two curves (B,C) seem to fit equally well to a biexponential. It should be pointed out that, despite the fact of a rising component in curve C, the onset of the fluorescence in its major part is essentially instantaneous. The difference function D(t) (fig. 1b) can be fit by bi-exponentials with lifetimes of 20 and 1300 ps, resp. The latter value corresponds to the slowest isotropic component, and can be explained by the electronic relaxation of "terminal", immobile chromophors, i.e. the allophycocyanins. Since the amplitude of the "100 ps"- component turns negative for long-wavelength monitoring, it can be assigned to the average energy-transfer time between phycocyanin and allophycocyanin.

Table 1 : Fit parameters derived from the fluorescence decay curves I(t) and D(t) of functionally intact phycobilisomes, isolated from Mastigocladus laminosus, by means of a least square fit procedure.

Exc.(nm)	/Obs.(nm)	T_1(ps)	/A	T_2(ps)	/A	T_3(ps)	/A
600	>620	63	27	137	30	1342	65
	>630			98	58	1346	81
I(t)	>645			158	13	1700	54
	>665			100	-7	1622	33
	>620	20	22			1284	4
D(t)	>630	23	15			1539	5
580	>590	25	29	129	16	1340	26
I(t)	>600	52	20	104	31	1414	38
	>610	42	18	114	11	1351	28
	>590	23	10			462	30
D(t)	>600	22	2			609	7

Because the transferring phycocyanin chromophors are mainly indirectly excited by energy-transfer from the primary absorbers, no additional depolarization is connected with this step.

With 580 nm light, relatively more of the ß2-chromophors of phycocyanin are excited (in accordance with [15] the phycocyanobilin bound to cystein 84 of the ß-chain is referred to as ß1-chromophor, the second, bound to cystein 155 is called ß2-chromophor). The isotropic decay curves (fig. 2a) show only decaying components (table 1): the long-lived allophycocyanin emission (~1350 ps), an intermediate component (~110 ps) and finally a fast component, whose apparent lifetime seems to depend on the observation window. Taking into consideration the parameters derived from the difference function D(t), one may suspect that the shortest component actually represents a mixture with two slightly different lifetimes, one of which is equal to 20 - 25 ps. This lifetime must characterize a very fast decay of directly excited chromophors, i.e. represent the transfer time from α-chromophors, which absorb also at wavelengths around 600 nm. The 50 - 60 ps component would then be related to the deactivation of the ß2-chromophors, which, at least in the PC-crystals [15], are located at the perimeter of the trimeric unit. The arrangement of the chromophors in the trimeric units is such that the preferential α-ß1 energy-transfer is one between neighbouring monomeric units. This fast

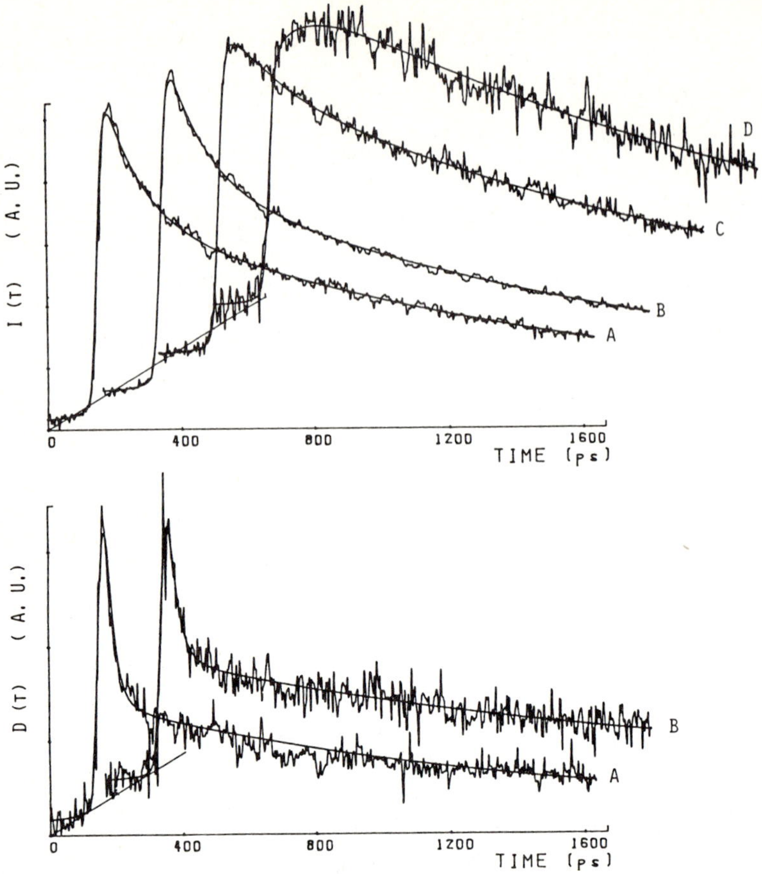

Figure 1: Dependence of fluorescence decay of functionally intact phycobilisomes isolated from Mastigocladus laminosus on observation window. Cut-off filters used are: 620 nm (A), 630 nm (B), 645 nm (C) and 665 nm (D). Excitation wavelength in all cases: 600 nm. The parameters of the fit curves are listed in table 1.

transfer route is absent in the monomeric unit and should, according to energetic criteria, be slowed down in phycoerythrocyanin (PEC). In the latter chromoprotein, the α-phycocyanobilin is replaced by a phycoerythrocyanobilin, which absorbs around 568 nm. The deactivation of the ß2-phycocyanobilin must necessarily occur by energy-transfer to the ß1- rather than to the α-chromophor.

In figures 3a and 3b, the functions I(t) amd D(t) of PEC-trimers are displayed for various combinations of excitation wavelength and observation window. For long-wavelength excitation (600 nm) the decay is governed by a ~350 ps and a ~1450 ps component. These values should correspond to the lifetimes of the ß1- (1450 ps) and the ß2- phycocyanobilin (350 ps),resp..

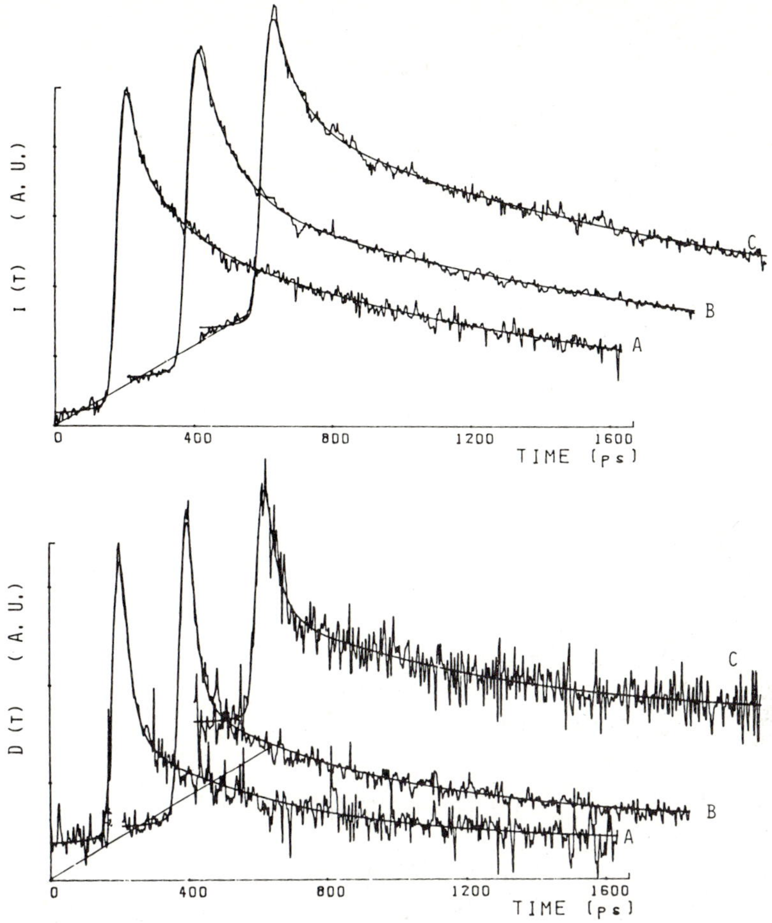

Figure 2: Dependence of fluorescence decay of isolated phycobilisomes on observation window. Cut-off filters used are: 590 nm (A), 600 nm (B), 610 nm (C); excitation wavelength is 580 nm. Fit parameters are listed in table 1.

Decay times of the same magnitude are found in D(t). For 545 nm excitation, a third component is detected both in I(t) and D(t). Its decay time of ~110 ps is assigned to the phycoerythrocyanobilin --> ß2- phycocyanobilin transfer time. It should be noted here again that going from 2 to 3-exponential fits, a splitting of one of the components in a faster and a slower one occurs. Therefore, the existence of a certain component must be established by measurements under conditions where that component dominates the decay.

The energetic degeneracy and the therefore possible energy-transfer between like species is, in our opinion, relevant for the interpretation of the decay curves of the trimeric unit

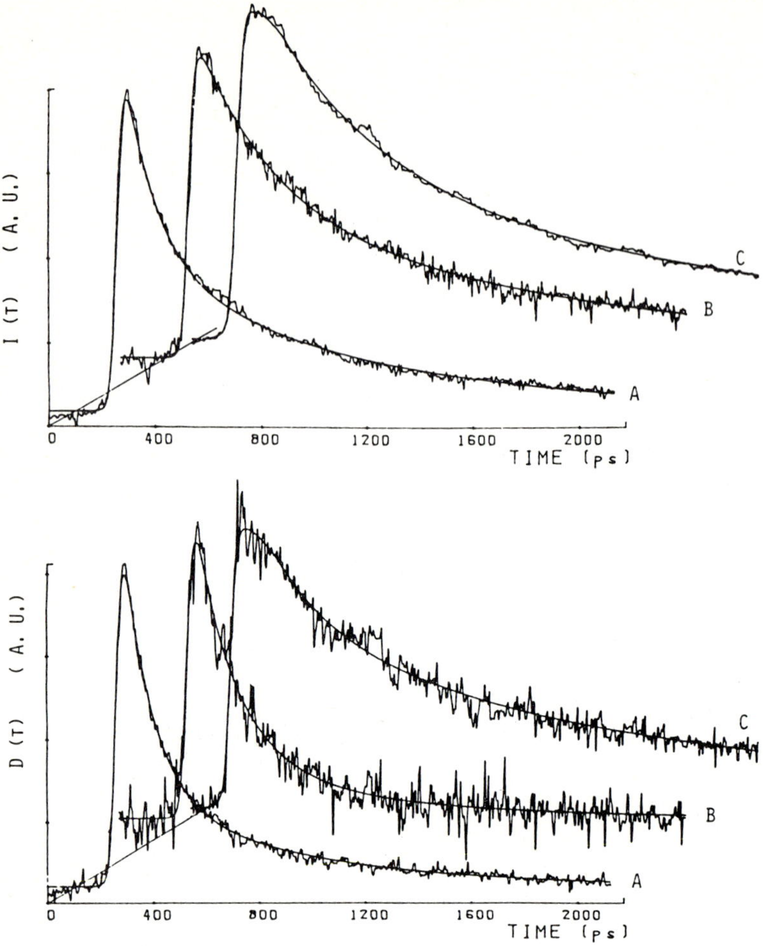

Figure 3: Dependence of fluorescence decay of isolated phycoerythrocyanin on excitation wavelength and observation window, characterized by the cut-off filter. Parameters are λ_{ex} / λ_{obs} τ_i in ps and A_i in paranthesis.

		I(t)	D(t)
(A)	545nm/ 570nm:	109(112)/ 470(58)/1500(32)	106(38)/ 580(13)
(B)	545nm/ 590nm:	270(32)/1400(28)	144(9)/ 520(3)
(C)	600nm/ 610nm	370(101)/1500(109)	260(12)/1530(13)

shown in figure 4. Under all experimental conditions, the major component is the longest-lived with a decay time of ~1400 ps. This value should characterize the "free" decay of the final energy accepting species, the ß1-chromophors. The 950 ps-component, which has a counterpart in the D(t) curve, must represent the decay of a rigidly held, but modified species, which is isolated in a sense that energy-transfer to ß1-chromophors is prohibited (a similar result and interpretation is given by

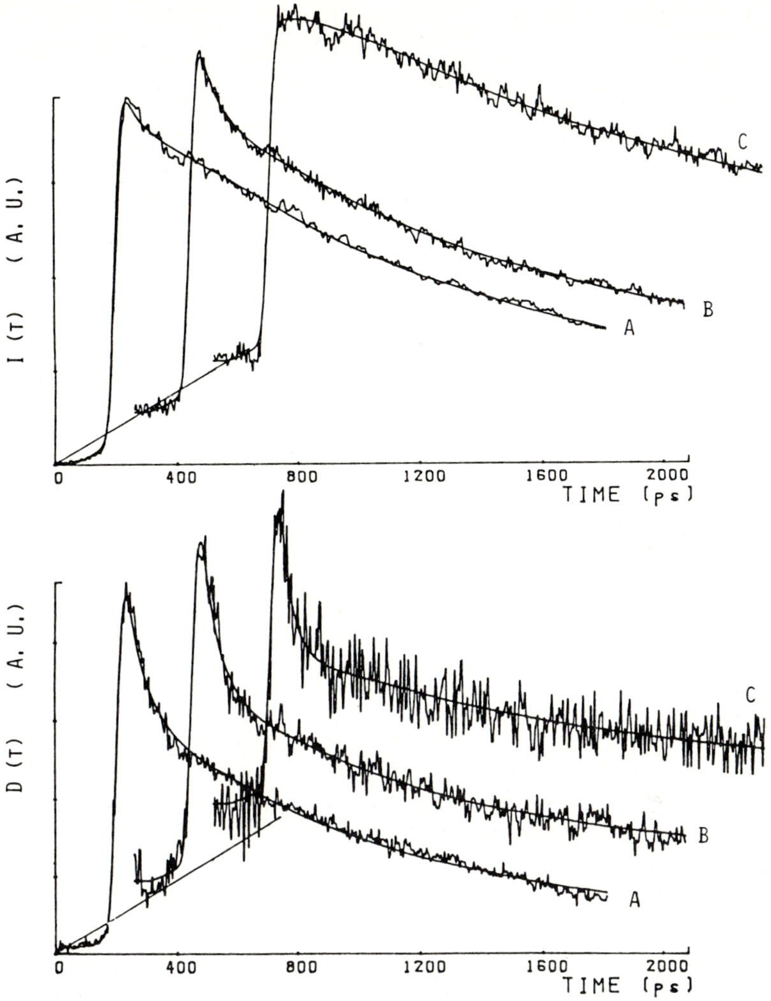

Figure 4: Dependence of fluorescence decay of phycocyanin trimers on excitation wavelength and observation window characterized by the cut-off filter. Parameters are λ_{ex} / λ_{obs}, τ_i in ps and A_i in paranthesis.

(A) 600nm/ 620nm: I(t):45(47)/1200(190) D(t):57(41)/ 850(40)
(B) 575nm/ 590nm: I(t):61(77)/1130(150) D(t):49(43)/ 800(25)
(C) 575nm/ 630nm: I(t):35(1)/1640(80) D(t):36(18)/1150(8)

Wendler et al. [16]). Upon short-wavelength excitation and observation the amplitude of the very fast component (40-60 ps) increases. This signals that this component is related to the ß2→ß1 energy-transfer; its value is close to that assigned to this type of transfer in the intact phycobilisomes. In contrast to the latter, the fast 20 ps component, seen mainly in the difference function D(t) is missing in the trimer. The explanation is probably connected to the fact that α→ß1 transfer is no longer irreversible in the trimer, since the energy-difference is of the order of kT only. Lacking the energy drain down the rod by very fast homo-transfer, the energy can circulate among the isoenergetic ß1-chromophors and eventually be back-transferred to α-chromophors. One may suspect that fluctuations in geometry result in variations of the excitation energy, which are sufficiently large to increase the back- and homo-transfer rates, resp.. For this reason, only energy-transfer from excited ß2- chromophors would be connected with a significant change of fluorescence polarization.

The above given interpretation could be in contrast to the interaction energies and transfer times derived from crystallographic data [17]. From those it is expected that deactivation of ß2-chromophors should be dominated by transfer to ß1-chromophors. The refinement of these data, however, is not yet good enough to give precise information on the complete chromophor structure, and therefore it is not possible to calculate the orientation factors of transition dipole moments and the overlap of donor emission and acceptor absorption spectra.

Acknowledgement. Financial support by Deutsche Forschungsgemeinschaft (SFB 143) and Fonds der Chemie is gratefully acknowledged. We also thank Dr. Hensel and Ms. Renz (München) for the ultracentrifuge measurements.

References

1 For a review see e.g.: H. Scheer, in: "Light Reaction Path of Photosynthesis", p. 7, ed. by F.K. Fong (Springer-Verlag, Berlin, 1982)
2 E. Mörschel, W. Wehrmeyer: Arch. Microbiol. 113, 83 (1977)
3 For a review see e.g.: A.N. Glazer: Ann. Rev. Plant Physiol. 32, 327 (1983)
4 For a recent review see e.g.: "Biological events probed by ultrafast laser spectroscopy", ed. by R.R.Alfano (Academic Press, New York, 1982)
5 J. Wendler, A.R. Holzwarth, W. Wehrmeyer: Biochim. Biophys. Acta 765, 58 (1984)
6 M. Seibert and J.S. Connolly: Photochem. Photobiol. 40, 267 (1984)
7 G.W. Suter, P. Mazzola, J. Wendler, A.R. Holzwarth: Biochim. Biophys. Acta 766, 269 (1984)
8 W. Haehnel, A.R. Holzwarth, J. Wendler: Photochem. Photobiol. 37, 435 (1983)
9 P. Hefferle, M. Nies, W. Wehrmeyer, S. Schneider: Photobiochem. Photobiophys. 5, 41 (1983)
10 P. Hefferle, M. Nies, W. Wehrmeyer, S. Schneider: Photobiochem. Photobiophys. 5, 325 (1983)
11 P. Hefferle, W. John, H. Scheer, S. Schneider: Photochem. Photobiol. 39, 221 (1984)
12 P. Hefferle, P. Geiselhart, T. Mindl, S. Schneider W. John, H. Scheer: Z. Naturforsch. 39c, 606 (1984)
13 G.R. Fleming, J.M. Morris, G.W. Robinson: J. Chem. Phys. 17, 91 (1976)

14 J. Papenhuizen and A.J.W.G. Visser: Biophys. Chem. 17, 57 (1983)
15 T. Schirmer, W. Bode, R. Huber, W. Sidler, H. Zuber: J. Molecular Biology, submitted; id. in: "Optical Properties and Structure of Tetrapyrroles", ed. by G.Blauer and H. Sund, (Walter de Gruyter, Berlin, 1985)
16 J. Wendler, W. John, H. Scheer, A.R. Holzwarth: submitted to Photochem. Photobiol.
17 T. Schirmer: PhD-Thesis, TU-München (1985)

Fluorescence Behaviour of Crystallized C-Phycocyanin (Trimer) from *Mastigocladus laminosus*

S. Schneider, P. Geiselhart, and C. Scharnagl
Institut für Physikalische und Theoretische Chemie,
Technische Universität München, D-8046 Garching, F.R.G.

T. Schirmer and W. Bode
Max-Planck-Institut für Biochemie, D-8033 Martinsried, F.R.G.

W. Sidler and H. Zuber
Institut für Molekularbiologie und Biophysik,
Eidgenössische Technische Hochschule, CH-8093 Zürich, Switzerland

1. Introduction

Of all light-harvesting biliproteins, C-Phycocyanin (C-PC) is certainly the most extensively studied one [1]. With respect to the investigation of energy-transfer between the phycocyanobilin chromophors, this biliprotein has the advantage that by now much is known about the structure of both the chromophor [1] and the protein [2,3]. It was furthermore established that the fluorescence properties depend sensitively on temperature and the state of aggregation (see e.g. [4]-[6] and references there). PRIESTLE et al. [4] found that the fluorescence behaviour changes significantly if phycobiliproteins are crystallized. In the case of C-PC from the cyanophyte Agmenellum quadruplicatum, the absorbance peak of a suspension of small C-PC crystals was red-shifted by only 7nm relative to the solution with hexamers, while the single crystal fluorescence was red-shifted by about 60 nm. The spectral distribution of the crystal fluorescence, whose lifetime was determined as ~0.5 ns, changed with excitation wavelength.

Recently trimers of C-PC from the cyanobacterium Mastigocladus laminosus Cohn were crystallized to prismatic hexagonal crystals of typical dimensions 0.2 mm x 0.2 mm x 0.4 mm and analyzed with 3 Å resolution by X-ray diffraction methods [3]. In agreement with earlier work [7,8] it was found that the trimers are disc-shaped (Ø ~110 Å) with a central hole of about 35 Å diameter (fig.1). In the crystal, the trimers of adjacent layers are rotated by 180° with the effect that the chromophors ß155 are arranged like a linear chain parallel to the six fold axis. The conformation derived for a pair of ß155 chromophors is shown as a stereo-picture in fig.2. Because there is a twist of 25° around all the methine bridges, the pyrrole rings deviate significantly from the average molecular plane, and show therefore a large circular dichroism (see below). Within one trimer the chromophors ß155 are located at the periphery and extend almost parallel to the trimer axis. The chromophors ß84 and α84 are topologically equivalent. The thioether bond and the first two pyrrole rings form a semi-circle around helix E [3], the electron density of the two last pyrrole rings could not be interpreted unambiguously so far. Nevertheless, one can state that the chromophor axis of α84 is lying almost in the trimer plane, whereas that of ß84 is inclined about 30°. The last pyrrole rings of ß84 reach into the

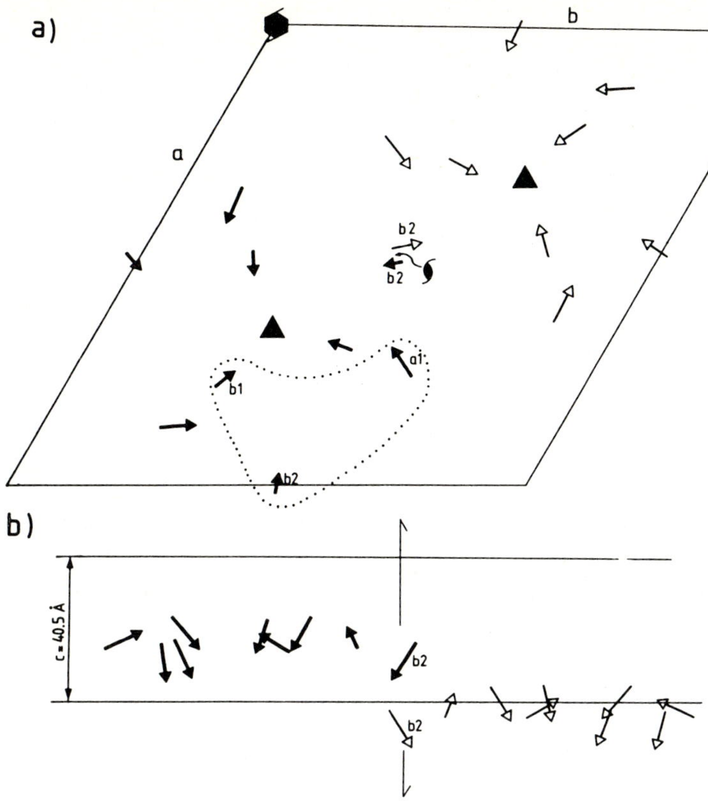

Fig. 1: Chromophor arrangement in C-PC crystals from Mastigocladus laminosus. Displayed are the least-squares fit-lines for the conjugated systems of the chromophors. The unit cell contains two trimers, the chromophors of which are marked by black and white arrows, resp..

a) View against the crystallographic c-axis, one monomer is outlined by dots.
b) View perpendicular to c-axis along the short diagonal of the ab - rhomb 110.

central hole and may be subject to specific interactions, e.g. with the linker peptides.

In this paper we report on measurements of the fluorescence decay of crystals of C-PC prepared identically as those used for the X-ray work [3]. The results are compared with solution data and interpreted on the basis of the above summarized structural information.

2. Experimental

Fluorescence emission and fluorescence excitation spectra of solution and crystal suspension were recorded with a home-built spectrometer. Fluorescence decay curves were recorded by means

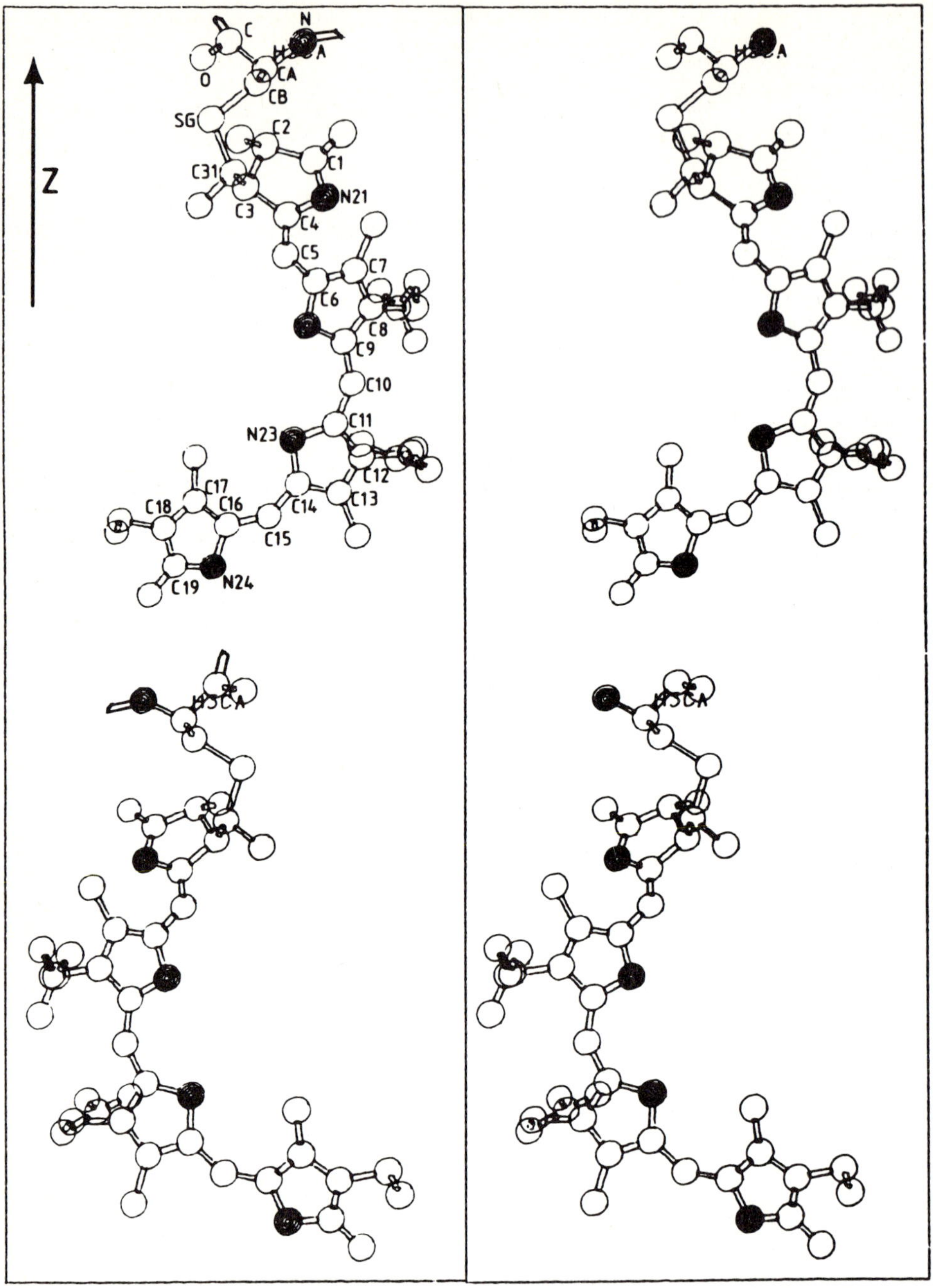

Fig. 2: Stereoscopic view of the spatial relation between ß155-chromophors of adjacent trimers in the M. laminosus C-PC crystals.

of a synchroscan streak camera at low-excitation light level to avoid non-linear processes [5,6,9]. Model calculations were performed in the π-electron approximation after PARISER, PARR and POPLE using "standard" sets of parameters [10,11]. The coordinates determined by X-ray analysis for chromophor ß155 (fig.2) were used in the model calculations.

3. Results and Discussion

The measurement of a single crystal absorption spectrum is made impossible by the fact that the crystals produced so far are too thick and the surface area is too small. We therefore must rely presently on fluorescence emission and excitation spectra of suspensions. As shown in fig.3, the fluorescence of a crystal suspension in 0.4 M phosphate buffer peaks around 750-760 nm (corrected for photomultiplier response). In contrast to PRIESTLE et al. no shoulder is found around 665 nm. The 815 nm-fluorescence excitation spectrum shows a long-wavelength maximum around 595 nm with a shoulder around 640 nm. At shorter wavelength, scattering by small particles in the buffer becomes significant, therefore, a strongly increasing signal is observed towards shorter wavelength. The existence of these small particles is also verified by their red fluorescence and in transmission measurements. Due to this scattering, the maximum in the excitation spectrum may appear somewhat hypsochromically shifted. For comparison, fig.3 includes also the spectra which were taken for crystals dissolved in phosphate buffer of lower concentration (0.1 M). The red absorption maximum is found at 610 nm; this could, however, indicate a monomeric rather than a trimeric state of aggregation. The steady-state fluorescence has its maximum at 650 nm, i.e. it shows a "normal" Stokes' shift.

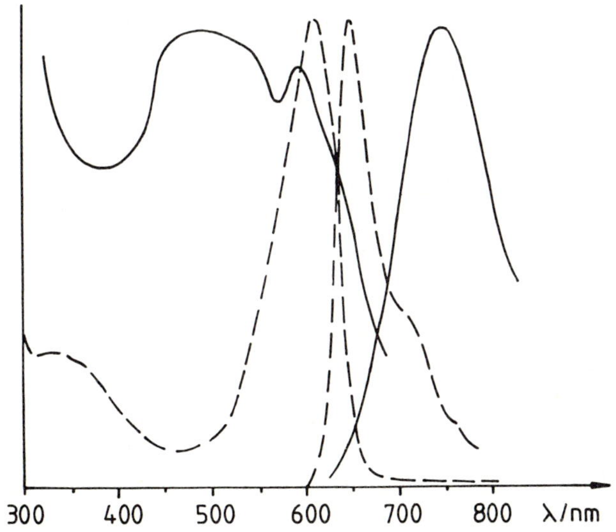

Fig. 3: Fluorescence excitation spectrum (det. 815nm) and emission spectrum (exc. 550nm) of microcrystal suspension (————). Absorption and fluorescence emission (exc. 550nm) spectrum of dissolved crystals (-------).

From the polarized fluorescence excitation spectra of the α- and β-subunits and the monomer, resp., Mimuro et al. [12] concluded that the absorption maxima of the chromophors α84, β84, and β155 in the monomer are around 618 nm (16.16 kK), 595 nm (16.03 kK) and 624 nm (16.80 kK), resp. Upon formation of isolated trimers, a strong interaction between chromophors α84 of one and β84 of the neighbouring monomer is introduced, due to the short center-to-center distance of ~20 Å. As a result of this interaction, the difference in excitation energy of the α84 and β84 chromophors should increase, and the long-wavelength absorbtion edge should shift bathochromic as it is actually observed (~6 nm ≙ 150 cm^{-1}). The interaction of the 1α84-chromophor with the other seven chromophors of the same trimeric unit is much smaller, the largest being probably the one with 2β155. For this pair the interaction should be weaker by about a factor 5, due to the increased distance, and another factor 5 due to the relative orientation (see table 1). The coupling strength among the three β155 chromophors should be of similar magnitude as that between β155 and β84 of the same monomer. The actual value of the interaction is dependent on the molecular geometry, since torsions around the methine bridges may significantly alter excitation energy and oscillator strength. As judged from the distances in an isolated trimer, an excited β155 chromophor should transfer its energy preferentially to the β84 chromophor of the same monomeric unit; the excited α84 chromophor should deactivate preferentially by transfer to β84 of the neighbouring monomeric unit. In monomers, the much longer transfer time to the β84 chromophor must become apparent under appropriate excitation conditions. In the crystalline sample, the distance between the β155 chromophors of neighbouring trimers (fig.1) is only about 21 Å, due to a nearly perfect alignment of the center lines of the molecules, the interaction energy is large.

Table 1: Center to center distances between chromophors of one trimeric unit in Å (top) and orientational factors κ (bottom) based on orientation of molecular frame (center line). After SCHIRMER [13]

	1α84	1β84	1β155	2α84	2β84	2β155	3α84	3β84	3β155
1α84		53.3	51.4	69.3	19.7	35.3	69.3	53.7	85.7
		.61	-.26	.62	-.99	-.14	.62	.044	.13
1β84			36.1	53.7	35.7	66.7	19.7	35.7	45.4
			.95	.44	1.30	1.16	-.99	1.30	.82
1β155				85.7	45.4	79.8	35.3	66.7	79.8
				.13	.82	1.11	-.14	1.16	1.11

A question raised generally in this context is whether the point-dipole approximation is still valid although parts of the molecules approach each other very closely. We have therefore performed MO-calculations using the coordinates derived for the arrangement of two ß155 chromophors to get net charge densities for all π-electron centers. Then the coulombic interaction between all centers of one with all centers of the second chromophor was calculated. For an additional cross-check, the MO calculation was performed for a "super-molecule" representing the two chromophors. As it is expected, the energy-splitting of the corresponding "plus"- and "minus"-linear combinations of the "super-molecule" is essentially equal to that calculated by summation of the point-point interactions, provided the same effective dielectric constant, as implicitly done in the coulombic integrals of the MO calculations, is taken into account.

Excitation energy and transition dipole moment of the $S_0 \rightarrow S_1$ transition of an individual chromophor vary significantly with the number of singly-excited configurations included in the CI-calculation. The results referred to in table 2 are based on SCF-CI calculations, including 14(occupied) x 9(unoccupied) excited configurations.

Due to the overall "extended" conformation, the oscillatory strength of the "red"-transition is very high ($f=1.22$, $D_{01}=12.4$ Debye) in agreement with estimates by e.g. SCATORDAY et. al. [14]. The transition dipolemoment is orientated roughly along the "long molecular axis", for which reason estimates of the orientation factor $\varkappa$ based upon molecular orientation (center line) are reasonably good approximations. The $\varkappa$ value derived from the calculated directions of the transition dipole moments is $\varkappa=-1.673$. In combination with a center to center distance of 21 Å and a transition dipole moment of 12.4 Debye, an interaction energy $V = 93/\varepsilon$ (cm^{-1}) is calculated in the point-dipole approximation. The explicit summation over all center-to-center interactions leads to $V = 203/\varepsilon$ (cm^{-1}). Assuming an effective dielectric constant of 2.5 [15] an interaction energy of 80 cm^{-1} would result.

Table 2: Excitation energies E_{oi} (wavelength λ_{oi}), transition dipolemoments D_{oi}, oscillatory strength f_{oi}, angle α_i between transition dipolemoments of $S_0 \rightarrow S_1$ and $S_0 \rightarrow S_i$ transition, resp., for a phycocyanobilin chromophor with the conformation derived from X-ray analysis for ß155.

	S_1	S_2	S_3	S_4	S_5	S_6	S_7
E_{oi} [kK]	17.6	21.8	24.1	27.4	28.3	32.1	32.8
λ_{oi} [nm]	567.5	458.8	414.5	367.2	352.7	310.8	304.9
D_{oi} [Debye]	12.4	7.8	3.3	5.4	2.9	4.0	1.4
f_{oi}	1.21	0.60	0.12	0.36	0.11	0.23	0.03
α_i [°]		88	79	49	89	62	60

A coupling strength of this seize between the ß155 chromophors of different trimers would not yet lead to perfect excitonic coupling of these chromophors. The situation should be rather described as intermediate case, since the bandwidth of the electronic transition must be assumed to be of the order of 900 cm^{-1} [14]. With respect to the energy-transfer kinetics, each chain of ß155-chromophors should appear as one donor only.

The rather strong deviation from planarity gives rise to a high rotatory strength of red (S_0-->S_1: R=5.5*10^{-38} cgs) and some higher electronic transitions (S_0-->S_2: R=-2.01*10^{-38} and S_0->S_4: R=-2.74*10^{-38} cgs).If in the native pigment the ß155 chromophor adopts similar large torsion angles, then this geometry could explain the observation that the CD-spectra of the ß-subunit show a rather high ellipticity at the short-wavelength side of the red absorption band.

Similar considerations as just made for the coupling of the ß155 chromophors should hold for the tightly coupled pairs of α84 and ß84. Due to the interaction between pairs in the same trimer (mainly via the ß84 chromophors), pairs in trimers laying above each other (distance ~40 Å), or in the neighbouring trimer above and below the excited states may no longer represent those of the isolated α84 and ß84 chromophors.

The appearance of the fluorescence decay curves monitored for 580 nm excitation with all fluorescence beyond 620 nm being collected is drastically different (fig. 4) for PC in microcrystal suspension and buffer solution. In the latter case, the isotropic

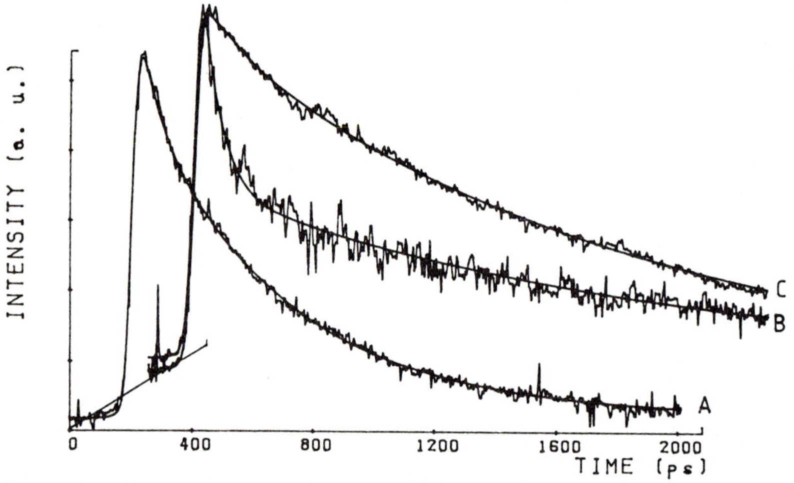

Fig. 4: Fluorescence decay of C-PC microcrystal suspension (A) and dissolved crystals (B):D(t)=I_p-I_s) and (C):I(t)=I_p+2*I_s). Excitation wavelength: 580 nm; cut-off filter: 600 nm

Parameters are τ_i and A_i in parenthesis:

(A) 59 ps (9) / 490 ps (25)

(B) 52 ps (28) / 1350 ps (13)

(C) 70 ps (22) / 1310 ps (99)

decay curve I(t) is dominated by a long-lived component with T_1= 1310 ps, (A_1=99). Superimposed is a short-lived component (T_2= 70 ps, A_2=22), which also has a counterpart in the difference function D(t). Since the corresponding lifetime is somewhat shorter and the amplitude larger, we must assume, that there is an additional energy-transfer step involved which is not resolved under our experimental conditions (see also discussion in [9]). By comparison with previous results and based on the arguments given above, we assign the 50 ps lifetime to the energy-transfer time between ß155 and ß84 chromophors. The 1300 ps component represents the free decay of the terminal acceptor chromophors ß84. In accordance with theory, the decay time found in D(t) is the same as derived from I(t), because the protein-bound chromophors do not undergo rotational relaxation.

For the microcrystal suspension we could not record polarized fluorescence decay curves; the decay curves were recorded with a polarization analyzer under magic angle (54°) to the polarization of the exciting light, i.e. they represent I(t) and therefore the excited state populations. Based on the discussion above, we furthermore assume that the short lifetime derived from the biexponential fit-procedure represents the transfer time between an excited ß155 chromophor coupled to the whole ß155- chromophor chain and the compound state made up from the strongly coupled α84-ß84 chromophors. The decay time of that compound state is obviously strongly decreased, due to that coupling (T_2= 500ps). In the picture usually drawn in connection with theory of radiationless transitions, this compound state could have a different geometry at the minimum of the excited state hypersurface. With this assumption, both the increased rate for radiationless deactivation and the unusual large Stokes´ shift could be made plausible. On the other hand, if the electronic coupling between the chromophors is strong enough, the excitation energy would migrate rapidly until it is trapped at suitable centers. In that case, the fluorescence decay pattern should vary with the chosen observation window. Unfortunately, the sensitivity of our streak-tube is too low beyond 700 nm to perform such a test. Further investigations using the single-photon timing technique are therefore in preparation.

References

1 For a review see e.g.: H. Scheer, in: "Light Reaction Path of Photosynthesis", p. 7-45, ed by F.K. Fong (Springer-Verlag, Berlin, 1982)

2 W. Sidler, J. Gysi, E. Isker, H. Zuber: Hoppe-Seylers Z. Physiol. Chem 363, 611 (1981)

3 T. Schirmer, W. Bode, R. Huber, W. Sidler, H. Zuber: submitted to J. Molecular Biology; id. in: "Optical Properties and Structure of Tetrapyrroles", ed. by G. Blauer and H. Sund, (Walter de Gruyter, Berlin, 1985)

4 J.P. Priestle, R.H. Rhyne, J.B. Salomon, M.L.Hackert: Photochem. Photobiol. 35, 827 (1982)

5 P. Hefferle, W. John, H. Scheer, S. Schneider: Photochem. Photobiol. 39, 221 (1984)

6 P. Hefferle, P. Geiselhart, T. Mindl, S. Schneider, W. John, H. Scheer: Z. Naturforsch. 39c, 606 (1984)

7 M. Dobler, S.D. Dover, K. Laves, A. Binder, H. Zuber: J. Mol. Biol. 71, 785 (1977)
8 E. Mörschel, K.P. Koller, W. Wehrmeyer, H. Schneider: Cytobiol. 16, 118 (1977)
9 S. Schneider, P. Geiselhart, T. Mindl, F. Dörr, W. John, H. Scheer: contribution in this volume
10 H. Scheer, H. Formanek, S. Schneider: Photochem. Photobiol. 36, 259 (1982)
11 C. Scharnagl, E. Köst-Reyes, S. Schneider, H.-P. Köst, H. Scheer: Z. Naturforsch. 38c, 951 (1983)
12 M. Mimuro and H. Zuber: contribution in this volume
13 T. Schirmer, PhD-Thesis, TU München (1985)
14 K. Csatorday, R. McColl, V. Csizmadia, J. Grabowski, C. Bagyinka: Biochem. 23, 6466 (1984)
15 J. Grabowski and E. Gantt: Photochem. Photobiol. 28, 39 (1978)

Energy-Transfer Kinetics in Phycobilisomes

A.R. Holzwarth

Max-Planck-Institut für Strahlenchemie, Stiftstr. 34-36
D-4330 Mülheim/Ruhr, F.R.G.

1. INTRODUCTION

Blue-green alga (cyanobacteria), red alga and cryptophyceae contain phycobiliproteins as major light-harvesting pigments which gather light in the wavelength region of low Chl absorption. The chromophores in these pigments are open chain tetrapyrroles which are bound covalently to apoproteins [1]. It has been shown that the energy absorbed by phycobiliproteins feeds the small pool of Chl in these algae [2]. In cyanobacteria and red alga the phycobiliproteins form large supramolecular antenna complexes, so-called phycobilisomes (PBS) [3,4] which are located at the outer surface of the thylakoid membrane. PBS are made up of two or three different types of phycobiliproteins which occur predominantly in hexameric aggregation. The first picosecond measurements of phycobiliprotein containing algae and isolated PBS were carried out by Porter et al. [5,6]. Phycobiliproteins and PBS are interesting objects for time-resolved studies for several reasons. Unlike the Chl protein complexes of higher plants, different phycobiliproteins have their absorption and emission spectra fairly well separated, which more easily allows the detailed sequence of energy-transfer steps to be explored. Furthermore, the single-step transfer times seem to be significantly longer than those of Chl complexes, which puts these processes in a time-range accessible to picosecond techniques.

The building blocks of PBS are disc-shaped aggregates (hexamers or trimers) of phycobiliproteins [7,8]. The discs composed of phycoerythrin and/or phycocyanin are stacked in rods, which are connected to a core of allophycocyanin discs. A single PBS may contain several hundred bilin chromophores. The structure with the high-energy pigments at the peripheral positions and the low-energy pigments at the center supports a cascade of directed down-hill energy-transfer [9]. The terminal emitters of PBS have their fluorescence maximum between 670 and 680 nm. The energy-transfer kinetics in PBS of the blue-green alga *Synechococcus* has been studied as a function of the rod length in order to arrive at a general model for the energy-transfer processes and kinetical laws governing this transfer in PBS. The rods of these PBS contain only C-phycocyanin, and their structure is well known [10].

2. RESULTS and DISCUSSION

The results of picosecond fluorescence and picosecond transient absorption studies on PBS of *S 6301* and *AN 112* [11-13] are summarized in Tables 1 - 3. Note that in the fluorescence study the PBS of *S 6301* had an average of 4 phycocyanin hexamers per rod [12], while in the picosecond absorption study they had only 3 hexamers [11,13].

Table 1: Analysis of the isotropic fluorescence decays from *Synechococcus 6301* in 0.75 M phosphate buffer, pH = 8.0, at room temperature.
The errors given are those obtained from the variations between several independent preparations. The statistical errors of the decay analysis are smaller than these values.

λ_{exc} [nm]	λ_{em} [nm]	τ_1 [ns]	R_1* [%]	τ_2 [ps]	R_2* [%]	τ_3 [ps]	R_3* [%]
570	590	1.80±0.05	14±2	300-600	11-17	63±15	73±3
570	600	1.83±0.02	16±2	270-400	10-15	80±12	71±2
570	620	1.85±0.02	19±2	300-450	8-12	94±12	70±2
570	640	1.85±0.02	28±2	270-480	4-15	100±20	65±5
570	680	1.90±0.02	86±4	1100±200	14± 4	120±20	-45±8
580	590	1.79±0.05	13±2	650±100	11±2	70±10	76±2
600	620	1.84	17	230	18	70	65
600	640	1.84	28	370	8	101	64
600	680	1.90	88	1200	12	131	-50
620	640	1.85±0.02	28±1	280±560	6-11	104±10	64±2
620	680	1.90±0.04	90±4	1100±250	10± 4	115±20	-52±13
645	680	1.89±0.01	88±1	1000±200	12± 1	120±10	-30±8
660	680	1.84±0.01	92±1	660±100	8± 1	45±10	-21±5
675	685	1.81	100				

* Normalized relative amplitudes.

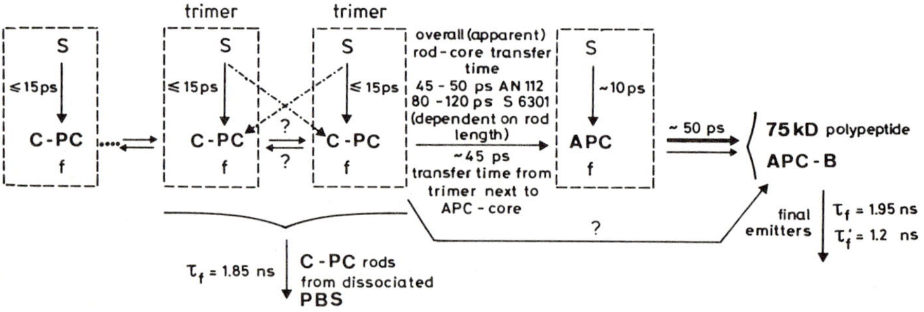

Fig. 1: Schematic model of the energy-transfer processes in PBS of *S 6301* and *AN 112* as deduced from picosecond fluorescence, absorption and anisotropy decay measurements.

Table 2: Absorption recovery ($\tau_{1,2}$) and anisotropy τ_r, $r(\infty)$ of phycobilisomes of *S 6301* and *AN 112* at different wavelength intervals.
Average values are given for all data. The λ_{exc} data were taken from isotropic two-wavelength experiment on *S 6301* phycobilisomes detection at λ_d = 650 nm.

	τ_1 [ps]	τ_2 [ps]	A_1/A_2 [a]	τ_r [ps]	$r(\infty)$ [b]
S 6301[c] 585 nm<λ<630 nm	12±3	84±8	1.0±0.5	10±2	0.07±0.02
λ_{exc} = 587 nm	-	86			
λ_{exc} = 618 nm	12	70			
AN 112					
609 nm<λ<630 nm	8.0±1.5	37±5.0	1.8±0.4	11±2	0.084±0.025
λ = 657 nm	9.9±1.4	51±5	0.8±0.2	11±2	0.084±0.025

[a] Amplitude ratio of short-lived and long-lived components.
[b] For evaluation and interpretation of the $r(\infty)$ value, see text.
[c] These phycobilisomes contained 3 hexamers of phycocyanin per rod on average.

The striking features of these studies are:

a) A fast isotropic and an anisotropic decay component with identical decay times ($\tau \sim$ 8-15 ps) is observed in both types of PBS independent of the rod length.
This fast decay is attributed mainly to s→f energy-transfer within hexameric phycocyanin in the rods, and in part to a similarly fast s→f transfer in the allophycocyanin core. The anisotropy decay data show that the s- and f-chromophores have non-parallel oriented transition dipoles.

b) The overall transfer time from the rod to the core increases with the rod length. It varies from ∿ 45 ps (1 hexamer per rod) to ∿ 90 ps (3 hexamers) and reaches ∿ 120 ps (4 hexamers). This overall transfer time has been observed both in the decay of phycocyanin fluorescence and in the corresponding rise of the allophycocyanin fluorescence, which were in excellent agreement to each other. The transfer kinetics of the rod→core transfer is close to mono-exponential in all cases. This transfer is related to an anisotropy decay, which is much slower than the corresponding isotropic decay. This shows that the energy is funnelled along the phycocyanin rods between f-chromophores with nearly parallel-oriented transition dipoles. The nearly parallel orientation ensures an efficient and fast transfer along the rods.

c) An energy-transfer time of 50-60 ps is found in the PBS core ($\lambda_{exc} \sim$ 650 nm). This transfer time is related to the energy-transfer from allophycocyanin to allophycocyanin B and/or the 75 kD linker peptide, which form the terminal emitters in these PBS.

Table 3: Fluorescence lifetimes (τ) and amplitudes (R, relative and B, absolute) of phycobilisomes of *AN 112* at different excitation (λ_{ex}) and emission (λ_{em}) wavelengths

λ_{exc} (nm)	λ_{em} (nm)	τ_1 (ps)	R_1 (%)	τ_2 (ps)	R_2 (%)	τ_3 (ns)	R_3 (%)	τ_4 (ns)	R_4 (%)
582	595	12	79	51	15	0.66	2	1.77	4
585	600			39	92	0.67	1	1.83	7
581	680			42[a]	-237[b]	1.44	31[b]	1.98	125[b]
620	680			51[a]	-311[b]	1.39	42[b]	1.98	176[b]
660	680			60[a]	- 49[b]	1.12	30[b]	1.96	228[b]
582,585[c]	595,600,620			32-39		0.61±0.12		1.85±0.04	
580,585,620[c]	640			43-45		0.65±0.19		1.87±0.04	
580,620[c]	680			45±6[a]		1.37±0.12		1.95±0.05	
650,660[c]	675,680			50±9[a]		1.13±0.47		1.97±0.07	

[a] Fluorescence risetime; negative amplitude.

[b] Absolute amplitude of the decay component.

[c] Average values.

d) The fluorescence of the terminal emitters shows two decay components with lifetimes of ∿ 1.1 ns and ∿ 1.9 ns. They should most probably be attributed to allophycocyanin B and the 75 kD linker peptide. Apparently both of them function as terminal emitters in PBS. The biological role of two different terminal emitters in PBS is presently unclear, but it might be relevant for the energy distribution between the two photosystems.

From these measurements of picosecond fluorescence, picosecond absorption and time-resolved anisotropy, the model shown in Fig. 1 has been proposed for the energy-transfer in homogeneous PBS rods [12,13]. The model involves a fast initial energy-transfer step of 10-15 ps assigned to s→f transfer within a hexameric unit of phycocyanin between non-parallel oriented dipoles. This transfer step gives the largest contribution to the anisotropy decay. Energy along the rods is then funneled between preferentially parallel oriented f-chromophores on the shortest possible path. The energies of the pigments ensure a slight asymmetry between forward and backward transfer. Parallel orientation of these f-chromophores in the different trimers and hexamers is inferred from the slow anisotropy decay related to transfer along the rods [11,13,14].

The assignment of the fast isotropic and anisotropic decay component with a lifetime of about 10 ps to s→f transfer in hexameric C-phycocyanin is supported by the time-resolved fluorescence spectra of isolated hexameric C-phycocyanin shown in Fig. 2. These spectra have been recorded by single-photon-

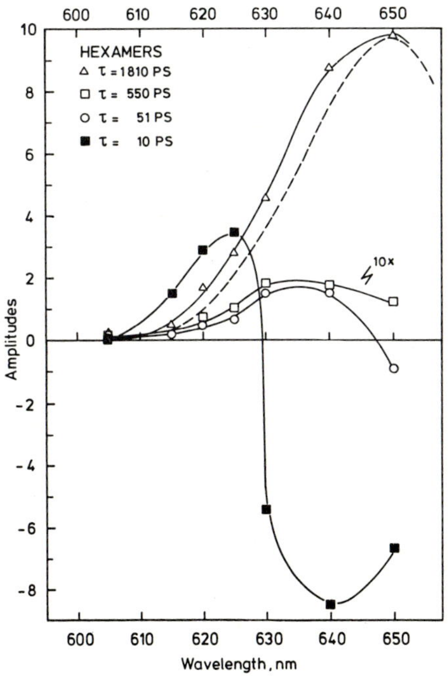

Fig. 2: Time-resolved spectra of the different decay components in hexameric C-phycocyanin from *Synechococcus 6301* as a function of the emission wavelength; λ_{exc} = 590 nm; pH = 5.5, 0.325 M phosphate buffer.

counting with high time-resolution using a channel-plate photomultiplier. For data analysis the technique of simultaneous decay analysis has been used. A sum of four exponentials was required for a good fit. The data shown in Fig. 2 are the amplitudes of the exponential components as a function of the emission wavelength upon excitation at 590 nm. These data clearly show a decay term (positive amplitude) of 10 ps lifetime at short wavelength, which turns into a rise term (negative amplitude) at longer wavelength. This behaviour is clearly indicative of internal s→f-transfer. The dashed curve indicates the steady-state fluorescence spectrum of the same sample. The long lifetime of about 1.8 ns is identical to the lifetime found at short wavelength in PBS of *S 6301* and *AN 112*. This component has been attributed to dissociated PBS rods [12].

The data of the PBS suggest that the transfer between hexamers/trimers along the rods occurs at a rate-constant comparable to fast s→f-transfer ($\leq$ 10 ps per step). The rate-limiting step is given by the transfer between the nearest neighbours of different pigment beds, i.e., phycocyanin/allophycocyanin core in the case of *S 6301*. This picture is close to a trap-limited random walk along a one-dimensional lattice of f-chromophores. Numerical simulations by Mimuro et al. [15] were also in agreement with this model. Theoretical considerations tell us that the number of steps before trapping, and thus the trapping time, should scale with N(N+1) in a one-dimensional array [16] when N is the number of lattice points. This dependence will be slightly modulated, however, by the spectral asymmetry of the rod components mentioned above.

Glazer et al. [17] have taken a different view. They also assumed a one-dimensional lattice, but attributed the rate-limiting step to disk-to-disk transfer. In the theoretical framework of Pearlstein and others, this model would correspond to a diffusion-limited trapping [18]. A strongly non-exponential kinetics is expected to result from a diffusion-limited trapping. Such a non-exponentiality is excluded by our own data, and does not seem to be supported either by the data of ref. [17].

For PBS of *Porphyridium cruentum* [6], as well as for the intact alga [5,19], the fluorescence of the different pigments has been analyzed in terms of a non-exponential $\exp(-2A\sqrt{t})$ transfer kinetics. This decay low was chosen mainly for empirical reasons, because it gave a good fit to the observed fluorescence kinetics with a minimal parameter set. Our studies showed that at the much higher signal/noise ratios that can be obtained by single-photon-timing, as compared to streak cameras, the non-exponential $\exp(-2A\sqrt{t})$ kinetics does not describe the data well [12,14]. Instead, a sum of exponential functions gives both a better fit and enables a more detailed interpretation of the results. The non-exponential kinetics cannot account properly for the energy-transfer cascade in PBS, because it averages over different kinetic components and different processes, which are clearly distinguished by their different associated spectra. Apart from that, the non-exponential kinetics is also not supported by theoretical considerations. The assumptions made for the derivation of the transfer kinetics for Förster energy-transfer in the case of randomly distributed donors and acceptors are not fulfilled in PBS [14,20]. A PBS rod may be described more closely as a linear array. In case of a one-dimensional energy diffusion, the exponent of the time τ would be different from 1/2 [21-23]. The reason why the kinetics for the irreversible transfer between different pigment beds in all PBS which we have studied so far is close to single exponential [12,14,20] is explained by the fact that the transfer is mainly trap-limited [18]. In other words, the transfer steps between trimers or hexamers of the same type of phycobiliprotein are faster than the irreversible transfer step to the next pigment in the series. This last step then becomes rate-limiting.

If a non-exponential function is used to describe the transfer kinetics of PBS, the physically distinct processes of fast $s \rightarrow f$ transfer, the intermediate transfer between pigment beds and the long terminal lifetimes of uncoupled phycobiliproteins are averaged by a single function. Such analysis leaves no room for the differential wavelength dependencies of these processes.

ACKNOWLEDGEMENTS

I like to thank Prof. K. Schaffner for his continuing interest and support of this work and the Deutsche Forschungsgemeinschaft for financial support.

REFERENCES

1 Scheer, H., (1982) in Molecular Biology Biochemistry and Biophysics 35, 7-45, Springer Verlag, Berlin

2 Wang, R.T., Stevens, C.L.R., and Myers, J., (1977) Photochem. Photobiol. 25, 103-108.

3 Wehrmeyer, W., (1983) in Photosynthetic Prokaryotes, pp. 1-22. Elsevier; Amsterdam, New York

4 Glazer, A.N., (1984) Biochim. Biophys. Acta 768, 29-51.

5 Porter, G., Tredwell, C.J., Searle, G.F.W., and Barber, J., (1978) Biochim. Biophys. Acta 501, 232-245.

6 Searle, G.F.W., Barber, J., Porter, G., and Tredwell, C.J., (1978) Biochim. Biophys. Acta 501, 246-256.

7 Koller, K.P., Wehrmeyer, W., and Schneider, H., (1977) Arch. Microbiol. 112, 61-67.

8 Glazer, A.N., (9182) Ann. Rev. Microbiol. 36, 173-198.

9 Gantt, E. and Lipschultz, C.A., (1973) Biochim. Biophys. Acta 292, 858-861.

10 Glazer, A.N., Lundell, D.J., Yamanaka, G., and Williams, R.C., (1983) Ann. Microbiol. 8134, 159-180.

11 Gillbro, T., Sandström, A., Sundström, V., and Holzwarth, A.R., (1983) FEBS Letters 162, 64-68.

12 Suter, G.W., Mazzola, P., Wendler, J., and Holzwarth, A.R., (1984) Biochim. Biophys. Acta 766, 269-276.

13 Gillbro, T., Sandström, A., Sundström, V., Wendler, J., and Holzwarth, A.R., (1985) Biochim. Biophys. Acta 808, 52-65.

14 Wendler, J., Holzwarth, A.R., and Wehrmeyer, W., (1984) Biochim. Biophys. Acta 765, 58-67.

15 Mimuro, M., Yamazaki, T., Yamazaki, I., and Fujita, Y., (1985) Photochem. Photobiol. in print.

16 Montroll, E.W., (1969) J. Math. Phys. 10, 753-765.

17 Glazer, A.N., Yeh, S.W., Webb, S.P., and Clark, J.H., (1985) Science 227, 419-423.

18 Pearlstein, R.M., (1982) Photochem. Photobiol. 35, 835-844.

19 Yamazaki, I., Mimuro, M., Murao, T., Yamazaki, T., Yoshihara, K., and Fujita, Y., (1984) Photochem. Photobiol. 39, 233-240.

20 Holzwarth, A.R., Wendler, J., and Wehrmeyer, W., (1982) Photochem. Photobiol. 36, 479-487.

21 Förster, T., (1949) Z. Naturforsch. 4A, 321-327.

22 Blumen, A., and Zumofen, G., (1982) J. Chem. Phys. 77, 5127-5140.

23 Zumofen, G. and Blumen, A., (1983) Chem. Phys. Lett. 98, 393-397.

Exciton State and Energy Transfer in Bacterial Membranes: The Role of Pigment-Protein Cyclic Unit Structures

R.M. Pearlstein

Physics Department, Indiana-Purdue University, 1125 East 38th Street, P.O. Box 647, Indianapolis, IN 46223, USA

H. Zuber

Institut für Molekularbiologie und Biophysik, Eidgenössische Technische Hochschule Zürich, CH-8093 Zürich, Switzerland

1. Introduction

Some of the recent findings on the structures of antenna pigment-protein complexes from a variety of photosynthetic organisms [1-7] invite a re-evaluation of the mechanism of electronic excited-state energy-transfer in these structures, and in the larger assemblies containing them. The clustering of pigments in these structures, and their overall rotational symmetries, have no obvious purpose in a Förster model of energy-transfer, but may play specific roles if the transfer mechanism is more coherent. The question of the role of these rotationally symmetric structures in the handling of electronic excitation energy is addressed here, in an admittedly somewhat speculative fashion, specifically for the antenna assemblies of purple bacterial membranes.

In section 2, the relation between antenna structure and energy-transfer by the Förster mechanism is discussed, and the relative insensitivity of this mechanism to structural details is noted. Section 3 summarizes the recent findings on antenna cyclic unit structures, and includes the results of calculations on the exciton energy levels of idealized structures having the same rotational symmetries as the actual ones. It is shown that the exciton angular momentum may be inverted (with respect to the order of exciton energy levels) in bacterial membrane structures. In section 4, the consequences of exciton angular momentum inversion are explored. It is argued that with greatly augmented exciton coherence times, which may also be a consequence of the protein structure, the momentum inversion might lead to singlet exciton storage and require a non-Förster mechanism for inter-unit-structure energy-transfer. The final section summarizes the results and presents concluding arguments.

2. Förster Transfer and the Arrangement of Pigments in a Photosynthetic Antenna

Theory suggests that energy-transfer by virtually any mechanism, including that of Förster, is in general more efficient in a regular structure than in an irregular one [8]. (By a "regular structure" is meant one in which the placement and orientation of pigment molecules is governed by definite rules of symmetry.) This view is in accord with the many recent analyses, which conclude that both intramembranal [1-6, 9-12] and extramembranal [7, 13, 14] pigment-protein unit structures are highly ordered, both internally and in their arrangements relative to one another. The Förster mechanism may be unique in that it appears to be rather insensitive to the details of the regular structure, at least in two dimensions [15]. Because, for various reasons, Förster transfer is frequently assumed to be the principal, if not the exclusive, means by which absorbed light energy ultimately migrates to reaction centers [8, 15-17], this point bears close examination.

KNOX [15] has clearly illustrated this structural insensitivity of Förster transfer with the example of 120 pigment molecules and an excitation-energy trap on a hypothetical square lattice. He showed that there is no significant difference in either the average trapping time, or the trapping time as a function of starting distance from the trap, in either of two structural situations: In one, the pigments are equally spaced, arrayed so as to occupy the individual lattice sites. In the other, the pigments are regularly clustered in groups of five (three at the corners), so that nearest neighbor distances are smaller within a cluster than between any two clusters, but such that the average pigment density is the same as in the first case. Knox (personal communication) conjectured, plausibly, that this result is general, i.e., as long as pigment density and two-dimensional lattice type (in his example, an isotropic square lattice) remain constant, diffusion-limited times for trapping of Förster-transferred excitation energy, also remain constant.

Knox's conjecture implies that many types of pigment arrangements would be compatible with the assumption of Förster transfer in a given pigment system. However, conversely, it leaves completely unexplained the fact of pigment clustering in the pigment-protein complexes of intramembranal photosynthetic antennas [1-6, 9, 11], which are fundamentally two-dimensional arrays [10, 12] of pigment molecules. Thus, if one accepts Knox's conjecture, and rejects the notion of "accidental" complexity in these antennas, one must consider the possibilities that intramembranal antennas have some function in addition to light absorption and (passive) energy transfer, the transfer does not proceed exclusively by the Förster mechanism, or perhaps both.

3. Cyclic Unit Structures and Their Characteristic Exciton Energy-Level Patterns

As a starting point for the consideration, on structural grounds, of the functions of photosynthetic antenna components and the mechanisms of energy-tranfer within and among them, let us examine the features of purple bacterial membrane complexes on the basis of a proposed structural model [5]. The model includes a pigment-protein unit structure that exhibits three features: (1) Pairing of polypeptides, with pair members designated as α and β, that are specifically but noncovalently linked. (2) An associated close pairing of pigments, one bound to each polypeptide. (3) Symmetric cyclic assemblies of the polypeptide pairs and their associated pigments. In bacterial membranes, each polypeptide has a central hydrophobic, alpha-helical region that spans the membrane. The pigments are bacteriochlorophylls that are either directly paired, i.e., within each α-β polypeptide pair, or cross paired, i.e., the pigment of one α-polypeptide is closest to the pigment of a β-polypeptide not paired to that α. The cyclic assembly is a hexameric (B800/850 complex) or dodecameric (B870/890 complex -- or B1015 in Rhodopseudomonas viridis) ring of α-β polypeptide pairs and their associated (noncovalently linked) bacteriochlorophylls. These cyclic structures, with their reaction-center cores (B870/890 or B1015), form regular two-dimensional arrays within the membranes. To the β-polypeptides of the B800/850 complex there is bound a second pigment of the same type that is generally assumed to interact more weakly with the remaining pigment system, and so is left out of consideration here.

With this precis of antenna composition, one can proceed to a consideration of the interactions among the pigments in each unit structure that fundamentally cause excited-singlet-state energy-transfer. In the simplest approximation, these are the Coulomb interactions between the differential charge distributions in the ground and lowest-excited-singlet states of the pigments, usually treated as point dipoles. The dipole-dipole interactions produce a splitting of the excited singlet level, leading to a set of quasi-stationary exciton states [8]. To describe the exciton states in detail requires precise information regarding transition dipole strengths and orientations, and inter-dipole (inter-pigment) distances. For the membrane complexes, the first and third of these are known approximately, but the second is more uncertain. For this reason, it is not worthwhile in this early description to attempt a complete analysis of the exciton states. However, it is possible to draw some useful conclusions from idealized structural models that incorporate the symmetry properties of the detailed structural models summarized above, and this is done here.

Consider first a set of 6 pigment molecules whose transition dipoles lie at the vertices of a regular hexagon. This model is too simple to represent even the B800/850 complex, but nonetheless demonstrates an important point: If the dipoles are symmetrically oriented, absorption into most of the exciton energy levels is symmetry-forbidden. Results for two examples with differing orientations are given in Table 1. (Formulas used for the calculation of exciton energies, total absorption, circular dichroism, and polarized absorptions are given in [8].) If the dipoles are oriented along the sides of the hexagon, it is seen (Table 1a) that only the two degenerate exciton levels just above the lowest level have any dipole strength for absorption. Conversely, for the geometry in which all the molecular dipoles point radially inward, it is the two degenerate levels just below the highest level that do all the absorption (Table 1b). Actually, these results can be viewed as a straightforward consequence of angular momentum conservation. In these geometries, exciton angular momentum parallel to the axis of rotational symmetry becomes a good quantum number and only those exciton states that have angular momentum = 1 (in units of Planck's constant h divided by 2π) can conserve the unit angular momentum of the photon upon absorption. In a geometry like that of Table 1a, where all of the dipole-dipole energies J are negative (attractive inter-

Table 1. Energies (relative to the unperturbed singlet transition energy), total dipole strengths (A), rotational strengths (CD), and dipole strengths polarized parallel ($A_{||}$) and perpendicular ($A_{\perp}$), to the rotational symmetry axis, that result from resonant transition dipole-transition dipole (exciton) interactions:

(a) For 6 dipoles oriented along the sides of a regular hexagon of radius R = 1.0 nm, each dipole having (squared) strength = $55D^2$

Exciton Energy (cm^{-1})	A	CD	$A_{\|\|}$	$A_{\perp}$
		(Relative Units)		
507.5	0	0	0	0
293.9	0	0	0	0
293.9	0	0	0	0
-240.8	0.5	0	0	0.25
-240.8	0.5	0	0	0.25
-613.7	0	0	0	0

(b) Same as (a), but with the 6 dipoles pointing radially inward

Exciton Energy (cm^{-1})	A	CD	$A_{\|\|}$	$A_{\perp}$
		(Relative Units)		
945.0	0	0	0	0
183.0	0.5	0	0	0.25
183.0	0.5	0	0	0.25
-369.0	0	0	0	0
-369.0	0	0	0	0
-573.1	0	0	0	0

actions), the angular momenta of the exciton states increase in more-or-less the same order as the energies of the states. For a geometry like that of Table 1b, i.e. positive or repulsive dipole-dipole interactions, the exciton angular momenta <u>decrease</u> as the exciton energies increase. The latter phenomenon, which may be termed "angular momentum inversion", is completely analogous to the well-known inversion of energy versus linear momentum in one-dimensional models of elementary excitations in crystals in the presence of repulsive interactions [18].

Exciton angular momentum remains a good quantum number in the case of 12 molecular transition dipoles placed at the vertices of a regular dodecagon. Exciton energy levels and absorptions for this geometry are given in Table 2a for dipoles directed radially inward, and (unlike Table 1) for polygon radius = 3.0 nm, a value consistent with the B800/850 structural model. Complete angular momentum inversion is, of course, maintained, while the energy of each absorbing exciton state is a greater fraction of the energy of the highest exciton level than in the case of the hexagon. (Absolute energy gaps are smaller, because the dipole-dipole separations are greater in <u>this</u> dodecagon.) The dodecagon can also be thought of as a hexagon of pairs of dipoles, as in the B800/850 model. However, the "paired hexagon" whose exciton levels are listed in Table 2a is probably too symmetric to represent B800/850 very well -- the J for an $\alpha-\beta$ pair equals that between the α-dipole and the β of the next pair. Table 2b gives the exciton properties of a paired hexagon in which the members of each pair are only 1.00 nm apart, rather than 1.55 nm as in the symmetric paired hexagon. Here, the energy of the absorbing states is even closer, fractionally, to the top of the exciton band, and the overall gap from the absorbing states to the bottom of the band is large in absolute terms, about 605 cm^{-1} or 2.9 k_BT at T = 300 K. However, in this asymmetric paired hexagon the exciton angular momentum operator is no

Table 2. Exciton energies and stick spectra (see Table 1 heading for definitions) for hexamers of dipole pairs:

(a) Symmetric case (regular dodecamer). Squared dipole strength = 55 D^2 for the monomer transitions, hexagon radius = 3.0 nm, nearest neighbor dipole separation = 1.55 nm; all dipoles directed radially inward

Exciton Energy (cm^{-1})	A	CD (Relative Units)	$A_{\parallel}$	$A_{\perp}$
209	0	0	0	0
138	0.500	0	0	0.250
138	0.500	0	0	0.250
57	0	0	0	0
57	0	0	0	0
-21	0	0	0	0
-21	0	0	0	0
-84	0	0	0	0
-84	0	0	0	0
-125	0	0	0	0
-125	0	0	0	0
-159	0	0	0	0

(b) Asymmetric case. Nearest neighbor dipole separation = 1.00 nm; other parameters as in (a)

Exciton Energy (cm^{-1})	A	CD (Relative Units)	$A_{\parallel}$	$A_{\perp}$
369	0	0	0	0
306	0.492	0	0	0.246
305	0.492	0	0	0.246
250	0	0	0	0
249	0	0	0	0
226	0	0	0	0
-269	0	0	0	0
-277	0	0	0	0
-277	0	0	0	0
-292	0.008	0	0	0.004
-292	0.008	0	0	0.004
-299	0	0	0	0

longer completely compatible with the dipole-dipole Hamiltonian, and as a result, two of the lowest-lying levels are slightly absorbing (less that 1% each).

While the asymmetric paired hexagon of Table 2b is reasonably realistic for B800/850 with regard to dipole strengths, inter-dipole distances, and overall symmetry, it is uncertain as to how well it represents the actual pigment dipole orientations. Other possibilities cannot be excluded, including orientational situations, for which there is no momentum inversion. Thus, we can at present only raise the possibility that exciton angular momentum inversion occurs in B800/850, or in the larger B870/890 or B1015 complexes. It has yet to be demonstrated experimentally. Moreover, it is by no means clear that other exciton properties of these structures are unimportant physiologically. Nonetheless, the momentum inversion concept is interesting, and is discussed further below.

4. Exciton Properties, Energy-Transfer, and Antenna Function

Exciton-level-structure calculations are useful for coordinating theory with experimental spectra and pigment-protein structural models. Whether the exciton-level-structure also has significance for antenna function, specifically energy transfer, depends on the degree of exciton coherence -- the persistence over time of definite phase relations among the individual localized terms in the superposition that constitutes each exciton wave function. If these phases become randomized in a time short compared to the time-scale for exciton migration and trapping, then the level-structure has minimal influence on the energy-transfer which in that case would proceed by the Förster mechanism [8].

One argument regarding exciton coherence is that its duration is of the order of Planck's constant divided by the width, in energy units, of the absorption band [19]. However, two of the assumptions that underlie this argument are now seen to be incorrect. The first is the assumption that the absorption band is homogeneous, i.e., the width encompasses the range of vibrationally-coupled states accessible to each pigment molecule. Photochemical hole-burning experiments have shown this argument to be invalid, at least for the phycobiliproteins [20, 21]. Indeed, the intrinsic, or homogeneous, linewidths of the photosynthetic pigments may be many orders of magnitude smaller than the observed, inhomogeneous, widths of the absorption bands (S. Voelker, personal communication). The second assumption is the idea that every interaction of the exciton with a vibrational quantum (phonon) is phase-randomizing for the exciton wave function. A new theory [22], the first to be inherently correct thermodynamically for the exciton-phonon interaction, shows that the dynamics of this interaction is much less randomizing than previously supposed. We conclude that the degree of exciton coherence in a photosynthetic pigment-protein structure remains to be determined, both theoretically and experimentally.

If, for the sake of argument, one assumes that the exciton coherence time can be long in a photosynthetic system, say at least as long as the exciton lifetime, it becomes meaningful to ask what effects angular momentum inversion might have long after absorption. One possible answer is the following: After initial absorption into an allowed, high-energy, exciton level, the exciton relaxes (presumably through sequential phonon-mediated processes) into the lower-lying exciton levels, according to the Boltzmann distribution. Because these levels have no dipole strength, they cannot radiate any more than they can absorb, i.e., the radiative lifetime of the exciton is greatly increased. (The magnitude of this effect depends in a complicated way on details of the exciton-phonon interaction.) If, as has been shown for the phycobiliproteins [13, 23], it is generally true for photosynthetic pigment-protein complexes that the quantum yields for deleterious nonradiative processes such as internal conversion and intersystem crossing are low, the increased radiative lifetime implies an overall substantial enhancement of exciton lifetime in the absence of energy-transfer and trapping -- provided only that the cyclic unit structure remains intact. This raises the possibility of exciton "storage", i.e., the stabilization of the exciton by the antenna structure for a time much longer than is necessary to trap and convert the energy (to that of a charge-separated state) at the reaction center. However, at present, one can only speculate as to the utility to a photosynthetic bacterium of exciton storage.

With long coherence times and angular momentum inversion, the lack of dipole strength in the low-lying exciton transitions (assumed populated) rules out the usual resonance energy-transfer mechanisms, either coherent or incoherent [8]. Exchange mechanisms are also unlikely, because of the large angular momentum differences and the inter-pigment distances involved. Of course, the excitation energy in a definite exciton state is already delocalized over the (central) pigments of a cyclic unit structure. The question is one of inter-unit-structure-transfer. A possible mechanism is transient exciton coupling, whereby two adjacent unit structures, initially uncoupled or very poorly coupled, may become strongly coupled for a brief time, after which they again become uncoupled. During the time of strong coupling, the exciton is delocalized over the combined system (two unit structures), and afterwards the exciton has a reasonable probability of being transferred from one to the other structure. Repetition of this process constitutes an inter-unit "random walk" of the excitation energy, as in the case of Förster transfer, but different in detail.

5. Conclusions

Emerging structural details of photosynthetic pigment-protein complexes, while consistent with the requirements of inter-pigment excitation energy-transfer by the Förster mechanism, are more complex than is necessary for this process. The existence of local 3-, 6-, and 12-fold rotational symmetries in these complexes that are not necessarily also translational symmetry elements (i.e., do not occur simply as a consequence of the point rotation group of a lattice structure with long-range order), are particularly puzzling in this respect. A possible explanation for the occurrence of such "cyclic unit structures" may be given if one assumes that the exciton states, which arise from the mutual interactions of the transition electric dipole moments of the individual pigment molecules, maintain the phase coherence of their (delocalized) wave functions for a much longer period of time than is customarily assumed. Such an assumption may be justified on the basis of recent findings, both theoretical and experimental [20-22]. From the apparent symmetry properties of the bacterial membrane cyclic unit structures [1-7, 10-12], one concludes that exciton angular momentum is a good quantum number, or nearly so, at least for a short time after exciton creation. With the exciton-coherence assumption, one further concludes that exciton angular momentum may be significant on a longer time-scale, including that of inter-unit-structure energy-transfer. In particular, depending on presently unknown details of bacteriochlorophyll binding to the protein, it is possible for angular momentum inversion to occur, i.e. the exciton state of highest angular momentum lies at lowest energy. Because none of the higher angular momentum states has any transition dipole strength, momentum-inverted excitons are particularly stable against radiation, and if deleterious nonradiative decay channels are also suppressed by the protein structure (as is certainly the case in phycobiliproteins [13]), momentum-inverted excitons would become excitation energy storage states. Inter-unit-structure energy-transfer would then require a mechanism such as transient exciton coupling rather than the Förster mechanism. The possible advantage to the organism of exciton storage is not obvious, but may be more related to photoprotection than to energy-transfer.

6. Acknowledgments

R.M.P. wishes to thank Profs. R. S. Knox, N. E. Geacintov, A. N. Glazer, and S. Voelker for valuable discussions. This work was supported by the Swiss National Science Foundation (project No. 3.286-0.82), by the ETH Zürich, and by IUPUI.

7. References

1. R. A. Brunisholz, P. A Cuendet, R. Theiler, and H. Zuber: FEBS Lett. 129, 150 (1981)
2. R. A. Brunisholz, F. Suter, and H. Zuber: Hoppe-Seyler's Z. Physiol. Chem. 365, 675 (1984)
3. R. Theiler and H. Zuber: Hoppe-Seyler's Z. Physiol. Chem. 365, 721 (1984)
4. H. Zuber: In Optical Properties and Structure of Tetrapyrroles, edited by G. Blauer and H. Sund (Walter de Gruyter and Co., Berlin 1985) p. 425
5. H. Zuber, W. Sidler, P. Füglistaller, R. Brunisholz, and R. Theiler: In Biology of the Photosynthetic Apparatus, edited by C. Arntzen, L. Bogorad, S. Boritz, and K. Steinback (Cold Spring Harbor Symposia, in press)
6. P. A. Loach, P. S. Parkes, J. F. Miller, S. Hinchigeri, and P. M. Callahan: In Biology of the Photosynthetic Apparatus, edited by C. Arntzen, L. Bogorad, S. Boritz, and K. Steinback (Cold Spring Harbor Symposia, in press)

7. T. Schirmer, W. Bode, R. Huber, W. Sidler, and H. Zuber: J. Mol. Biol. (in press)
8. R. M. Pearlstein: In Photosynthesis: Energy Conversion by Plants and Bacteria, Vol. 1, edited by Govindjee (Academic, New York 1982) p. 293
9. J. M. Kramer, R. van Grondelle, C. N. Hunter, W. H. J. Westerhuis, and J. Amesz: Biochim. Biophys. Acta 765, 156 (1984)
10. W. Stark, W. Kühlbrandt, I. Wildhaber, E. Wehrli, and K. Mühlethaler: EMBO J. 3, 777 (1984)
11. D. Gulen and R. S. Knox: Photobiochem. Photobiophys. 7, 277 (1984)
12. W. Kühlbrandt: Nature 307, 478 (1984); J. Li: Proc. Natl. Acad. Sci. USA 82, 386 (1985)
13. A. N. Glazer: Biochim. Biophys. Acta 768, 29 (1984)
14. J. M. Olson, P. D. Gerola, and G. H. van Brakel: In Antennas and Reaction Centers of Photosynthetic Bacteria -- Structure, Interactions and Dynamics, edited by M. E. Michel-Beyerle (Springer-Verlag, Berlin, in press)
15. R. S. Knox: In Topics in Photosynthesis, Vol. 2, edited by J. Barber (Elsevier, Amsterdam, 1977) p. 55
16. R. M. Pearlstein: Photochem. Photobiol. 35, 835 (1982)
17. R. M. Pearlstein: In Advances in Photosynthesis Research, Vol. I, Part 1, edited by C. Sybesma (Martinus Nijhoff/Dr. W. Junk, The Hague, 1984) p. 13
18. C. Kittel: Introduction to Solid State Physics, 4th edition (John Wiley and Sons, New York, 1971) p. 293
19. V. M. Kenkre and R. S. Knox: Phys. Rev. Lett. 33, 803 (1974)
20. J. Friedrich, H. Scheer, B. Zickendraht-Wendelstadt, and D. Haarer: J. Chem. Phys. 74, 2260 (1981)
21. J. Friedrich, H. Scheer, B. Zickendraht-Wendelstadt, and D. Haarer: J. Am. Chem. Soc. 103, 1030 (1981)
22. B. J. West and K. Lindenberg: J. Chem. Phys. (in press)
23. J. Grabowski and E. Gantt: Photochem. Photobiol. 28, 39 (1978)

Carotenoid-Bacteriochlorophyll Interactions

R.J. Cogdell

Department of Botany, University of Glasgow, Glasgow G12 8QQ, U.K.

Carotenoids perform two major functions in bacterial photosynthesis [1,2]. They act as photoprotective agents, preventing the harmful photodynamic reaction, and as accessory light-harvesting pigments, extending the spectral range over which light drives photosynthesis. Most bacterial carotenoids are rather hydrophobic molecules, and they are typically found associated with photosynthetic membranes. However, they are not freely mobile within the lipid interior of these membranes, but are rather non-covalently bound to the photochemical reaction centres and the light-harvesting complexes [3]. It is important to emphasise that it is the specific, close, well-ordered interactions between the carotenoids and the bacteriochlorophyll molecules within these pigment-protein complexes that allows the carotenoids, efficiently, to fulfil their photoprotective and light-harvesting roles.

Figure 1 compares the relative energy-levels of the first excited singlet and triplet states of both bacteriochlorophyll a and a typical bacterial carotenoid, spheroidene, which has ten conjugated double bonds. The most striking feature depicted in Fig. 1 is that on the one hand the carotenoid has a relatively energetic first excited singlet state, while on the other it has a rather low-lying first excited triplet state. The functional

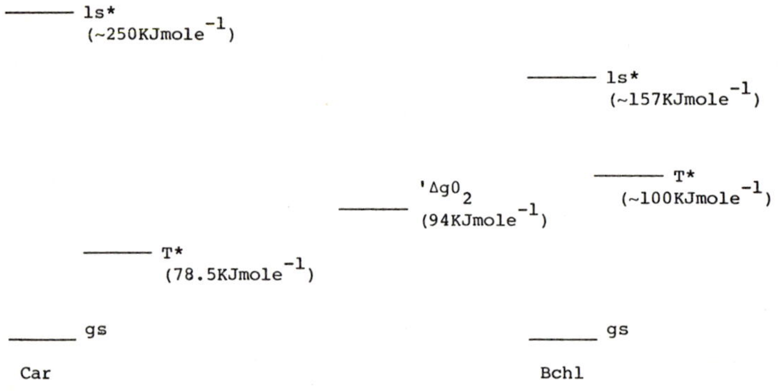

Fig. 1 The energy levels of the 1st excited singlet and triplet states of the carotenoid, spheroidene, and bacteriochlorophyll a

gs - ground state
1s* - 1st excited singlet state
T* - 1st excited triplet state
The value for the bacteriochlorophyll triplet state is only approximate

consequence of this is that the carotenoid's first excited singlet state is energetically able to donate energy to the bacteriochlorophyll's first excited singlet state (this is the antenna function), while the low-lying carotenoid triplet state is able to quench both triplet bacteriochlorophyll and singlet oxygen (this is the photoprotective function).

In this short paper, due to the limited space available, I shall concentrate upon the antenna role of the carotenoids in the light-harvesting complexes.and use this to illustrate the main features of how carotenoids and bacteriochlorophylls are thought to interact *in vivo*. Singlet-singlet energy-transfer from carotenoids to bacteriochlorophylls within isolated, purified antenna complexes is rather easy to demonstrate experimentally, by sensitised fluorescence [cf. 4]. The efficiency of this process is quite variable, but can in some cases approach 100% (Table 1). It is also worth noting that, where it has been investigated, this singlet-singlet energy-transfer has been shown to be temperature-independent, even down to 4 K[5,6]. Although the antenna function of carotenoids has been recognised for over thirty years,the details of the molecular mechanism by which this process occurs still remains to be defined.

In organic solution carotenoids are essentially non-fluorescent,and show no significant degree of direct intersystem crossing to the triplet state [7,8]. Indeed,their excited state lifetime has been estimated to be no greater than 1 psec [8,9]. These findings would seem to make singlet-singlet energy-transfer by a conventional Förster (long-range) dipole-dipole exchange reaction extremely unlikely. If this is a correct analysis then what possible molecular mechanisms remain that can explain the efficient carotenoid to bacteriochlorophyll singlet-singlet energy-transfer reaction? So far there have been two alternative reaction mechanisms proposed [10,11]. THRASH et al. [10] have suggested that energy-transfer may proceed by way of a forbidden, low-lying excited singlet state (the ^{1}Ag state). This

Table 1 A comparison of the efficiences of carotenoid to bacteriochlorophyll singlet-singlet energy-transfer in some isolated antenna complexes

Type of Complex	Energy Transfer Efficiency	Ref.
Rps. sphaeroides B800-850-complex	≥ 90%	4.5
Rps. acidophila B880-complex	~ 25%	6
Rps. acidophila type I B800-850-complex	50-55%	6
Rps. acidophila type II B800-850-complex	70-75%	6
Rps. acidophila B800-820-complex	70-75%	6

state has been studied in shorter polyenes by biphotonic spectroscopy and its energy-level in carotenoids with longer than five conjugated double bonds has been calculated [8 and SCHULTEN and co-workers, personal communication]. Even though this is a 'low-lying' excited singlet state, it is still sufficiently energetic to serve as a potential energy donor to bacteriochlorophyll a. However, since the cross-section for absorption and fluorescence of the ^{1}Ag state can be expected to be rather small (it is after all a forbidden state), if this mechanism is to operate efficiently then the distance between the donor carotenoid and acceptor bacteriochlorophyll a molecule must be very small i.e. less than 7 Å [7,8]. A second possible mechanism has been proposed by Ravi-Naqvi [11]. This author has suggested that the energy-transfer could be mediated by an electron-exchange mechanism. Interestingly, it is also a strict requirement of this second, proposed mechanism that the carotenoid and bacteriochlorophyll molecules must be in very closs proximity to each other. Possibly even so close that their respective Π electron clouds interact directly. This requirement for a close apposition of the donor carotenoid to its acceptor in order for singlet-singlet energy-transfer to occur has been elegantly demonstrated by DIRKS et al. and MOORE et al. [12,13]. These workers have synthesised a range of covalently linked carotenoporphyrin dimers. In these model compounds, singlet-singlet energy-transfer from the carotenoid to the porphyrin only occurs when the two molecules are held in very close contact i.e. when the carotenoid is caused to fold back over the porphyrin plane so that the two Π systems interact directly.

It must now be asked whether there is any direct experimental evidence for the close ordering of the pigments within the isolated antenna complexes, and then whether it is possible to try and test the two proposed mechanisms of the singlet-singlet energy-transfer process.

It is now well documented, based upon a variety of experimental observations, that carotenoids are bound into the antenna complexes in a highly specific and well-ordered fashion. When antenna complexes are isolated and purified, analysis of their composition reveals that there is a fixed ratio of carotenoids to bacteriochlorophylls present. A typical antenna complex consists of a small number of bacteriochlorophyll and carotenoid molecules non-covalently bound to two low-molecular weight, hydrophobic apoproteins [14]. In the B890-complex from Rhodospirillum rubrum, for example, the carotenoid:bacteriochlorophyll a ratio is 1:2 [15,16] while in the B875-complex from Rhodopseudomonas sphaeroides it is 1:1 [17]. When the carotenoid is bound to the antenna complex, its binding is accompanied by a well-defined red-shift in its absorption spectrum [18]. Table 2 illustrates this point. Binding of the carotenoid to the antenna complex also induces optical activity [3]. Most bacterial carotenoids are symmetric molecules and therefore optically inactive in their main absorption bands in the visible region. However, when the carotenoids are bound to the antenna complex, a large induced CD-signal is seen in the visible absorption bands, indicating a rather specific, pigment-protein interaction within the complex. Denatration of an antenna complex reverses both the red-shift in the carotenoid's absorption spectrum and the induced CD-signal. However, perhaps the most dramatic demonstration that the carotenoids are well-ordered in the antenna complexes comes from recent studies upon crystals of an antenna complex. If crystals of the B800-850-complex from Rhodopseudomonas acidophila strain 7750 are viewed under a polarising microscope they show a high degree of dichroism (see Fig. 5 in [19]. The colour of the crystals seen by eye is mainly due to the strong visible absorption bands of the carotenoids. Therefore, this

Table 2 The absorption maxima of spheroidene in different environments [18]

Environment	Absorption Maxima (nm)		
Petroleum Spirit (40-60°)	427	452	482
Phosphatidylcholine	436	464	497
Bovine Serum Albumen	440	466	500
Rps. sphaeroides [(a)] B800-850-complex	450	475	510

(a) Only in the case of the antenna complex does spheroidene show any CD-signals in the visible absorption bands

strong dichroism shows directly that the carotenoids are arranged, within the antenna complex, in a highly ordered way. While the ordering of the carotenoids within the antenna complexes is quite clear, it must be said quite openly that, at present, there is no data that indicates how closely packed the carotenoids and bacteriochlorophyll molecules are within the antenna complexes. However.crystals of antenna complexes have now been obtained by three groups [19-21], and it is to be hoped that, soon, a high-resolution structure determined by X-ray diffraction will be able to answer this point. There is, though, one piece of evidence that suggests that in some complexes the bacteriochlorophyll molecules are able to 'sense' the presence of the carotenoids [22]. There is a well known carotenoidless mutant of R. rubrum called $G9^+$. The antenna complex from this mutant is identical to that from the wild-type strain,except that it lacks the carotenoid, spirilloxathin. When the absorption spectra of the two complexes are compared in the bacteriochlorophyll Qy absorption bands in the NIR, it can be seen that the wild-type complex absorbs ~10 nm further to the red than the mutant complex. The origin of this extra red-shift in the presence of the carotenoid is not clear. It could be due to a direct carotenoid-bacteriochlorophyll interaction, or an indirect interaction due to a change in protein structure on binding the carotenoid. However,it does indicate some measure of carotenoid-bacteriochlorophyll interaction.

Now to the second question. Is it possible to test the two proposed mechanisms of the carotenoid to bacteriochlorophyll singlet-singlet energy-transfer by direct experiment? This energy-transfer process must be a very fast process,and it will therefore only be amenable to study by psec and sub-psec spectrophotometry. There is one experimental system that looks to be the best first choice for this type of study. Four different types of antenna complex can be isolated and purified from Rps. acidophila [23]. These complexes show a range of efficiences for the carotenoid to bacteriochlorophyll singlet-singlet energy-transfer (see Table 1). It is to be hoped that this range of efficiences will be reflected in varying rate-constants for the energy-transfer process,and that these will be able to be measured (both by ground-state recovery of the carotenoid absorptions and by monitoring the rise time for bacteriochlorophyll fluorescence). If this can be achieved, then the involvement of carotenoid radicals can be looked for to see if the electron-exchange mechanism operates.

References

1 M. Griffith, W.R. Sistrom, G. Cohen-Bazire and R.Y. Stanier: Nature, Lond. 176, 1211 (1955)
2 R.J. Cogdell: Pure and Applied Chem. 57, 723 (1985)
3 R.J. Cogdell: Phil. Trans. R. Soc. Lond. B. 284, 569 (1978)
4 R.J. Cogdell, M.F. Hipkins, W. MacDonald and T.G. Truscott: Biochim. Biophys. Acta 634, 191 (1981)
5 R. van Grondelle, H.J.M. Kramer and C.P. Rijgersberg: Biochim. Biophys. Acta 682, 208 (1982)
6 A. Angerhofer, R.J. Cogdell and M.F. Hipkins: Biochim. Biophys. Acta in press
7 P.-S. Song and T.A. Moore: Photochem. Photobiol. 19, 435 (1974)
8 B.S. Hudson, B.E. Kohler and K. Schulten: Excited States 6, 1 (1982)
9 R.F. Dallinger, W.H. Woodruff and M.A.J. Rodgers: Photochem. Photobiol. 33, 275 (1981)
10 R.J. Thrash, H.L.B. Fang and G.E. Leroi: J. Chem. Phys. 67, 5930 (1978)
11 K. Ravi-Naqvi: Photochem. Photobiol. 31, 523 (1980)
12 G. Dirks, A.L. Moore, T.A. Moore and D. Gust: Photochem. Photobiol. 32, 277 (1980)
13 A.L. Moore, G. Dirks, D. Gust and T.A. Moore: Photochem. Photobiol. 32, 691 (1980)
14 J.P. Thornber, R.J. Cogdell, B.K. Pierson and R.E.B. Seftor: J. Cell. Biochem. 23, 159 (1983)
15 R.J. Cogdell, J.G. Lindsay, J. Valentine and I. Durant: FEBS Lett. 150, 151 (1982)
16 R. Picorel, G. Bélanger and G. Gingras: Biochemistry 22, 2491 (1983)
17 R.M. Broglie, C.N. Hunter, P. Delepelaire, R.A. Niederman, N.-H. Chua and R.K. Clayton: Proc. Natl. Acad. Sci. U.S.A. 77, 87 (1978)
18 E. Davidson and R.J. Cogdell: Biochim. Biophys. Acta 635, 295 (1981)
19 R.J. Cogdell, K. Woolley, R.C. Mackenzie, J.G. Lindsay, H. Michel, J. Dobler and W. Zinth: this volume
20 W. Welte, T. Wacker, M. Leis, W. Kreutz, J. Shiozawa, N. Gad'on and G. Drews: FEBS Lett. 182, 260 (1985)
21 J.P. Allen, R. Theiler and G. Feher: Biophys. J. 47, 4a (1985)
22 J.P. Thornber, T.L. Trosper and C.E. Strouse: In Clayton R.K. and Sistrom W.R. (eds.): 'The Photosynthetic Bacteria' Plenum Press, New York and London pp. 133-160 (1978)
23 R.J. Cogdell, I. Durant, J. Valentine, J.G. Lindsay and K. Schmidt: Biochim. Biophys. Acta 722, 427 (1983)

Bacteriochlorophyll *a*- and *c*-Protein Complexes from Chlorosomes of Green Sulfur Bacteria Compared with Bacteriochlorophyll *c* Aggregates in CH_2Cl_2-Hexane

J.M. Olson, P.D. Gerola, and G.H. van Brakel*

Institute of Biochemistry, Odense University, Campusvej 55, DK-5230 Odense M, Denmark

R.F. Meiburg and H. Vasmel

Department of Biophysics, Huygens Laboratory, State University, P.O. Box 9504, NL-2300 RA Leiden, The Netherlands

1. Bacteriochlorophyll c--Binding Proteins

The two groups of green bacteria, Chlorobiaceae and Chloroflexaceae, contain bacteriochlorophyll (BChl) c or a related pigment as the main light-harvesting pigment [1]. In both groups all of the BChl c is located in chlorosomes, oblong bodies appressed to the inner surface of the cytoplasmic membrane as shown in Fig. 1. Inside the chlorosome are proteinaceous structures called rod elements which are thought to be made up of BChl c--binding proteins. These BChl c--binding proteins have recently been extracted from chlorosomes of Chloroflexus aurantiacus (Chloroflexaceae) [2] and Chlorobium limicola f. thiosulfatophilum (Chlorobiaceae). The Chloroflexus protein contains 51 amino acid residues, of which 4 are Asn and 4 are Gln. Similarly the Chlorobium protein contains ca. 55 residues, of which ca. 9 are Asn + Gln. Neither protein contains any Cys. The Chloroflexus protein contains no Lys and only one His, while the Chlorobium protein appears to contain little or no Arg, Trp, Tyr and His. From the amino acid sequence of the Chloroflexus protein it has been deduced that the protein can form an α-helix in which the amide

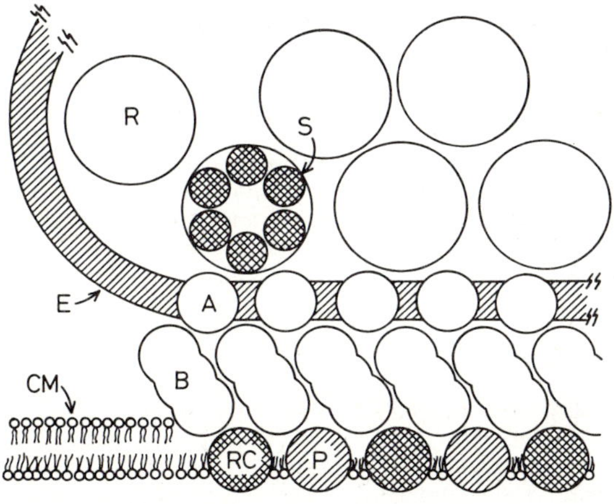

Fig. 1. Model of a chlorosome and associated cytoplasmic membrane (CM) in Cb. limicola. A = BChl a--anchor protein, B = BChl a--baseplate protein, E = chlorosome envelope, P = pigment complex, R = rod element, RC = reaction center complex, S = 15 kd subunit = BChl c--antenna protein.

* Present address: Sezione Fisiologia Vegetale-Fotosintesi, Dipartimento di Biologia, Università di Milano, Via Celoria 26, Milano (Italy)

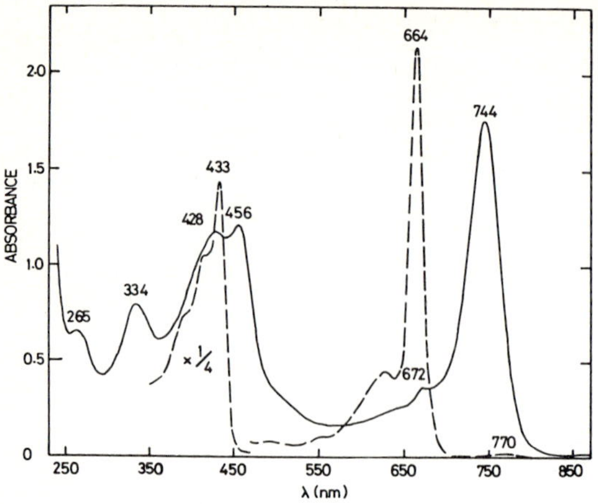

Fig. 2. Chlorosome fraction (——) prepared in 2 M Na SCN and its 90% methanol extract (---).

groups (Asn and Gln) are located predominantly on one side of the helix. WECHSLER et al. [2] have proposed that 7 BChl c molecules are bound to each α-helix via interactions with 7 amide groups on the protein. Although the Chlorobium protein apparently shows little homology with the amino acid sequence of the Chloroflexus protein, we suppose that BChl c might be bound to the Chlorobium protein by a similar interaction. However, we do not yet know the complete sequence and probable conformation of the polypeptide.

The absorption spectrum of chlorosomes from Cb. limicola f. thiosulfatophilum (Fig. 2) is very similar to that of chlorosomes from Cf. aurantiacus [3], but the circular dichroism (CD) spectra for the Q_y-bands are sign-reversed in the two species. In Chlorobium the CD pattern (Fig. 4) is 724 nm (-) and 753 nm (+), whereas in Chloroflexus, it is 725 nm (+) and 757 nm (-) [3]. It seems probable, in Chloroflexus at least, that the optical properties of a chlorosome depend more on the BChl-BChl interactions between BChl c--proteins than on BChl-BChl interactions between chromophores on the same protein. If this is correct, the difference between the chlorosome CD spectra of the two species may reflect mainly differences in the arrangement of the BChl c--proteins in the rod elements (5.2 nm diameter in Chloroflexus vs. 10 nm diameter in Chlorobium) resulting from differences in the secondary structures of the two BChl c--proteins.

2. A New Bacteriochlorophyll a--Protein

When low or medium density chlorosomes are prepared from Chlorobium in the absence of 2 M Na SCN, they are unstable. The ca. 750-nm absorption band gradually diminishes, and a new band at ca. 673 nm (probably due to bacteriopheophytin c) takes its place. When such degraded chlorosomes are purified on a sucrose gradient, two fractions with absorption bands at 794 nm (Fig. 3) appear, one at the top of the gradient and one at about 18% sucrose. The 794-nm pigment is BChl a, and the major protein associated with the fraction is a ca. 5 kDa polypeptide. There is no evidence for the presence of the 42-kDa polypeptide known to be associated with the BChl a--baseplate protein (809 nm).

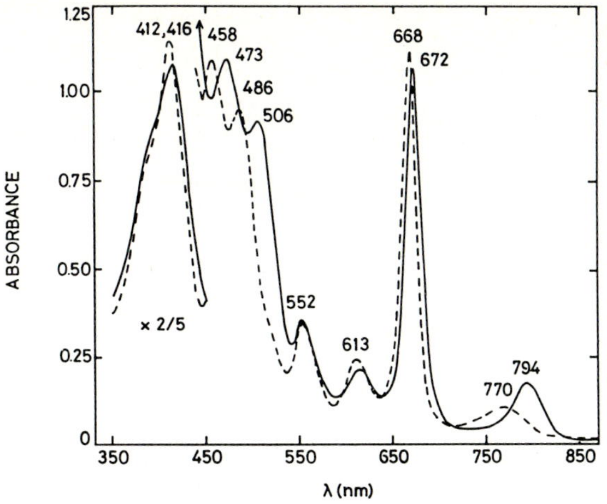

Fig. 3. Absorption spectra of the 794-nm fraction (——) and its 90% methanol extract (---).

When stable chlorosomes are prepared in the presence of 2 M Na SCN (Fig. 2) the ratio of BChl c to BChl a is ca. 90 (about 2 or 3 x the value for Chloroflexus chlorosomes). At 5 K the absorption spectrum shows a peak at 754 nm (BChl c) and a minor shoulder at ca. 800 nm (BChl a). Upon excitation at 460 nm the 5 K fluorescence emission spectrum shows peaks at 784 nm (BChl c) and 825 nm (BChl a) (R. J. van Dorssen and J. Amesz, unpublished). This represents a high energy-transfer efficiency from BChl c to BChl a in the chlorosome at 5 K. This new BChl a--protein (794 nm) associated with the chlorosome in Chlorobium appears to play a similar role to that of the 792-nm BChl a--protein in the chlorosome of Chloroflexus [3]. We suppose that the BChl a--proteins (794 nm) receive excitation energy from the BChl c--proteins in the rod elements, and then transfer this excitation to the BChl a--baseplate proteins as shown in Fig. 1.

3. Bacteriochlorophyll C Aggregates in CH_2Cl_2-Hexane

When concentrated solutions of BChl c in CH_2Cl_2 or CCl_4 are diluted in hexane, the BChl c molecules form aggregates [4-6] whose Q_y-absorption maximum (Fig. 4) closely matches that of BChl c in chlorosomes. The CD spectrum in the Q_y region is also similar to that for the chlorosomes, except that the rotational strengths are sign-reversed and somewhat smaller in magnitude (see Fig. 4). We decided to compare this in vitro system without protein to the in situ system with protein (i.e. the Chlorobium chlorosome), because we thought that understanding the in vitro aggregate might help us to understand the interactions of the chromophores in situ.

Two models have been proposed for the aggregate in vitro. In the SKF model (Fig. 5A) the only link between neighboring BChl c molecules is the non-covalent bonding between the 2-(1-hydroxyethyl) oxygen atom of one BChl c and the Mg atom of the other [6]. The BChl c rings lie in parallel planes 3.4 Å apart, and the Mg-Mg distance between nearest neighbors is calculated

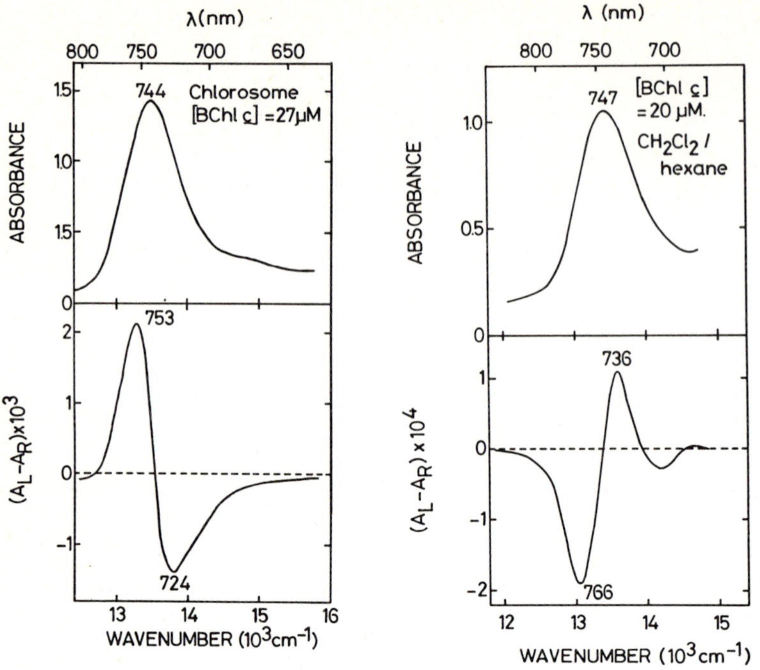

Fig. 4. Absorption and CD spectra for the BChl c Q_y-band in chlorosomes (left panel) and in CH_2Cl_2/hexane, 1:200 (right panel).

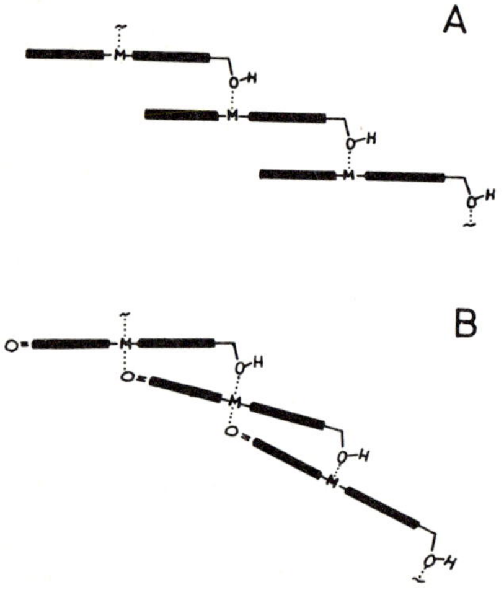

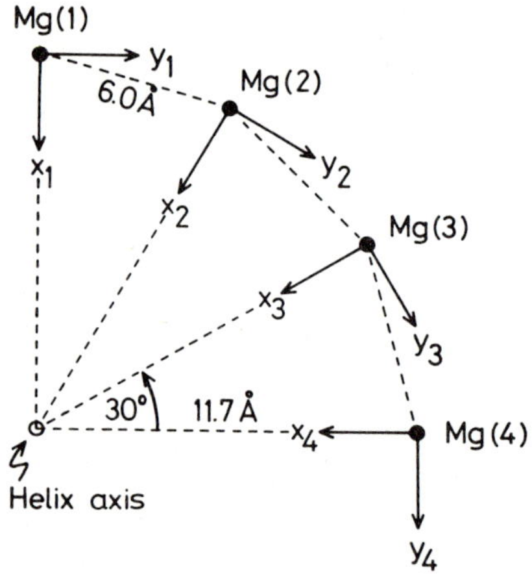

Fig. 5. Models of the BChl c aggregate in (CH_2Cl_2 or CCl_4)-hexane. A. SKF-model [6]; B. BMK-model [4, 5].

Fig. 6. Helical arrangement of BChl c molecules in the SKF-model with $\Theta = 30^0$.

to be 6.9 Å. In order to introduce optical activity into the SKF model, we arranged the BChl c molecules in a helix as shown in Fig. 6. In the BMK model (Fig. 5B) there is an additional link between neighboring molecules through the keto oxygen atom on ring V of one molecule [4, 5]. We have assumed that the keto oxygen on one BChl c is located in the 6th position relative to the Mg atom of another BChl c molecule at a distance of 2.06 Å. The strength of the interaction is not considered. This gives an angle of 12.5^{0} between the planes of neighboring BChl c molecules and a Mg-Mg distance of 6.6 Å.

The optical properties (exciton splittings, dipole strengths, and rotational strengths) were calculated for several versions of each model with the computer program HERMIT, which calculated the interactions between transitions with the point-dipole approximation and the assumption that the one-electron term and the electric magnetic coupling term are both zero. The dipole strength of the BChl c monomer Q_y transition was assumed to be the same as that for chlorophyll a: $2{,}56 \times 10^{-58}$ $(\text{coulomb-meter})^2$ [7]. The optical characteristics for the SKF model were calculated for N (number of BChl c molecules) = 2 to 20 with $\Theta = 30^{0}$. The results showed that most of the dipole strength (D) and rotational strength (R) was concentrated in 3 or 4 exciton components. The main exciton-component was red-shifted from 660 nm to 695 nm for N = 10, to 698 nm for N = 12, and to 692 nm for N = 14. This meant that the red-shift was not very sensitive to changes in N in this range. On the other hand, the splitting between the main exciton-components was sensitive to changes in N, becoming noticeably smaller as N increased. The SKF model was also tested for more tightly-wound helices ($\Theta = 35^{0}$ and 40^{0}) and less tightly-wound helices ($\Theta = 20^{0}$ and 25^{0}) with N = 10. In no case did the wavelength of the lowest energy exciton component exceed 704-705 nm. Thus we may conclude that the SKF model cannot generate a red-shift beyond ca. 700 nm.

We calculated the optical properties of the BMK model for 9 to 11 BChl c molecules (Fig. 7) taking into account that interaction of the keto oxygen (ring V) with an electrophile (i.e. Mg) can shift the Q_y transition in the BChl c monomer to ca. 676 nm [8]. In this model most of the dipole strength and rotational strength is concentrated in the two lowest energy

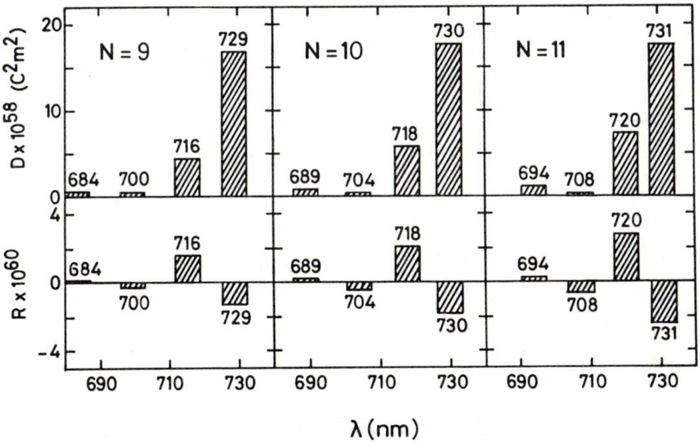

Fig. 7. Values of dipole strength (D) and rotational strength (R) for a number (N) of BChl c molecules in the BMK model.

Table 1. Wavelengths of the two major exciton components as a function of the number of BChl c molecules in the BMK model (λ_m = 676 nm except for one end molecule where $\bar{\lambda}_m$ = 660 nm)

Number	λ_1 (nm)	λ_2 (nm)	$\lambda_2-\lambda_1$ (nm)
8	712.1	728.0	15.9
9	715.6	728.2	13.6
10	718.3	730.1	11.8
11	720.5	730.8	10.3
12	722.3	731.4	9.1

exciton components. The main exciton-component is shifted from 676 nm to 730 nm for N = 10, to 732 nm for N = 14, and to 734 nm for N = 20. Again the redshift is not very sensitive to changes in N. However, the exciton splitting between the main bands is sensitive to changes in N as shown in Table 1.

For comparison with the SKF and BMK models, the experimental absorption and CD spectra for the BChl c aggregate in CH_2Cl_2-hexane (Fig. 4) were deconvoluted simultaneously [9] to yield the values of D and R shown in Fig. 8. The exciton splitting of 12 ± 1 nm corresponds to 10 ± 1 BChl c molecules in the BMK model and ca. 9 molecules in the SKF model. The SKF model (N = 9, data not shown) gives a poor fit to the experimental spectra, but the BMK model is somewhat better. Since the total dipole strength of the Q_y band in the BChl c aggregate is about 70% higher than that of the BChl c monomer, all the D- and R-values for the BMK model should be increased by 70% for comparison to the experimental data. Even with this correction the theoretical R-values are noticeably smaller than the experimental values. Also the 21-nm difference between the experimental and theoretical values for the red-shift remains to be accounted for.

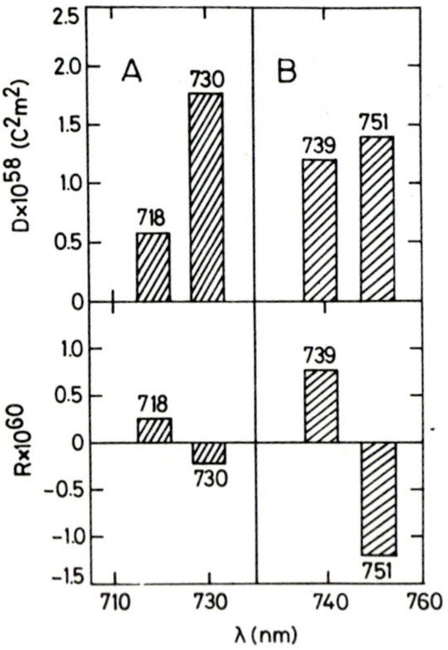

Fig. 8. Values of dipole strength (D) and rotational strength (R) normalized to one BChl c molecule. (A) BMK-model (N = 10). (B) BChl c aggregate in CH_2Cl_2-hexane (data from Fig. 4).

4. Conclusions

1) The BChl *c* aggregates in CH_2Cl_2-hexane may consist of about 10 BChl *c* molecules in a configuration something like the BMK model. 2) The BChl *c* in chlorosomes of *Chlorobium* probably is attached to a 55-amino acid polypeptide in the rod element. From the 10 ± 1 nm splitting of the Q_y-band, we estimate strong exciton interactions between 10-12 BChl *c* molecules in the chlorosome. 3) It is not yet clear how the structure of the BChl *c* aggregate in CH_2Cl_2-hexane might be related to the organization of BChl *c* in chlorosomes.

5. Acknowledgement

This work was supported by grant GM 32109 from the U.S. National Institute of General Medical Sciences.

6. References

1. J.M. Olson: Biochim. Biophys. Acta 594, 33 (1980)
2. T. Wechsler, F. Suter, R.C. Fuller, and H. Zuber: FEBS Lett. 181, 173 (1985)
3. J.A. Betti, R.E. Blankenship, L.V. Natarajan, L.C. Dickinson, and R.C. Fuller: Biochim. Biophys. Acta 680, 194 (1982)
4. M.I. Bystrova, I.N. Mal'gosheva, and A.A. Krasnovskii: Mol. Biol. 13, 440 (1979) [translated from Molekulyarnaya Biologiya 13, 582 (1979)]
5. A.A. Krasnovsky and M.I. Bystrova: BioSystems 12, 181-194 (1980)
6. K.M. Smith, L.A. Kehres, and J. Fajer: J. Am. Chem. Soc. 105, 1387 (1983)
7. K.D. Philipson, S. Cheng Tsai, and K. Sauer: J. Phys. Chem. 15, 1440-1445 (1971)
8. J.J. Katz and J.C. Hindman: *In Biological Events Probed by Ultrafast Laser Spectroscopy*, edited by R.R. Alfano (Academic, New York 1982) p. 119
9. L.L. Shipman, T.M. Cotton, J.R. Norris and J.J. Katz: J. Am. Chem. Soc. 98, 8222 (1976)

Reverse-Phase High-Performance Liquid Chromatography of Antenna Pigment- and Chlorosomal Proteins of *Chloroflexus aurantiacus*

R. Feick

Botanisches Institut, Universität München, D-8000 München 19, F.R.G.

INTRODUCTION

During the past few years, reverse-phase high-performance liquid chromatography (RP-HPLC) has emerged as one of the techniques best suited for the isolation and separation of low molecular weight proteins (M_r 5000) and polypeptide fragments generated by enzymic or chemical cleavage of proteins (1,2). The distinct advantages of RP-HPLC, speed, sensitivity, reproducibility and resolution, have contributed to great advances in protein structure analysis. Despite the success in isolating hydrophilic polypeptides of even considerable size (3), it was not until recently that RP-HPLC was employed as a final purification step in the isolation of hydrophobic membrane proteins. Examples are the separation of subunitsof chloroplast coupling factor (4), of mitochondrial cytochrome c-oxidase (5) and various structural envelope proteins of viral origin (sendai-, polio-, influenza-virus; 6,7,8).

As shown in a previous study (9), five membrane-associated polypeptides (apparent M_rs: 3700, 5300, 5800, 11000, 18000, referred to as A, B, C, D, E) are constituents of the antenna system in the green photosynthetic bacterium Chloroflexus aurantiacus. While the 5800-protein and bacteriochlorophyll a form the antenna pigment protein complex located on the cytoplasmic membrane, the other polypeptides and bacteriochlorophyll c are specific for the chlorosome, where the bulk of the antenna pigment is housed. The 5800-polypeptide is believed to be part of the chlorosome-cytoplasmic membrane attachment site, providing a functional linkage between the chlorosome-antenna and the antenna-reaction center complexes on the cytoplasmic membrane.

Due to similarities in the protein molecular weights, one might anticipate problems in the isolation of these polypeptides if traditional chromatographic or electrophoretic techniques are employed. Hence, the objective of this work was to investigate whether RP-HPLC can be successfully applied as an one-step isolation technique for purifying some of the photosynthetic membrane proteins of Chloroflexus aurantiacus.

RESULTS

The RP-HPLC was performed on a large pore (30 nm), silica-based C-3 column. Among all the solvents and ion-pairing agents tested, the best results were obtained by employing a gradient from water to acetonitrile/propanol for elution. Trifluoroacetic acid (TFA) and triethylamine (TEA) were present in both solutions as ion-pairing agents (a more detailed description will be given elsewhere). To effect solubilization, the membrane suspension was first subjected to a 10% TFA-precipitation. The resultant pellet was routinely dissolved in 67% formic acid-33% 2-propanol and injected

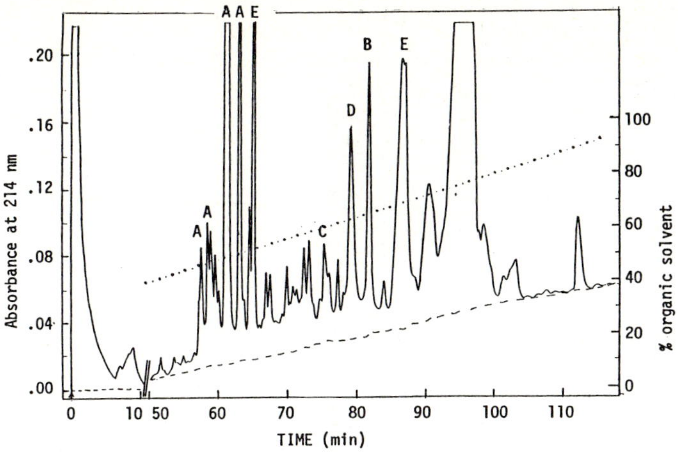

Fig. 1a

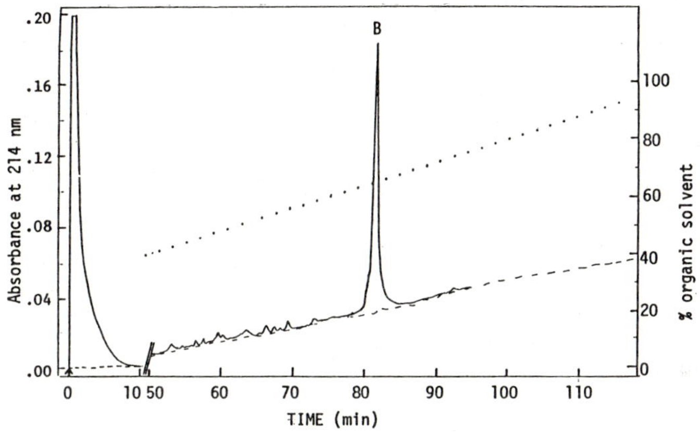

Fig. 1b

Fig. 1a: RP-HPLC elution profile of a membrane fraction containing chlorosomes and cytoplasmic membranes. Flow rate: 1 ml/min; 0.8 % organic solvent /min; room temperature.
— — — blank gradient run
• • • • • % organic solvent

Fig. 1b: Rechromatography of peak fraction B. Collected eluent was concentrated by vacuum evaporation and rechromatographed under identical conditions.

immediately onto the column. The RP-HPLC elution profile of a membrane fraction containing chlorosomes and cytoplasmic membranes is shown in Fig. 1. As judged by SDS-PAGE most of the proteins, including the light-harvesting-specific polypeptides, eluted within the 45 - 75% organic solvent range while lipids, pigments and (presumably) highly aggregated proteins are contained in the later peak fractions. Without any loss in separation efficiency, the time required to complete the chromatography could be reduced by applying a 30 - 70% gradient over a total of 50 min, beginning 10 min after sample injection.

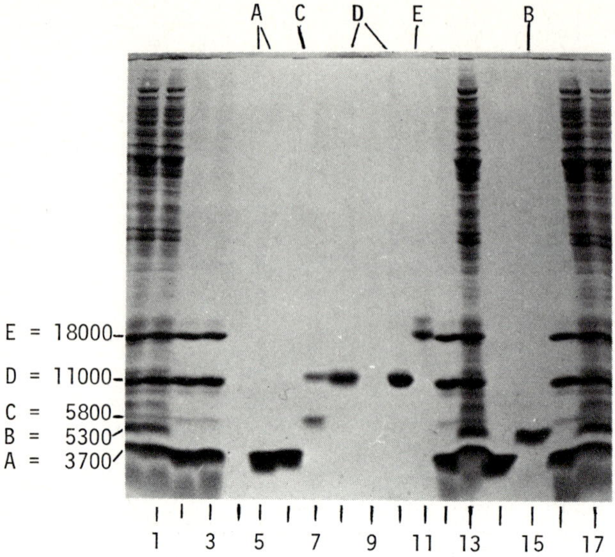

Fig.2: Polypeptide patterns after SDS-PAGE in a continous 11.5 - 16.5% gel.

lane 1,13,17:	membrane fraction containing cytoplasmic membranes with attached chlorosomes. This fraction was subjected to RP-HPLC.
lane 2,3,12,16:	isolated chlorosomes
lane 5,6,14:	peak fraction A
lane 7	peak fraction C
lane 8,10	peak fraction D
lane 11:	peak fraction E
lane 15:	peak fraction B

A comparison of the sample elution profile with the blank gradient profile revealed some degree of peak overlap, suggesting incomplete separation. However, rechromatography of the dominant peak material (e.g. peak B, Fig. 1a) yielded a single component with unaltered retention time. Nearly complete recoveries were obtained upon rechromatography.

The identification and purity of fractionated polypeptides were evaluated by sodium dodecyl sulfate-polyacrylamide gel electrophoresis (SDS-PAGE). As shown in Figure 2, the isolation and purification of four polypeptides (A=3700, B=5300, D=11000, E=18000) was accomplished in a single RP-HPLC run. The purification of the 5800-polypeptide which coeluted with a contaminating polypeptide is still in progress. Gel electrophoresis of the RP-HPLC fractions clearly demonstrated that caution is advised in attempting to assign a polypeptide to a peak. It was found that up to four peak fractions contained the 3700-polypeptide with nearly the same degree of purity. The 18000-protein eluted at both 52 and 72% organic solvent, though in a purer but lower yield in the earlier fraction.

These findings verify results reported in the literature that RP-HPCL can cause abnormal elution behavior of proteins: Single polypeptides can be eluted as multiple peaks and, conversely, one single peak may represent more than one protein. Such microheterogeneity might indicate partial chemical modification such as oxidation, esterification, or deamidation of amino acid residues (10). However, different solubilization conditions (eg. 8 M urea or 8 M guanidine hydrochloride in the presence of 1% 2-mercaptoethanol) did not significantly alter the elution profile from that shown in Figure 1.

ACKNOWLEDGEMENT

The author would like to thank Dr. Hugo Scheer for his interest in this work and for financial support (SFB 143). Special thanks to R.C. Fuller in whose laboratory this work was performed.

REFERENCES

1. Walsh, K.A., Ericsson, L.H., Parmelee, D.C. and Titani, K. (1981) Ann. Rev. Biochem. 50, 261.
2. Hearn, M.T.W. (1980) J. Liq. Chromatogr. 3, 1255.
3. Fallon, A., Lewis, R.V. and Gibson, V.D. (1981) Anal. Biochem. 110, 318.
4. Tandy, N.E., Dilley, R.A. and Regnier, F.E. (1983) J. Chromatogr. 266, 599.
5. Power, S.D., Lochrie, M.A. and Poyton, R.O. (1983) J. Chromatogr. 266, 585.
6. Phelan, M.A. and Cohen, K.A. (1983) J. Chromatogr. 266, 55.
7. Heukeshoven, J. and Dernick, R. (1982) J. Chromatogr. 252, 241.
8. Van Der Zee, R., Welling-Webster, S., Welling, G. (1983) J. Chromatogr. 266, 577.
9. Feick, R. and Fuller, R.C. (1984) Biochemistry 23, 3693.
10. Heukeshoven, J. and Dernick, R. (1984) Chromatographia 19, 95.

Fluorescence-Detected Magnetic Resonance of the Antenna Bacteriochlorophyll Triplet States of Purple Photosynthetic Bacteria

A. Angerhofer, J.U. von Schütz, and H.C. Wolf

Physikalisches Institut, Teil 3, Universität Stuttgart, Pfaffenwaldring 57, D-7000 Stuttgart 80, F.R.G.

1. Summary

Fluorescence-detected magnetic resonance (FDMR) measurements in zero-field have been performed on the antenna bacteriochlorophyll (BChl) complexes of the purple photosynthetic bacteria Rhodopseudomonas (Rps.) sphaeroides strains R 26.1 and R 26 S (original mutant) and Rps. capsulata A1a$^+$. In every case, at least two distinct triplet states are observed.

2. Introduction

The whole fluorescence emission spectrum of the photosynthetic membrane of purple photosynthetic bacteria can be divided into fluorescence from BChl-biosynthesis precursor pigments [1] and the fluorescence of the antenna (AT) BChl pigments. In reduced chromatophores, one observes the RC triplet by magnetic resonance via the AT fluorescence emission. This is due to the efficient energy-transfer between AT and RC. Because of the rather strong RC-triplet signal, the AT-triplets have been observed until now only in a RC-less mutant A1a$^+$ pho$^-$ of Rps. capsulata [2]. By application of high-resolution electron-electron double resonance (EEDOR), Beck et al. found different triplet states which differed mainly in their E-values.

In this contribution we report on the first FDMR experiments performed on the AT system of the carotenoidless mutants R 26.1, R 26 S [3] of Rps. sphaeroides and A1a$^+$ of Rps. capsulata. In order to measure these signals we had to get rid of the overlapping RC triplet signals. This was obtained by appropriate preparation of the chromatophores, oxidation or soft dilution with detergent in low concentration.

3. Materials and Methods

The chromatophores from Rps. sphaeroides R 26.1 and Rps. capsulata A1a$^+$ were kind gifts from Prof. Dr. G. Drews, Freiburg. Dr. Richard Cogdell supplied us with chromatophores from Rps. sphaeroides R 26 S. The concentrated samples had an optical density (OD) of about 50 at the long wavelength maximum (1 cm cuvette).

Oxidized chromatophores were obtained by the addition of ferricyanide in tris buffer (pH 7,5) to a final concentration of 50 mM. Detergent-treated chromatophores were obtained by dropwise addition of 1 % Triton TX 100 up to a final concentration of 1/2 % in dim light at room temperature while stirring.

Dilution was performed with tris buffer (pH 7,9) and glycerol in 30 %/70 % v-v ratio to yield an OD of about 1 at the long wavelength maximum (1 cm cuvette).

FDMR measurements were performed with the apparatus described in [4].

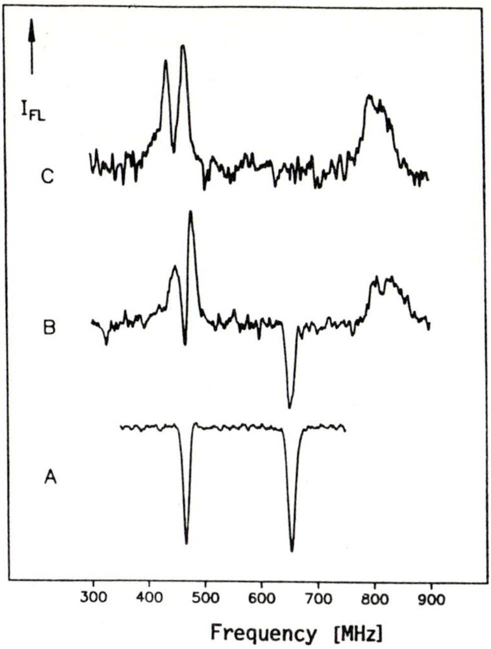

Fig. 1 FDMR-spectra of chromatophores from Rps. sphaeroides R 26.1
A: reduced with dithionite
B: without chemical changes
C: oxidized with ferricyanide

4. Results and Discussion

Fig. 1 demonstrates the effect of oxidation or detergent treatment on the FDMR-signals of Rps. sphaeroides R 26.1. While the RC triplet (negative sign) comes up strongly in reduced chromatophores (reduced with dithionite, spectrum A), a complex spectrum is seen in untreated chromatophores (spectrum B). Clearly visible are the negative peaks of the RC-triplet overlapped by additional signals with positive sign, which we attribute to the AT triplet states. By oxidation or detergent treatment (data not shown) the RC is energetically decoupled from the AT pigments,and its triplet state spectrum is no more visible on the AT emission. Remaining are two signals at 457 MHz, 477 MHz,and a broad,sometimes structured band around 820 MHz. Different samples of the same species and the same preparation procedure show different intensity ratios of the two signals at 457 and 477 MHz, whereas at the same time the peak of the 820 MHz band varies between 810 and 835 MHz. Because 2E-signals do not appear, we want to attribute the two signals around 470 MHz to the resolved D-E-transitions of two distinct triplet states, whereas the 820 MHz band shows the concomittant unresolved D+E signals. Coordination of the D-E and D+E signals can be done by viewing different samples with different intensity ratios of the D-E signals,and attributing the maximum of the D+E signal to the most intense D-E signal. The so calculated D- and E-values are tabulated in Table 1.

Because of the relatively broad linewidth of 12 MHz for the D-E-signals and 30 MHz for the D+E-signals, the D- and E-values have an uncertainty of

Table 1 Zero-field splitting values of the triplet states in AT BChl and wavelength of detection of the AT fluorescence in *Rps. sphaeroides* R 26.1, R 26 S and *Rps. capsulata* $A1a^+$.
A: monomeric BChl in MTHF (hexacoordinated). Values derived from [5].
B: monomeric BChl in the bacterial membrane (pentacoordinated). Values derived from [6].

	|D| $[10^{-4}cm^{-1}]$	|E| $[10^{-4}cm^{-1}]$	λ_{Emis} [nm]
Rps. sphaeroides R26.1	215	56	903
	216	63	
Rps. sphaeroides R 26 S	217	51	905
	220	61	
Rps. capsulata $A1a^+$	214	56	910
	209	67	
monomeric BChl A	230	58	773
monomeric BChl B	232	69	783

about 5 %. The D-variations of the two distinct triplet states lie within this error margin, whereas the variations of E are greater than 11 % and have to be taken into account.

A similar difference of the E-value is reported in the literature for BChl in penta- or hexacoordinated form [5,6]. This suggests that we find the triplet states of AT BChl in penta- and hexacoordinated form. This hypothesis is strengthened by the fact that the Q_x-absorption of AT-BChl lies at 596 nm, just between the location of pentacoordinated BChl (583 nm) and hexacoordinated BChl (612 nm) [7].

5. References

1 J. Beck and G. Drews, Z. f. Naturforsch. 37c, 199 (1982).
2 J. Beck, J.U. von Schütz and H.C. Wolf, Chem. Phys. Lett. 94, 147 (1983).
3 E. Davidson and R.J. Cogdell, FEBS Lett. 132, 81 (1981).
4 A. Angerhofer, J.U. von Schütz and H.C. Wolf, Z. f. Naturforsch. (in press).
5 H.J. den Blanken and A.J. Hoff, Chem. Phys. Lett. 96, 343 (1983).
6 J. Beck, J.U. von Schütz and H.C. Wolf, Z. f. Naturforsch. 38c, 220 (1983).
7 T.A. Evans and J.J. Katz, Biochim. Biophys. A 396, 414 (1975).

High-Resolution ^{1}H NMR of Light-Harvesting Chlorophyll-Proteins

C. Dijkema, G.F.W. Searle, and T.J. Schaafsma

Agricultural University, Department of Molecular Physics, De Dreijen 11, NL-6703 BC Wageningen, The Netherlands

The structure of the monomeric (29 kD) protein subunit of the chlorophyll-a and -b containing light-harvesting complex (LHC) from barley (*Hordeum vulgare*) has been investigated in D_2O by 500 MHz ^{1}H NMR, using a DASWEFT pulse sequence [1] to suppress HDO resonances. The results are presented as an illustration of the potential of this method for similar bacterial proteins.

Chl-a/b particles, isolated from cation-depleted thylakoids and fractionated on a Triton X-100 containing sucrose-gradient according to the method of BURKE *et al* [ ], were obtained from Drs. D. von Wettstein and B.M. Møller (Carlsberg Laboratories, Copenhagen). In order to prevent aggregation and, consequently, spectral broadening, Triton was replaced by perdeuterated SDS, up to 1% by weight.

We observed resolved resonances assigned to Chl-a α(9.1 ppm), β(9.5 ppm) and δ(8.5 ppm) methine protons and the Chl-b formyl proton (11.0 ppm). The corresponding in vitro values are found to be 9.4, 9.7 and 8.5 ppm for Chl-a and 11.3 ppm for Chl-b [3] . Note that there are no large ligand-shifts. No resonances were observed of Chl-b α- and β-methine protons, presumably due to heterogeneous broadening. These results are consistent with Raman-data [4] indicating at least two different Chl-b populations, both penta-ligated at Mg and hydrogen-bonded at the formyl- and/or ring V keto-groups, and a single Chl-a population. Upon ageing, pheophytinization occurs, giving rise to α- and β-methine resonances of pheophytin-b at 10.3 and 9.7 ppm, and changes in the absorption- and fluorescence spectra. At the same time, very sharp resonances appear in the methyl- and methylene region, presumably due to mobile phytol-chains. Nevertheless, the protein spectrum remains unaffected, in agreement with the rigidity and high stability of the protein.

These results demonstrate that ^{1}H NMR provides an option for structural studies on isolated chlorophyll-proteins, when no crystals are available.

References:

1. C.A.G. Haasnoot: J. Magn. Reson. 52, 153 (1983)
2. J.J. Burke, C.L. Ditto and C.J. Arntzen: Arch. Biochem. Biophys. 187, 252 (1978)
3. G.L. Closs, J.J. Katz, F.C. Pennington, M.R. Thomas and H.H. Strain: J. Am. Chem. Soc. 85, 3809 (1963)
4. M. Lutz, J.S. Brown and R. Remy: "Resonance Raman Spectroscopy of Chlorophyll-Protein Complexes" in Ciba Foundation Symposium 61 (1977) Chlorophyll Organization and Energy Transfer in Photosynthesis, pp. 105-125.

Crystallization and Linear Dichroism Measurements of the B800–850 Antenna Pigment-Protein Complex from *Rhodopseudomonas sphaeroides* 2.4.1

J.P. Allen, R. Theiler, and G. Feher‡

University of California, San Diego, Department of Physics, B-019, La Jolla, CA 92093, USA

1. Purification

Crude chromatophore membranes of *R. sphaeroides* 2.4.1 were purified by differential centrifugation and suspended in buffer (20 mM Tris-HCl, pH 8.0, 2mM Tris - EDTA) to a final absorbance $A_{1cm}^{850}=100$. An equal volume of 1.6% (w/v) β-D-octylglucopyranoside (βOG) in water was added at room temperature, and the resulting solubilizate centrifuged at $\sim 2\times 10^5$ g for 60 min. at 4° C. The supernatant was brought to room temperature, and 10% nonanoyl-N-methyl-glucamide (Nga) was added to a final concentration of 1%. The B800-850-complex was isolated by column chromatography on DEAE-cellulose in the presence of 1% Nga and buffer (10 mM Tris-HCl, pH 8.0, 1mM EDTA-Tris, 1 mM β-mercaptoethanol). The B800-850-complex was eluted with 200 mM NaCl in Nga and buffer. Further purification was accomplished by repeating the DEAE chromatography. For the crystallization, 1% βOG was substituted for Nga by repeated concentration and dilution in an Amicon concentration cell and a final dialysis.

The purified complex was characterized by its optical spectrum which exhibited the characteristic absorption bands of the B800-850-complex. The absorbance ratio $A_{1cm}^{800}/A_{1cm}^{845}$ was 0.63 indicating that the labile B800 chromophore was retained. SDS-polyacrylamide gel electrophoresis showed two prominent bands at the positions of α and β polypeptides of the B800-850-complex with minimal contamination by other proteins.

2. Crystallization

Crystallization was performed by vapor diffusion in the presence of polyethyleneglycol 4000 (PEG) and NaCl as previously reported [1]. The procedure was adapted from the technique used by GARAVITO and ROSENBUSH to crystallize matrix porin from *E. coli* [2]. Included with the protein solution was an amphiphile, 1,2,3 heptane-triol (m.p.$\sim$80° C), which had been used by MICHEL in the crystallization of reaction centers from *R. viridis* [3]. A variety of crystal forms were observed after 2 to 4 weeks or room temperature and under dark conditions. In Fig. 1 four representative forms are shown. Crystal forms A and B grew under phase-separation conditions, with the phase boundary apparent in frame B. When the initial protein concentration was decreased from $A_{1cm}^{845}=18$ to $A_{1cm}^{845}=9$, crystal form C was observed. By decreasing the initial concentration of PEG (9.0% to 7.5%) and heptane triol (3.0% to 2.5%) form D was obtained.

3. Linear Dichroism

Linear dichroism experiments were performed on crystal form B using a microspectrophotometer as previously described [4]. The optical spectrum from 450 nm to 870 nm is

‡Work supported by the National Science Foundation (GM-13191) and the National Institute of Health (PCM 82-02811-AO3).

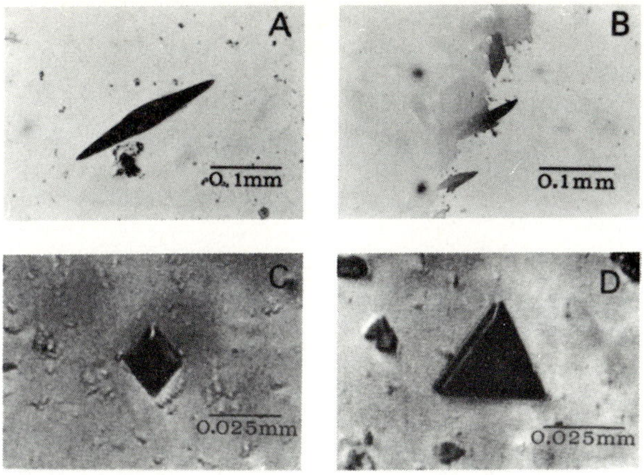

Figure 1. *Different crystal forms of the B800-850- complex from R. sphaeroides 2.4.1.*

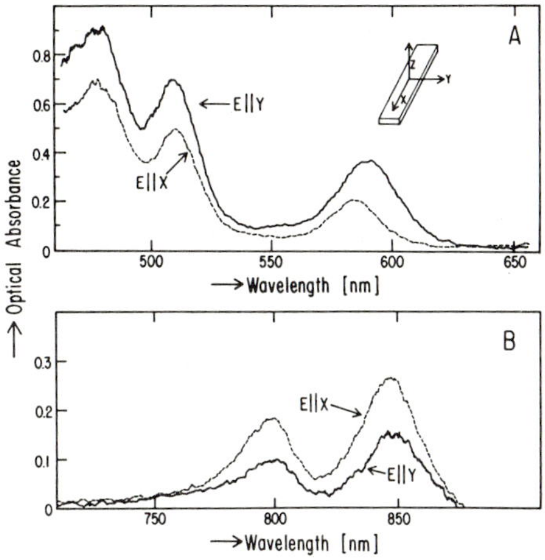

Figure 2. *Linear dichroism spectra of crystal form B of the B800-850-complex from R. sphaeroides 2.4.1. The measurements in the visible region (frame A) were per- formed on a thicker (~×7) crystal, but of the same form as those in the infrared region (frame B).*

presented in Fig. 2. The characteristic absorption bands of the native B800-850-complex show a strong dependence on the polarization of the incident light. The data indicate a preferential alignment along the x-axis of the transition moments (Q_y) associated with the absorption bands in the near infrared. The greater absorbance for E || y at the high wavelength side of the 580 nm peak indicates a preferential alignment along the y-axis of the Q_x transition moments associated with the B850 chromophores. A preferential alignment along the y-axis is suggested for the carotenoid transition moments (absorption bands at 477 and 510 nm) along the y-axis. An interpretation of the spectra with respect to other data, including linear dichroism measurements of the complex from *Rhodopseudomonas capsulata,* is provided in the paper by MÄNTELE, *et al.* [5].

A more precise assessment of the relative orientation of the chromophores will have to await the results of linear dichroism measurements of tilted crystals or different crystal forms.

1. J. P. Allen, R. Theiler, and G. Feher *Biophysical Journal* **47,** 4a (1985).
2. R. M. Garavito and J. P. Rosenbush *J. Cell. Biol.* **86,** 327-329 (1980).
3. H. Michel *J. Mol. Biol.* **158,** 567-572 (1982).
4. J. P. Allen and G. Feher *Proc. Natl. Acad. Sci. USA* **81,** 4795-4799 (1984).
5. W. Mäntele, K. Steck, T. Wacker, W. Walte, B. Levoir, and J. Breton, presented in this volume.

Crystallization of the B800–850-complex from *Rhodopseudomonas acidophila* Strain 7750

R.J. Cogdell[1], K. Woolley[1], R.C. Mackenzie[1], J.G. Lindsay[2], H.Michel[3], J. Dobler[4], and W.Zinth[4]

[1] Department of Botany, University of Glasgow, Glasgow G12 8QQ, U.K.

[2] Department of Biochemistry, University of Glasgow, Glasgow G12 8QQ, U.K.

[3] Max-Planck-Institut für Biochemie, D-8033 Martinsried, F.R.G.

[4] Physik-Department, E11, Technische Universität München, D-8000 München, F.R.G.

Recently, methods have become available for producing good quality three-dimensional crystals of integral membrane proteins, in the presence of detergents [1,2]. We were especially encouraged in this respect by the successful determination of the structure of the reaction centre from Rhodopseudomonas viridis [3], which has been so well described at this meeting.

We adopted a 'shotgun' approach to try and get crystals from a bacterial antenna complex. A range of antenna complexes were prepared and tested to see whether they would form crystals, using the vapour diffusion method [2]. Crystals were obtained in two cases, the B800-850-complexes from Rhodopseudomonas sphaeroides and Rhodopseudomonas acidophila. The B800-850-complex from Rps. acidophila forms crystals much more readily, and so we have concentrated our efforts upon this antenna complex. The B800-859-complex from Rps. acidophila contains bacteriochlorophyll a and carotenoids non-covalently bound to three low molecular-weight polypeptides [4].

We have experimented with the conditions for crystal formation, varying the detergent (lauryl-dimethylamine-N-oxide (LDAO), N-octyl-rac-2,3 dioxypropyl-sulphoxide or β-ocytl-glucoside), the precipitant (ammonium sulphate, potassium phosphate or polyethylene glycol), the pH (pH7.0 to 10.0), the temperature (4°-20°C) and the type of small amphiphile present (1,2,3 heptane-triol, piperidine-2-carbonic acid and benzamidine hydrochloride).

So far, the best results have been obtained with phosphate as the precipitant and at pH values above pH9.0. The crystal form seems to be very dependent upon the type of small amphiphile present. This is illustrated in Figs. 1-4.

Crystallisation in the presence of heptane-triol/piperidine-2-carbonic acid mixtures gives rise to long needles or large, flat, plate-like crystals, while use of piperidine-2-carbonic acid alone yields rather square crystals. So far our largest crystals have been obtained in the presence of benzamidine hydrochloride, and in this case the crystals are usually wedge-shaped or rhomboid.

The degree of order of the pigment molecules within the crystals has been investigated optically, by looking for dichroism. Figure 5 shows the result of photographing some crystals with polaroid light. The colour of the crystals in the visible region is dominated by the absorption bands of the carotenoids. A comparison of the two pictures in Fig. 4 shows quite clearly that the carotenoids, within these crystals, exhibit a high degree of dichroism, and are therefore well-ordered within the crystals.

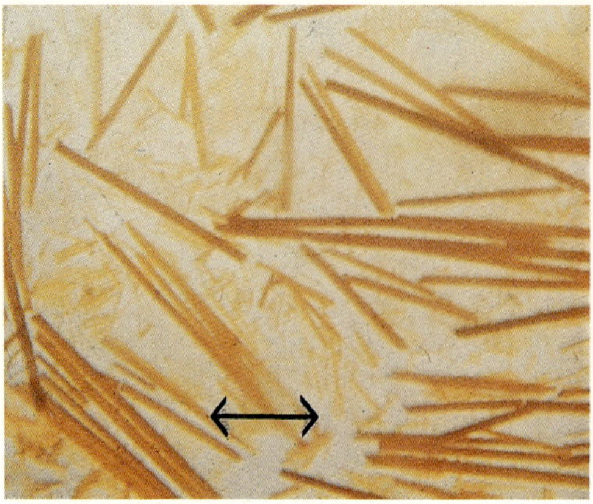

Fig. 1 Crystals grown in the presence of heptane-triol and piperidine-2-carboxylic acid

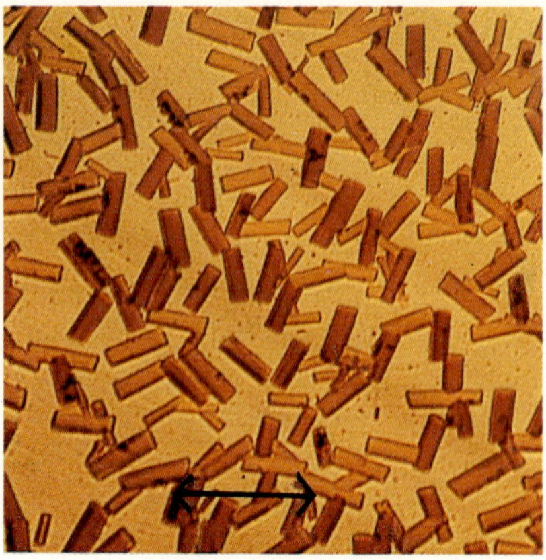

Fig. 2 Crystals grown in the presence of heptane-triol and piperidine-2-carboxylic acid

In each case the bar represents 100 μ

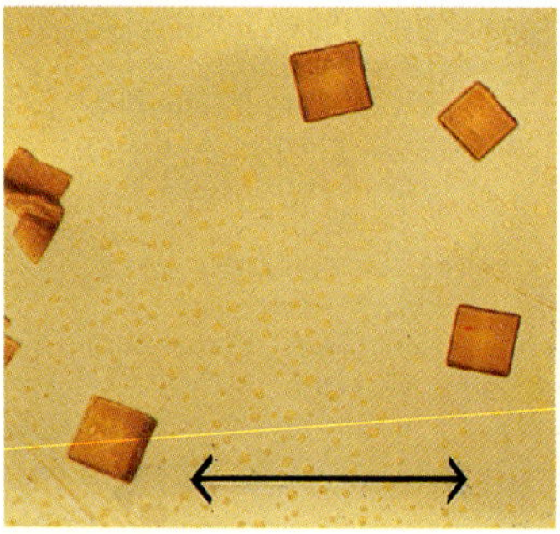

Fig. 3 Crystals grown in the presence of piperidine-2-carboxylic acid

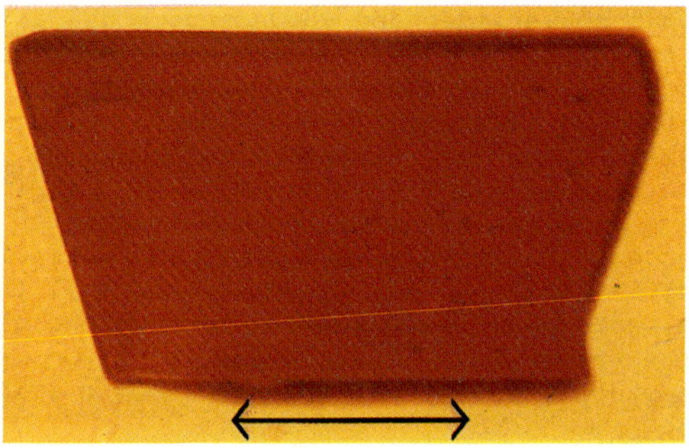

Fig. 4 Crystal grown in the presence of benzamidine hydrochloride

In some preliminary studies we have looked at this phenomenon in more detail, by determining the linear dichroism spectra of some thin needle-shaped crystals (Fig. 6). Even though the crystals were, optically, rather too dense in the region of the 800 and 850 nm absorption bands, the linear dichroism spectra do indicate a remarkable degree of order in these bacteriochlorophyll a Qy absorption bands. In these bands, the dichroic ratio is clearly ⩾20x. In the Qx band at ~590 nm the sign of the dichroism is reversed, and the dichroic ratio is about 3:1. These initial results are very encouraging and indicate well-ordered crystals.

We are now continuing with this project, trying to get larger crystals that will be suitable for X-ray diffraction studies.

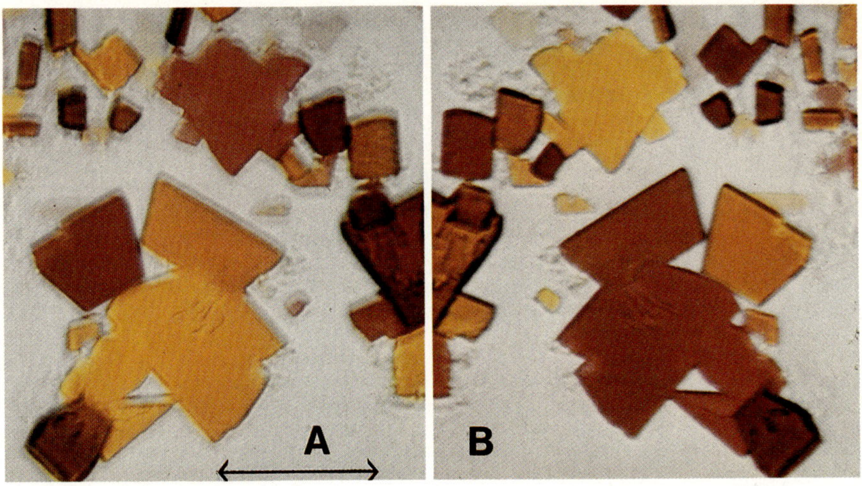

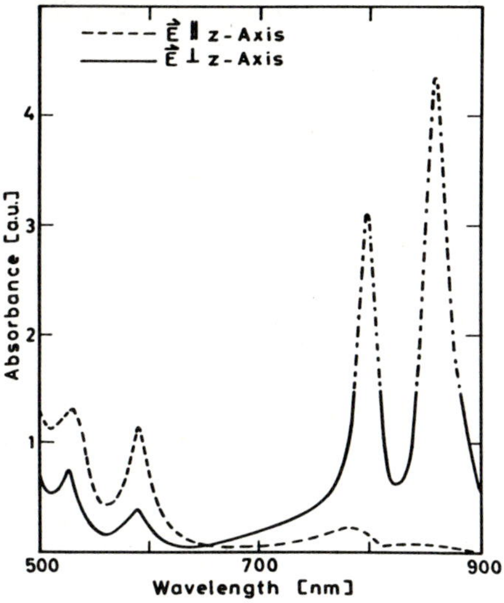

Figs. 5 A/B Crystals photographed under polarised light. The polarised light was 90° out of phase in the two photographs. The bar represents 100μ.

Fig. 6 The linear dichroism spectrum of some long, thin, crystals.

Where the line goes above 1.5 absorbance units the spectrum has been computed by reference to the absorption spectrum of the antenna complex in solution.

Acknowledgements

This work was supported by grants from the SERC and the Sonderforschungsbereich 143. We would like to thank Ms Lynne Roberts for expert technical assistance.

References

1 R.M.Garavito and J.P. Rosenbusch: J. Cell. Biol. 86, 327 (1980)
2 H. Michel and D. Oesterhelt: PNAS 79, 1283 (1980)
H. Michel: J. Mol. Biol. 158, 567 (1982)
3 J. Deisenhofer, O. Epp, K. Miki, R. Huber and H. Michel: J. Mol. Biol. 180, 398 (1984)
4 R.J. Cogdell, I. Durant, J. Valentine, J.G. Lindsay and K. Schmidt: Biochim. Biophys. Acta 722, 427 (1983)

Linear Dichroism (LD) and Absorption Spectra of Crystals of B800–850 Light-Harvesting Complexes of *Rhodopseudomonas capsulata*

W. Mäntele, K. Steck, T. Wacker, and W. Welte
Institut für Biophysik und Strahlenbiologie der Universität Freiburg
Albertstrasse 23, D-7800 Freiburg, F. R. G.

B. Levoir and J. Breton
Service de Biophysique, Départment de Biologie, C.E.N. Saclay,
F-91191 Gif-sur-Yvette Cédex, France

The B800-850 light-harvesting complex of Rps. capsulata was crystallized in the presence of detergents. The crystals were obtained by vapour diffusion technique using polyethylene glycol 4000 (Merck, for gas chromatography). A technique to produce large and thick crystals suitable for X-ray diffraction is described elsewhere (1). For spectroscopic work, due to the high optical density of the crystals, thin extended crystalline plates are advantageous. The crystallization technique was therefore modified. The best yield of thin crystals was found when ß-D-octylglucopyranoside (1% w/v) or octyltetraoxyethylene (1%) plus heptanetriol (3%) were used as detergents and sodium chloride (0.5 M) was present. Other conditions were similar to the ones given in (1). Similar crystallization methods were applied by us to the B800-850 complex of Rps. palustris, and small crystals were obtained. Crystallization of the B800-850 complex of Rps. spaeroides has also been reported by Allen et al. (see J.Allen, R.Taylor and G.Feher, this volume). Thin crystalline plates of the B800-850 complex of Rps. capsulata, along the shortest dimension, can have optical densities as low as 0.3 at 850 nm. They have the form of parallelograms with edges being 0.05 to 0.2 mm in length. These thin crystals have similar geometry to the triclinic ones used for X-ray diffraction (1). We therefore assume that both lattices are identical. Spectra were obtained both from thin crystals grown together with large crystals for X-ray studies and from thin crystals grown with the modified procedure as described above. Crystals were carefully pipetted with some mother liquor into a microcell, formed by two glass windows and a spacer of 20 µm thickness. The crystal was thus aligned with its largest surface parallel to the microcuvette walls.

Absorbance spectra were generally taken with the crystal perpendicular to the beam. For LD spectra, the crystal could be rotated around the beam axis (rotation angle α) and tilted by up to $\pm$ 40° with respect to the beam direction (tilt angle ß). For a definition of the angles see fig.1.

Spectra were taken on a single-beam instrument described in (2). Two microscope objectives were used to condense the beam to a spot size of about 100µm x 100µm and to enlarge the image of this spot. An iris diaphragm was used to cut off false light around the crystal. An absorbance spectrum was obtained by taking a reference spectrum next to the crystal (I_0), a sample spectrum through the crystal (I) and forming $\log(I_0/I)$ on an instrument computer. Linear dichroism spectra were either obtained from spectra with fixed polarization, or by using a photoelastic modulator with phase-sensitive detection as described in (2). Figure 2a and 2b show absorbance spectra of two crystals in the region of 450nm-750nm and 750nm-950nm.

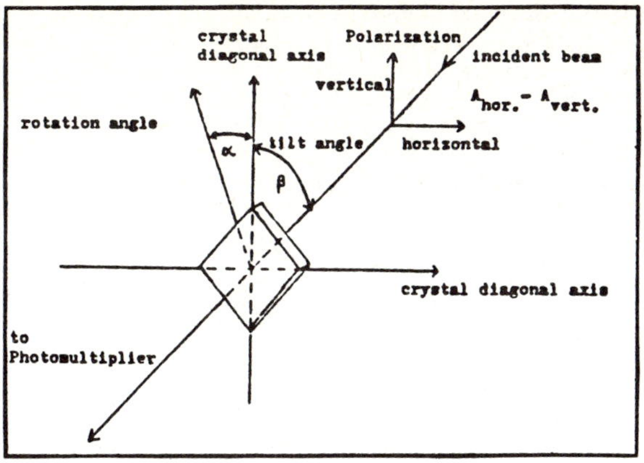

Fig. 1

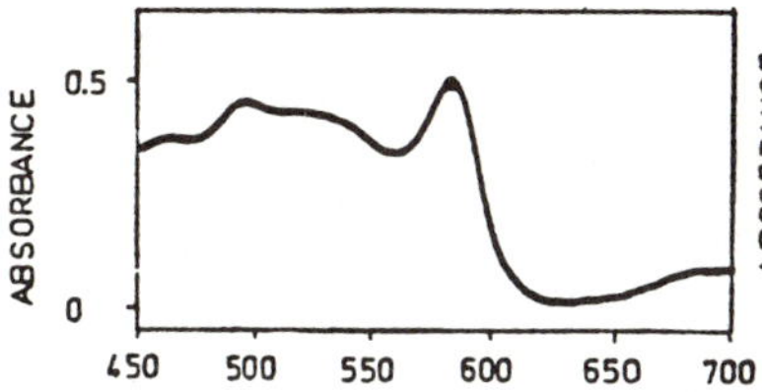

Fig. 2a

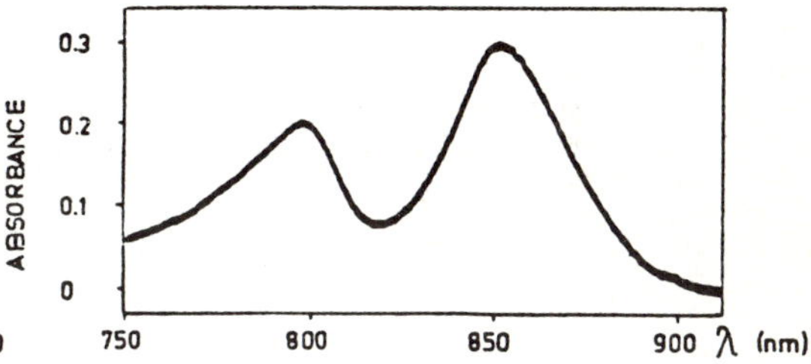

Fig. 2b

Carotenoid absorption and absorption from Bchl Q_x and Q_y transitions closely correspond to the ones observed for this chlorophyll-protein complex solubilized in detergent. At 680nm in the spectrum in fig. 2a, a very weak absorption indicates the presence of only a small amount of oxidation products. The stability of the crystals appears rather high, since even prolonged exposure of the crystal to white light of high intensity in the instrument did not significantly increase this band. A LD spectrum in the 450nm-750nm region is shown in fig.3.

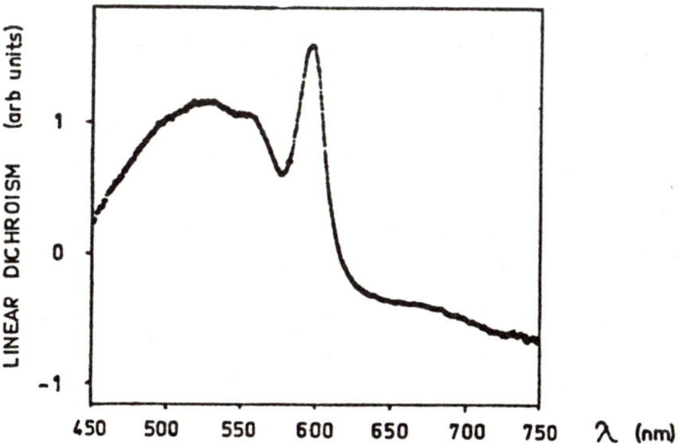

Fig. 3

A positive LD signal is observed both for the carotenoid and the Q_x transition. The maximum of the Q_x LD is shifted to longer wavelengths with respect to the absorbance maximum, in agreement with previous reports (**3-5**). For the 800 and 850nm bands, a distortion of the LD signal, due to the low photomultiplier output, is present in our spectra. However, the sign of the LD is always opposite to that of the Q_x band (data not shown) as previously reported for uniaxially oriented B800-850 complexes of Rps. sphaeroides. This is also demonstrated by the LD spectra from crystals of the B800-850 complex of Rps. sphaeroides obtained by Allen et al. (see J.Allen, R.Taylor and G.Feher, this volume). Rotating the crystal around the beam direction by 90° leads to sign change for both the carotenoid and Q_x LD, and maximum, minimum and zero crossovers for both LD signals are found at the same angular positions (see fig.4).

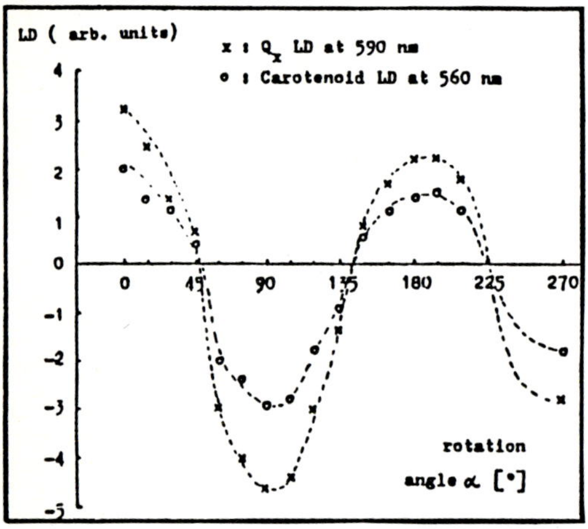

Fig. 4

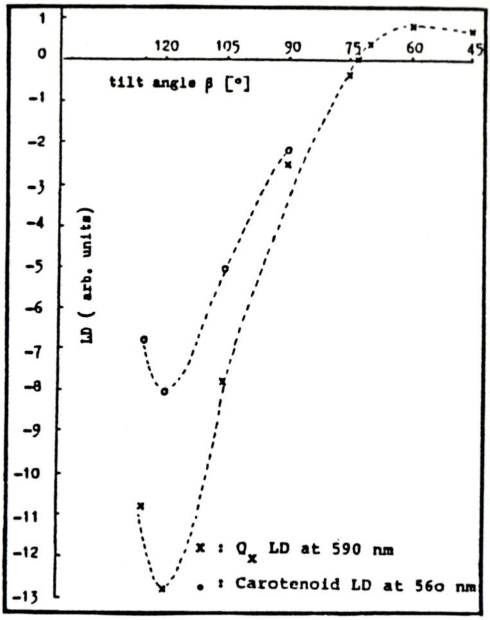

Fig. 5

By considering the number of zero crossovers of the LD signals, we can exclude a threefold, fourfold and sixfold rotation axis, no rotation symmetry axis or a twofold axis are in accord with our results. This finding supports our initial assumption that the thin crystalline plates belong to the triclinic spacegroup. The dichroism signal was found to change drastically when the crystal was tilted out of the plane perpendicular to the measuring beam. For a given rotation angle α, maximum Q_x and carotenoid dichroism was found when the crystal plane formed an angle of 120° to the beam (see fig.5 for the dependence of LD on the tilt angle β). The transition moments are thus tilted out of the crystal plane. A detailed determination of the angles between the transition moments and the crystal plane and crystal axes, however, with a more accurate positioning of the crystals with respect to the polarization axis and beam direction, is in progress.

Acknowledgements: We wish to thank Prof. W.Kreutz for continuous support and interest. We are grateful to Nasser Gad'on and Prof. G.Drews for providing us with the purified light-harvesting complex used for crystallization.

References:

(1) Welte,W. et al. (1985) FEBS Lett. 182, 260-264
(2) Breton,J. (1977). Thèse, Université Paris-Sud
(3) Bolt,J. and Sauer,K. (1979) Biochim. Biophys. Acta 546, 54-63
(4) Breton,J. et al. (1981) in "Photosynthesis III. Structure and Molecular Organisation of the Photosynthetic Apparatus" (Akoyunoglou,G. ed.) Balaban Int. Sci. Philadelphia, pp.445-459
(5) Kramer, H.H.M. et al. (1984) Biochim. Biophys. Acta 765, 156-165

Part II

Reaction Centers: Structure and Interactions

The Crystal Structure of the Photosynthetic Reaction Center from *Rhodopseudomonas viridis*

J. Deisenhofer and H. Michel

Max-Planck-Institut für Biochemie, D-8033 Martinsried, F.R.G.

The X-ray analysis at 3 Å resolution of well-ordered crystals (1) of the photosynthetic reaction center (RC) from the purple bacterium Rhodopseudomonas viridis allowed the calculation of an electron density map. Interpretation of this map led to the construction of atomic models of the prosthetic groups, and of the protein subunits in the RC complex. In this article we briefly summarize results of the X-ray work which have been described in more detail elsewhere (2),(3),(4).

Fig. 1 shows the four bacteriochlorophyll-b (BChl-b) molecules, two bacteriopheophytin-b (BPh-b) molecules, one menaquinone, one non-heme iron, and four heme groups located in the electron density map. In Fig. 2 these molecules are shown, together with the polypeptide chains of the RC's protein subunits L, M, H, and cytochrome. The heme groups are covalently linked to the cytochrome subunit. The remaining chromophores are associated with the subunits L and M in the central part of the RC. In this region, the structure shows a high amount of local twofold symmetry, which relates large parts of the polypeptide chains of L and M, and also the pyrrole ring systems of corresponding BChl-bs and BPh-bs. The non-heme iron sits on this L-M local diad.

Close to the L-M local diad, two BChl-bs are in intimate contact, and form the "special pair", the primary electron donor. Their pyrrole rings I are stacked with a distance of about 3 Å; the acetyl groups at these rings interact with the Mg^{2+} ions of the other BChl-bs. On either side of the L-M local diad an "accessory" BChl-b is in contact with the special pair. The BPh-b are located next to these accessory BChl-bs. So, we find in the RC two strings of chromophores, which both originate at the special pair, and appear to be suitable as electron pathways.

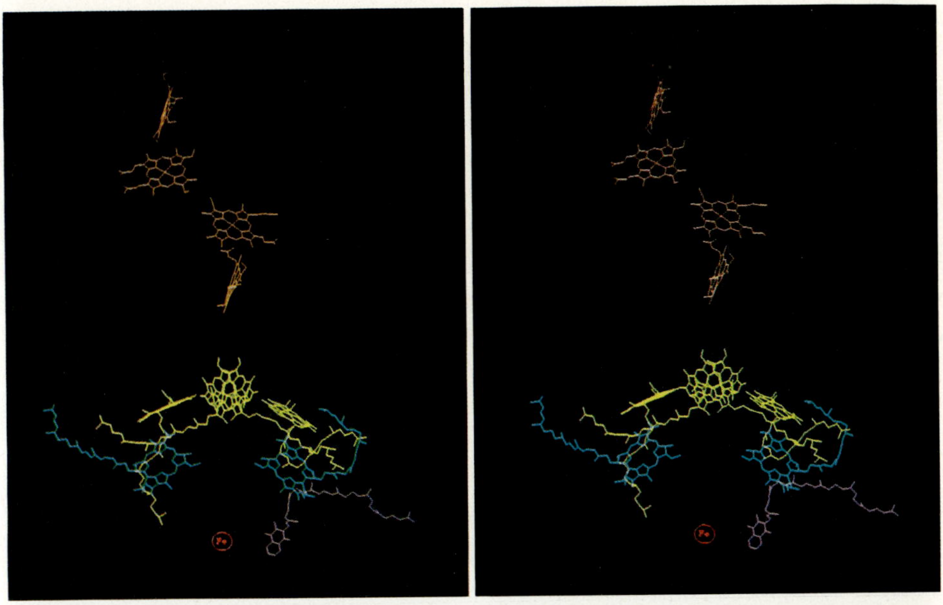

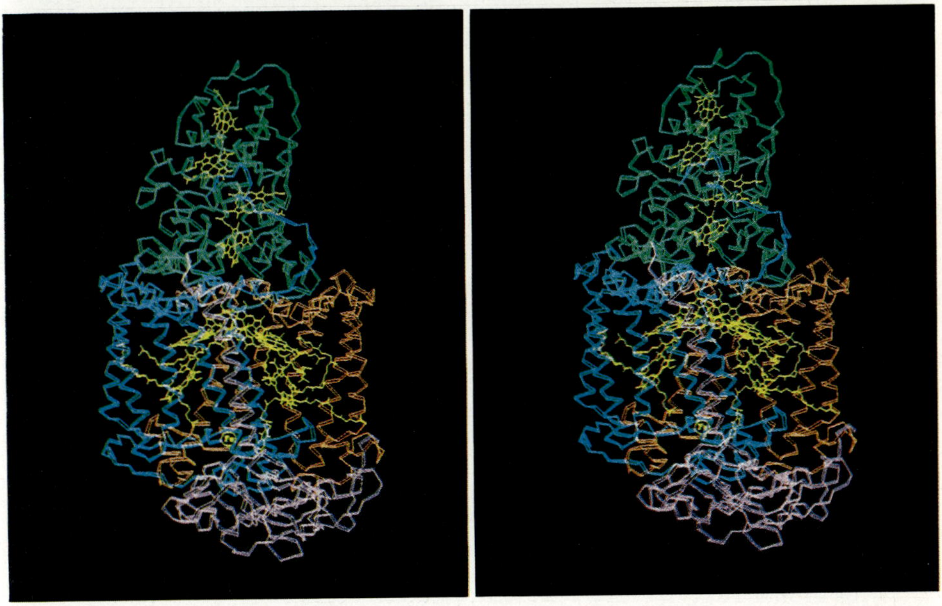

Fig.1. Stereo pair showing the arrangement of the chromophores in the RC. (Colours: *yellow*: BChl-b, *light blue*: BPh-b, *purple*: quinone, *red*: non-heme iron, *brown*: heme groups)

Fig.2. Stereo pair showing the structure of the protein subunits (represented as ribbons), together with the chromophores in the RC. (Colours: *green*: cytochrome, *brown*: L, *blue*: M, *purple*: H. *yellow*: chromophores)

The local twofold symmetry ends at the level of the quinones: Only Q_A was found in the electron density map near the BPh-b positioned on the L-side of the L-M complex. Most of the second quinone, Q_B, was lost during preparation of the RCs (Sinning and Michel, unpublished). Binding studies in RC crystals with o-phenanthroline, terbutryn, and ubiquinone-1 revealed a common binding site for these three compounds, which we believe to be the Q_B binding site. This Q_B site is related to the Q_A site by the L-M local diad. It remains an intriguing question why the electron pathway in the RC leads almost exclusively from the special pair first to Q_A, and then to Q_B. The four heme groups of the cytochrome form an almost linear string, leading from the tip of the subunit towards the special pair. Two pairs of heme group are related by another local diad; about one third of the polypeptide chain also obeys this internal local symmetry of the cytochrome subunit.

Outstanding structural features of each of the subunits L and M are five helices running in the general direction of the L-M local diad, with lengths between 24 and 30 residues. From the hydrophobic character of these helical segments we conclude that they span the bacterial membrane. On both sides of the helix region of L and M the polypeptide segments connecting the transmembrane helices and the terminal segments form flat surfaces perpendicular to the L-M local diad. The cytochrome is bound to the surface close to the special pair; the carboxy-terminal globular domain of the H-subunit is in contact with the surface that is near the non-heme iron and the quinone. Near its amino-terminus, the H-subunit forms another transmembrane helix; the remainder of this subunit is the only part of the RC with a significant amount of beta-sheet.

References:

1 H. Michel: J. Mol. Biol. 158, 567-572 (1982)

2 J. Deisenhofer, O. Epp, K. Miki, R. Huber, and H. Michel: J. Mol. Biol. 180, 385-398 (1984)

3 J. Deisenhofer, H. Michel and R. Huber: Trends Biochem. Sci. 10, 243-248 (1985)

4 J. Deisenhofer, O. Epp, K. Miki, R. Huber and H. Michel: submitted

Single Crystals from Reaction Centers of *Rhodopseudomonas viridis* Studied by Polarized Light

W. Zinth, M. Sander, J. Dobler, and W. Kaiser

Physik Department, E11, der Technischen Universität München, D-8000 München, F.R.G.

H. Michel

Max-Planck-Institut für Biochemie, D-8033 Martinsried, F.R.G.

The spectral properties and the pigment composition of isolated reaction centers (RC) have been studied extensively in the past /1-3/. Recently the reaction centers from Rhodopseudomonas viridis (Rps. viridis) were crystallized /4/ and an X-ray structure analysis yielded the pigment arrangement /5/. From the latter publication we know that the four bacteriochlorophyll b (BC) and the two bacteriopheophytine b (BP) molecules are arranged in two branches (see Fig. 1). They are related pairwise by a twofold local symmetry axis (broken line). The pigments of the branch containing the quinone, probably a menaquinone, MQ, are closer to the L protein subunit, the other pigments are closer to the M subunit. Consequently we index them by L and M. It was assumed /5/ that BC_{LP} and BC_{MP} form the special pair (P) and that BC_{LA} and BC_{MA} are the accessory bacteriochlorophylls.

In this Letter we present the results from optical investigations of the crystallized reaction centers and combine the optical and the structural information. The optical properties of the reaction centers in the single

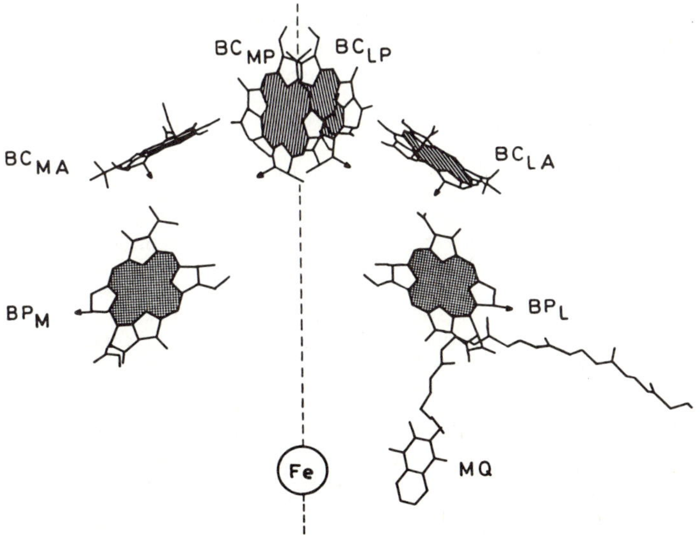

Fig. 1 Model of the central part of the reaction center of Rps. viridis containing the four bacteriochlorophyll b (BC), the two bacteriopheophytine b (BP) molecules and the menaquinone (MQ) after Ref.5. The molecules are arranged in two branches, L and M.

crystals are characterized and the influences of oxidation, reduction and photooxidation on the absorption spectra are shown. We make use of the ordered arrangement of the RC in the crystals to interpret the optical spectra: We assign absorption bands to particular pigments in the molecular model of the reaction center.

The reaction centers were isolated and the crystals were grown as described recently /6,4/. The space group of these crystals is $P4_32_12$. The tetragonal unit cell has the dimension of $223\times223\times114$ A^3 and contains eight asymmetric units. Each asymmetric unit is occupied by one RC molecule yielding a RC-concentration of 2.35×10^{-3} mol/l. In most cases the crystals grow in the form of tetragonal columns with the preferential growth direction parallel to the z-axis. Fortunately, the crystals can be obtained in the form of thin platelets of several hundered micrometers in length and width and with a uniform thickness of a few micrometers. The surfaces of the platelets are parallel to the xy, xz, or yz planes. The orientation of the crystallographic axis can be determined upon microscopic inspection. During the investigations the crystals were kept in closed cells which contained a salt solution of 2.7 M ammonium sulfate, 20 mM sodium phosphate, 0.1% N,N-dimethyldodecylamine N-oxide, 1% triethylammonium phosphate, and 1% heptane-1,2,3-triol, pH 6.5 (chemically reduced crystals were prepared by adding 20 mM sodium ascorbate to the salt solution, whereas oxidation of the RC was accomplished by adding $K_3Fe(CN)_6$ (100 mM).

The transmission spectra were recorded with a self-constructed microscope set-up: Light from a tungsten lamp was monochromized by a first grating spectrometer of 3nm resolution. The light beam of a certain linear polarization was focussed onto the crystal via a microscope objective (illuminated area 70 μm × 70 μm). The intensity of the measuring light in the crystals was kept at the lowest possible level,in order to avoid absorption changes due to photooxidation of the RC. We have confirmed that less than 3% of the RC were photooxidized due to the action of the measuring light. Part of the incoming measuring light was split-off by a beamsplitter and used as a reference. The light transmitted through the crystal was imaged by another objective onto the slit of a second spectrometer, and detected by a suitable photomultiplier (photocathodes S 20 or S 1). The second spectrometer shielded the photomultipliers against stray light. The electrical signals were processed by a phase-sensitive detection system.

Light-induced absorbance changes were measured in the following way: At certain wavelengths the transmitted light power was recorded prior to (I_{dark}) and after illumination ($I_{illum.}$) of the RC by actinic light (for 30 s). The logarithm of the ratio $I_{illum.}/I_{dark}$ yielded the absorbance change ΔA. Actinic light was obtained from a Xe-arc-lamp used in conjunction with a spectrometer transmitting radiation between 960 nm and 980 nm. The actinic light was linearly polarized and travelled in the crystals collinearly to the measuring beam. In order to avoid saturation of the absorbance changes, the level of the actinic light was kept so low, that the relative absorbance changes were always smaller than 10%.

Fig. 2 shows absorption spectra of the RC crystals in different surroundings. The same geometry of observation was used when recording the three spectra: the crystals were oriented with the tetragonal axis perpendicular to the propagation direction (k) of the measuring light. The polarisation (vector E) was parallel (broken curves) and perpendicular (solid curves) to the tetragonal axis.

The absorbance spectrum of the native RC crystals (Fig 2a) displays all features known from RC in solutions. The Q_y absorption bands of the BC and

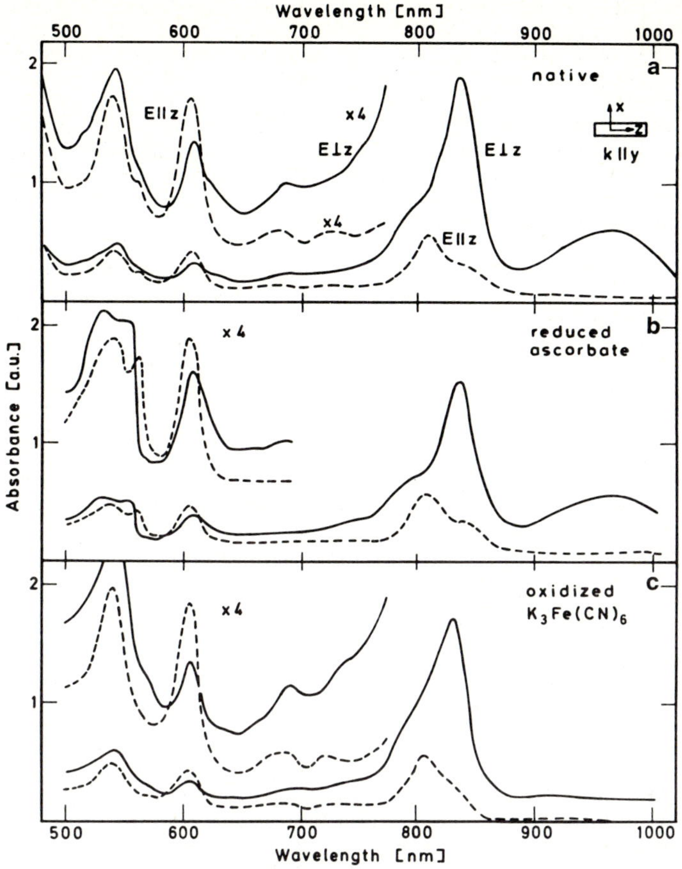

Fig. 2 Polarized absorbance spectra of single crystals of reaction center preparations from Rps. viridis: (a) native crystals, (b) crystals treated with ascorbate, (c) crystals treated with $K_3Fe(CN)_6$.

BP are between 750 nm and 850 nm . Q_y of the special pair (P) is around 960 nm. The Q_x absorption band of the BC is close to 605 nm, whereas the Q_x absorption of the BP and the bands of the cytochrome c are seen near 550 nm. Of special interest is the comparison of the extinction coefficients determined for RC solutions and RC crystals. Using the transmission data for λ = 960 nm, the measured crystal thickness, the known RC concentration and taking into account that the transition moment of the 960 nm band is polarized perpendicular to the tetragonal axis, we obtain the value for the extinction coefficient of ε = 110,000 M^{-1} cm^{-1}. In order to compare the data from RC crystals and RC solutions, the extinction coefficient ε was calculated for isotropically arranged RC. The value determined for the crystallized RC agrees with the published extinction coefficient of ε = 125,000±25,000 $M^{-1}cm^{-1}$ for RC in solutions /2/. This finding gives a strong indication that the RC crystals contain intact RC where the strength of the special pair transition at 960 nm is not influenced by crystallisation.

The polarized absorption spectra allow us to deduce qualitatively the orientation of the transition moments within the RC crystals. The combination with the structural data allows us to assign absorption bands to particular pigments. Detailed information can be obtained for the BP molecules: It is known from the literature that the BP absorb around 800 nm /1/. In the polarized absorption spectra we find one band at 805 nm polarized parallel to the tetragonal z axis, whereas a strong shoulder is seen in the $(E \perp z)$ spectrum at 790 nm. Since BP_L has a considerable absorption for light polarized parallel to the z axis, we can assign the absorption at 805 nm to BP_L and the shoulder at 790 nm to BP_M.

RC crystals reduced by ascorbate show changed absorption spectra (see Fig. 2b): Between 550 nm and 560 nm, in the range of the cytochrome c molecules absorption increases considerably. It is known that the RC of Rps. viridis in total contain four cytochrome c molecules. The X-ray structure analysis shows that the four cytochrome c molecules are positioned in the RC roughly along a line pointing away from the special pair. There are two different types, two cytochrome 558 and two cytochrome 553 molecules named according to the position of their long-wave absorption peak. The observed absorbance increase suggests that in the native crystals the cytochrome c molecules are mainly oxidized,having negligible absorption around 555 nm. In the reduced crystals the additional absorption due to the cytochromes has a pronounced dichroism: A peak is seen in the $E \| z$ spectrum at 558 nm whereas the $E \perp z$ spectrum displays a shoulder at 558 nm and a peak at 553 nm. The different polarisation properties of the two absorption bands in combination with the results of the structure analysis indicate that the cytochrome molecule which is closest to the special pair is a cytochrome 558.

In Fig. 2c the absorption spectra of RC crystals oxidized by $K_3Fe(CN)_6$ are shown. The 960 nm transition of the special pair disappears and the 830 nm band is shifted to shorter wavelengths. When inspecting more carefully the absorption spectra, one realizes a weak absorbance decrease around 605 nm. In addition, the broad shoulder around 620 nm in the $E \perp z$ spectrum is reduced (as compared with Fig. 2a). The complete lack of any absorption peak between 550 nm and 560 nm suggests that all the cytochrome molecules are oxidized. Finally there is a very broad absorption background extending throughout the visible in the $E \| z$ spectrum. This background may be tentatively assigned to the oxidized special pair P^+ (see below).

The action of oxidation on the crystallized RC is seen more clearly when the absorbance changes induced by photooxydation are studied. The relative polarisation of actinic and measuring light in the crystals allows to obtain additional information. In tetragonal crystals different geometries are needed to deduce the complete spectral information. In principle the spectra of Fig. 3 allow to determine the direction fo the transition moments of the pigments in the crystals /7/. As an example we have calculated from the light-induced absorbance changes the direction of the special pair transition at 960 nm. The transition is polarized mainly in the xy-plane (within $\pm 7^\circ$) and at an angle of $30^\circ \pm 5^\circ$ relative to the x-direction. It is interesting to note that this value agrees within the experimental accuracy with the direction determined by excitonically coupling the Q_y-transitions of the BC molecules forming the special pair (BC_{LP} and BC_{MP}). This finding supports the interpretation that: (i) BC_{LP} and and BC_{MP} form indeed the special pair and that (ii) the interaction between the special pair and other pigments does not influence the direction of the 960 nm transition.

The quantitative analysis of the directions of the transition moments in the Q_y range of the BP and the accessory BC is made difficult by the strong overlap of the absorption bands. Nevertheless,some interesting features are readily determined from the absorption spectra of Fig. 3. Around 850 nm the

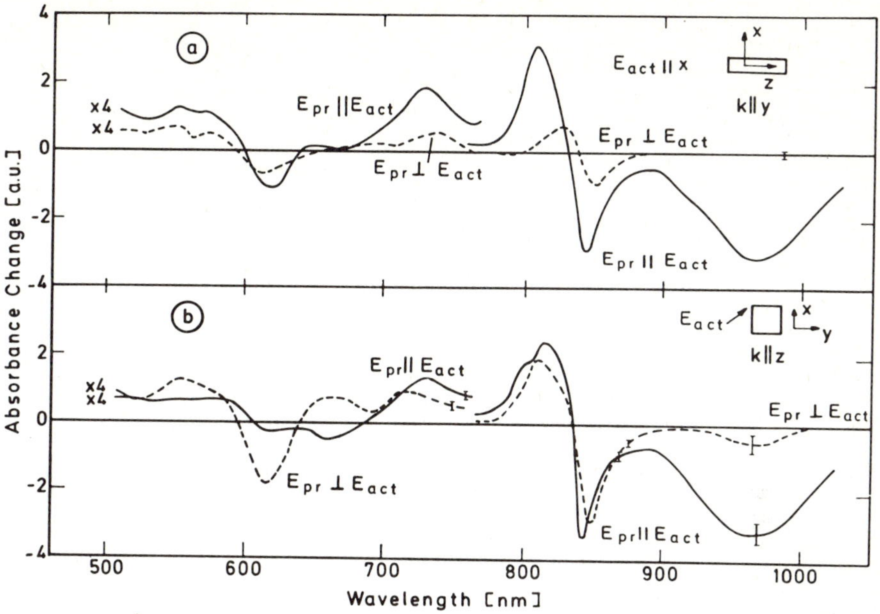

Fig. 3 Absorbance changes of the crystallized reaction centers induced by illumination with actinic light at 970 nm. The propagation direction k of actinic and measuring light is parallel to the y-axis (Fig. 3a) or parallel to the z-axis (Fig. 3b). The electric - field vector of the actinic light (E_{act}) is parallel to the x-axis and at 45° to the x-axis in Fig. 3a and Fig. 3b respectively. The solid and the broken curves were taken with the electric-field vectors of measuring (E_{pr}) and actinic light (E_{act}) parallel and perpendicular to each other, respectively.

difference spectra have a dispersion-type shape. A more careful inspection of the experimental data reveals that the points of zero-crossing, as well as the position of the minima and maxima, do not coincide in the four spectra of Fig. 3. This indicates that the absorption bands show shift and reorientation upon photooxidation. This can be caused by the oxidation of the special pair changing the excitonic interactions and leading to Stark-shifts of the resonance frequencies. In the spectral region of the Q_x bands of the BC one finds a pronounced absorbance decreases at 618 nm with a width of 30 nm. The polarization dependence of that absorbance decrease supports the assignment that the Q_x band of the special pair is at 618 nm. The apparent red-shift of the special pair Q_x transition relative to the absorption band of the accessory BC at 605 nm allows a selective excitation of the special pair by light with $\lambda \approx 620$ nm. This finding is important for the interpretation of very recent time-resolved experiments on the RC /8/.

The detailed optical investigations of the crystallized reaction centers from Rhodopseudomonas viridis give valuable information: (i) The reaction center crystals contain intact reaction centers which are photochemically active. There is quantitative agreement of the absorption cross-section of the special pair in solution and in the crystal. (ii) The reaction centers can be reduced and oxidized in the crystalline form. This allows structural studies from chemically treated reaction centers. (iii) The assignment of absorption bands to particular pigment molecules is achieved.

References
1 J.P. Thornber, P.L. Dutton, J. Fajer, A. Forman, D. Holten, J.M. Olson, W.W. Parson, R.C. Prince, D.M. Tiede and M.W. Windsor in Proceedings of the 4th International Congress on Photosynthesis, 1977 (D.O. Hall, J.Coombs and T.W. Goodwin, eds. The Potochemical Society, London, 1978), pp. 55-70
2 R.K. Clayton and B.J. Clayton: Biochim. Biophys. Acta 501, 478 (1978)
3 J.P. Thornber, R.J. Cogdell, R.E.B. Seftor and G.D. Webster: Biochim. Biophys. Acta 593, 60 (1980)
4 H. Michel: J. Mol. Biol. 158, 567 (1982)
5 J. Deisenhofer, O. Epp, K. Miki, R. Huber and H. Michel: J. Mol. Biol. 180, 385 (1984)
6 W. Zinth, W. Kaiser and H. Michel: Biochim. Biophys. Acta 723, 128 (1983)
7 E.W. Knapp, W. Zinth, M. Sander, S.F. Fischer, W. Kaiser, J. Deisenhofer and H. Michel: PNAS (1985)
8 W. Zinth, M.C. Nuss, M.A. Franz, W. Kaiser and H. Michel, same book.

On the Analysis of Optical Spectra of *Rhodopseudomonas viridis* Reaction Centers

E.W. Knapp and S.F. Fischer

Physik-Department, Technische Universität München, D-8046 Garching, F.R.G.

1. Introduction

Since the structure of the reaction center (RC) Rhodopseudomonas viridis is now known /1,2/, it is an interesting challenge to make model calculations for optical spectra and to compare these with the experiments /3-5/. We choose a simple exciton model for the six pigments of the RC /6/. At present such model calculations cannot be entirely free of adjustable parameters, if one wants to achieve good agreement, with line positions and intensities of the observed spectra. However, by starting from the structural data and by treating simultaneously several spectroscopic data, like polarized absorption spectra /3,4/, light-induced absorbance changes /5,6/ and circular dichroism (CD) spectra /7,8/, we are able to determine the model parameters quite well; in particular we learn about the excitonic interaction within the special pair and its inherent and solvent-induced spectral shifts. Moreover, we can predict the admixture of the excitonic wavefunctions between the pigments. This information may help to clarify questions concerning the dynamics of the light-induced charge separation.

2. Theoretical Methods

The central part of the RC of Rhodopseudomonas viridis contains six pigments: four bacteriochlorophyll b (BC) and two bacteriopheophytin b (BP) molecules. The pigments are arranged in two branches (M and L), which exhibit a nearly twofold rotation symmetry. Two BC one of each branch interact most closely. They form the so-called special pair (P) and are labeled BC_{MP} and BC_{LP}. The two accessory (A) BC's (BC_{MA}, BC_{LA}) are next to the special pair BC's. The two BP's (BP_M, BP_L) are situated at the corresponding BC_A's opposite to the special pair BC's.

To calculate the spectra, only the lowest electronic excitation forming the Q_y bands of the BC and BP pigments are considered. The higher electronic excitations forming the Q_x and the Soret bands B_x and B_y have an energy, which is about 4600 cm^{-1}, 11800 cm^{-1} and 14400 cm^{-1} above the energy of the Q_y band respectively. The influence of these bands on an exciton spectrum of the RC in the regime of the Q_y bands can be calculated by perturbation theory. It is largest for the special pair pigments, where the excitonic interactions estimated from the structure are 300 cm^{-1}, -130 cm^{-1}, 1100 cm^{-1} between the Q_y band and the Q_x, B_x, B_y band respectively. Hence in all three cases the resulting admixture to the exciton states of the Q_y bands is below 8 %. Within our model /6/ the state of the RC corresponding to a Q_y band excitation localized at the pigment n is denoted by $|n\rangle$ its energy by E(n). Any two states $|n\rangle$ and $|m\rangle$ are coupled via interaction matrix elements V(n,m). The Hamiltonian of the RC containing six pigments can be written as

$$H = \sum_{n=1}^{6} | n \rangle E(n) \langle n | + \sum_{n \neq m=1}^{6} | n \rangle V(n,m) \langle m |. \quad (1)$$

The theoretical spectra are calculated by use of the transition moment vectors $\vec{\mu}^{(n)}$ of the individual pigments and the solution of the eigenvalue problem of the Hamiltonian, Eq. (1), describing the excitonic system /6/.

3. Model Paramaters

The parameters of the exciton model are the excitonic interaction energies V(n, m) between different pigments, the diagonal energy-shifts E(n) and the transition moment vectors $\vec{\mu}^{(n)}$ of the individual pigments. In principle, the calculation could be done free of adjustible parameters. However, the solvent shifts, the dipolar transition moments, and even some interaction energies are not given with sufficient accuracy. The transition moment vectors of the Q_y bands of BC and BP molecules are along the line connecting the nitrogen atoms of the pyrrole rings I and III. Their orientation in the RC is determined by the structure. The absolute value of the transition moment vectors $|\vec{\mu}^{(n)}|^2$ given by experiments are 30D^2 and 19D^2 /8/ for BC and BP monomers respectively. Recent experiments /9-11/ point to much larger values. To learn more about the sensitivity of the results upon these parameters, the values 45D^2 and 30D^2 are also considered. The excitonic interaction energies V(n,m) are calculated with an extended dipole model /6/ using the known transition moment vectors of the pigments. The dipole model is appropriate for pigments at a distance not closer than 3 Å /12/. Since this condition is not fulfilled for the special pair BC's the corresponding interaction parameter $V_P = (BC_{MP}, BC_{LP})$ is varied together with the diagonal energy parameters E(n) under the constraint that the position of the low-energy transition is fixed by the experiment. Minor adjustments are also introduced to other diagonal energy parameters in order to fit the polarized absorption spectra /6/. To facilitate a comparison with experiments, the excitonic transition in the calculated spectra are represented by Gaussian line profiles. The widths of these lines is taken from the experimental absorption spectrum (Fig.1).

4. Results and Discussion

The absorption spectrum of the Q_y bands of the RC contains a broad peak at 965 nm, which is shifted about 1750 cm^{-1} below the more intense peak at 830 nm. The peak at 965 nm can be bleached /5/, it corresponds to an exciton state that is formed by the special pair. The large energy-shift ΔE of this excitonic special pair state is determined by the diagonal energy shift parameter E_P of the special pair pigments relative to the other pigments, and by the excitonic interaction V_P between the BC's of the special pair. The excitonic interaction leads to a splitting of the special pair exciton states into a lower band state P(-) and a upper band state P(+), whose energy-shifts are roughly given by $\Delta E_{+}^{-} = E_P \mp V_P$ relative to the energy of the other pigments. The energy-shift is kept fixed at ΔE = -1750 cm throughout this work to account for the relative position of the lower special pair band P(-).

The contribution of the dipole-dipole interaction to the excitonic interaction of the special pair is 600 cm^{-1} and 900 cm^{-1} assuming a transition dipole strength of 30D^2 and 45D^2 for the BC molecules respectively. Fig. 1 displays absorption and CD spectra, where this interaction energy is varied between 600 cm^{-1} and 1200 cm^{-1}. Thereby, the energy of the upper special pair band P(+) changes from 865 nm to 780 nm. Nevertheless, we find

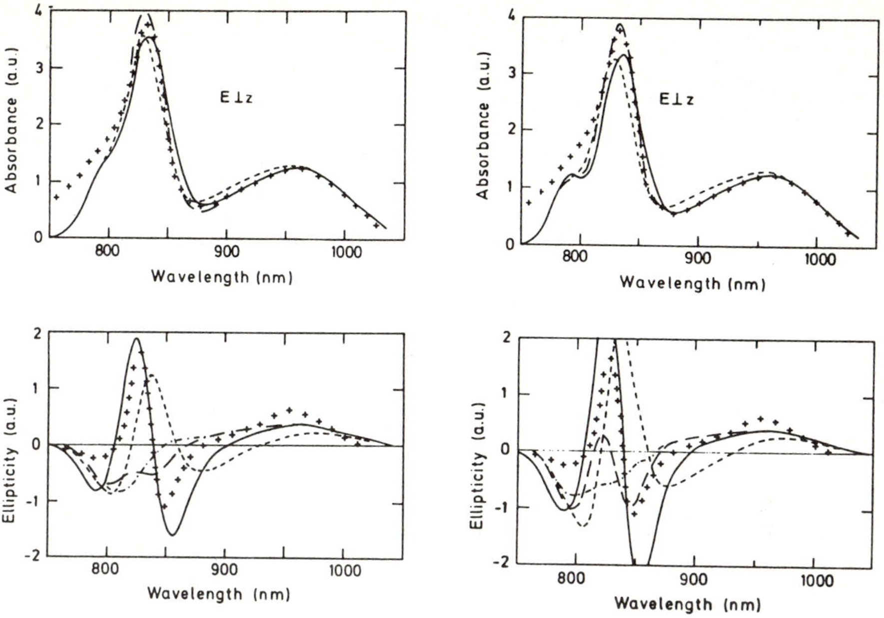

Fig. 1. Polarized absorption and CD spectra /7/ of RC's. Experiments are represented by crosses. The special pair interaction energy V_P is varied. The parameter values of V_P /cm^{-1}/ are:- - - - -600; ——— 800; — — 1000; -·-·-· 1200. In the left part $30D^2$ and $19D^2$ in the right part $45D^2$ and $30D^2$ have been used for the dipolar interaction strength of the BC and BP molecules respectively. In each part a single intensity adjustment has been made.

that the absorption spectra are not very sensitive to variations of the excitonic interaction V_P. Only minor changes appear in the absorption spectra for V_P = 600 cm^{-1}. These are due to the upper special pair state P(+), whose energy falls for this value of V_P below the energy of the excitonic states of the accessory BC and the BP pigments,and borrows now some intensity from them.

The CD spectra are much more sensitive to the position of the upper special pair state. Only the spectra calculated with the special pair interaction parameter V_P = 800 cm^{-1} provide an acceptable agreement with the experimental spectra. In each part of Fig.1 the intensities of the calculated spectra are fitted only for one parameter set to the lower band of the special pair. The main difference between the calculated spectra with the larger and the smaller transition dipole moments is found in their absolute value of absorption intensity and rotational strength. For the lower special pair band P(-) the calculated rotational strength is 15 % above (40 % below) the experimental value of 3.2 Debye Bohr magneton /8/ if one uses the large (small) transition dipole moment values respectively.
The best fits (Fig.2) are obtained using the excitonic interaction V_P = 850 cm^{-1} (900 cm^{-1}) of the special pair pigments for the small (large) values of the transition dipole moments of the pigments. The first value is just 250 cm^{-1} larger than the dipolar contribution to the excitonic interaction and may be due to enhanced exciton interaction /12/ of the special pair pigments. The latter value is just the amount of pure dipolar interaction of the special pair pigments for the larger transition dipole moments.

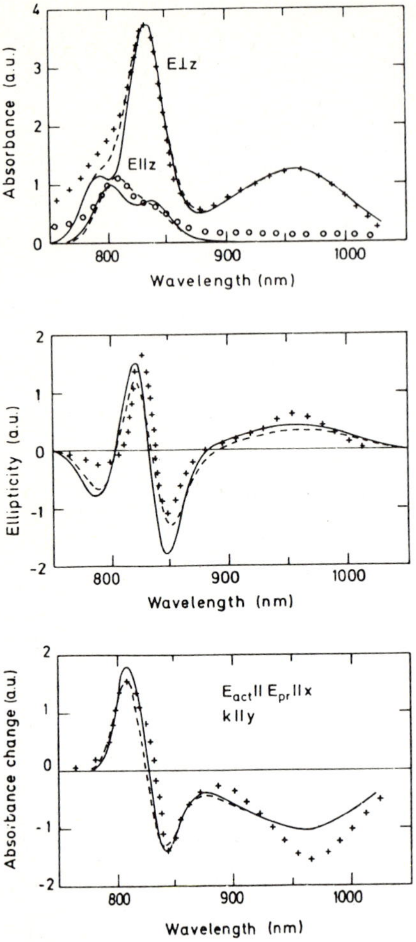

Fig. 2

Polarized absorption, CD and light induced absorbance change spectra. Polarization and geometry are indicated in the figure. The symbols represent the experimental spectra. The calculations are best fits. The solid lines are spectra with a special pair interaction of V_P = 905 cm^{-1} and a dipolar strength of 45 and 30D^2 for the BC and BP molecules respectively. The dashed lines are calculated spectra with V_P=850 cm^{-1} and a dipolar strength of 30D^2 and 19D^2 for BC and BP respectively. Intensities have been adjusted to the experimental values in the special pair band at 965 nm.

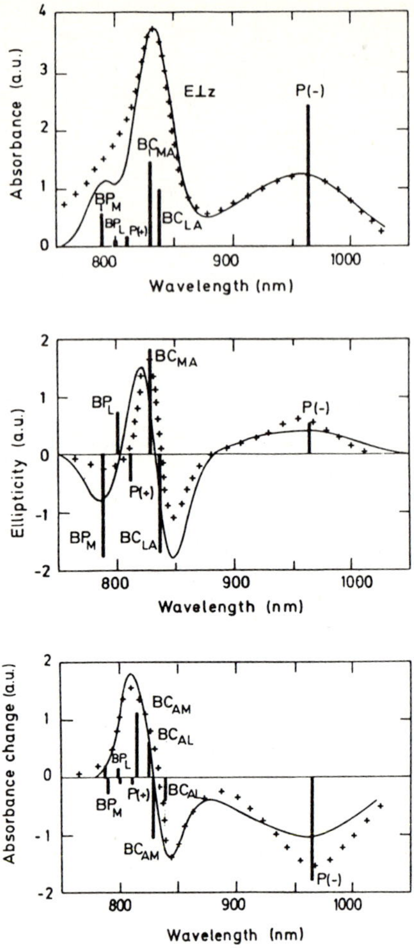

Fig. 3

Same as Fig.2. Best fit with a dipolar strength of 45 and 30D^2 for BC and BP respectively. The individual excitonic transitions are given by bars, whose lengths indicate the corresponding intensities.

Comparing the best fits of the absorption spectra (Fig.2) for the large and small transition dipole moments, the only discrepancy appears at 800 nm, where the BP molecules absorb. Apparently the larger dipolar interaction of the BP's with the neighbour BC's leads to a loss of intensity of the BP's in favour of the BC's as compared to the BP monomer intensity. For the large and small transition dipole moments the upper special pair band P(+) is situated between the excitonic states from the BP and BC molecules. Figure 3 displays the decomposition of the calculated spectra in the individual excitonic states. The lower special pair band contains only contributions from the special pair BC's. Also the exciton state with the largest energy contains nearly a single pigment contribution (BP_M). All other excitonic states have large contributions from a single pigment, but also considerable admixture from other pigments. For instance, the upper special pair band P(+) is strongly mixed with the BP_L and BC_{MA} pigments. Also the lowest excitonic state from the accessory BC molecules corresponding to BC has a considerable admixture (35 %) from the special pair BC's. The upper special pair band has often been identified with this state /6/. Our analysis provides now a refined interpretation.

In contrast to absorption spectra, where the upper special pair band P(+) carries only little intensity, the rotational strength of the P(+) state in the CD spectrum is considerable (Fig.3). This is one reason for the sensitivity of the CD spectrum on the energy position of this state, which varies with the excitonic interaction V_P. The largest deviation of the calculated and the measured CD spectra are in the regime of the BP molecules at 800 nm. These discrepancies, which also appear in the absorption spectra, indicate that the calculated excitonic interactions of the BP molecules with the pigments in the neighbourhood are too large. Another source of discrepancy relates to the fact that the experimental CD spectrum is non-conservative (positive and negative ellipticity do not cancel) in contrast to the simple theoretical description used /6/.
The absorbance change spectra (Fig.3) exhibit a negative lobe at 965 nm due to the bleaching of the lower special pair band and a negative (845 nm) and positive (820 nm) lobe. The latter relates to the Stark shift of the accessory BC excitonic states to higher energies after illumination. This shift is due to the presence of a negative charge at the special pair BC's induced by illumination of the RC.

Conclusions

Starting from the structural information, the optical spectra can be interpreted in a consistent way. The contributions of the pigments to the various excitonic states of the spectra are obtained. The analysis shows that the accessory BC and the BP molecules of the two pigment branches are rather inequivalent. The energy position of the upper exciton band P(+) of the special pair has been determined to be at 811 nm. It has, however, a considerable admixture from the accessory BC of the M branch (20 %) and from the BP molecule of the L branch (27 %) /6/. If this P(+) state is involved in the charge separation process, it should favour the electron transfer to the BP of the L branch. This is consistent with recent findings from time-dependent fluorescence data /13/.

Acknowledgement

The analysis has been done in close contact with W. Zinth, W. Kaiser, and H. Michel, who have given the experimental results in the preceding article. The authors like to thank also Prof. M.E. Michel-Beyerle and Prof. H. Scheer for valuable discussions. The work was supported by the Deutsche

Forschungsgemeinschaft under the project "Primary Processes in Bacterial Photosynthesis (SFB 143).

References

/1/ J. Deisenhofer, O. Epp, K. Miki, R. Huber, and H. Michel: J. Mol. Biol. 180, 385 (1984).
/2/ J. Deisenhofer and H. Michel: This volume.
/3/ W. Zinth, W. Kaiser, and H. Michel: Biochim. Biophys. Acta 723, 128 (1983).
/4/ W. Zinth, M. Sander, J. Dobler, W. Kaiser, H. Michel: This volume.
/5/ W. Zinth, E.W. Knapp, S.F. Fischer, W. Kaiser, J. Deisenhofer, and H. Michel: Chem. Phys. Letters (1985) in press.
/6/ E.W. Knapp, W. Zinth, M. Sander, S.F. Fischer, W. Kaiser, J. Deisenhofer, and H. Michel: Proc. Natl. Acad. Sci. US (1985) in press.
/7/ K.D. Philipson and K. Sauer: Biochemistry 12, 535 (1973).
/8/ V.A. Shuvalow and A.A. Asadov: Biochim. Biophys. Acta 545, 296 (1979).
/9/ A. Scherz and W.W. Parson: Biochim. Biophys. Acta 766, 653 (1984).
/10/ W.W. Parson, A. Scherz, and A. Warshel: This volume.
/11/ A. Scherz: Private communication.
/12/ A. Warshel: J. Am. Chem. Soc. 101, 744 (1979).
/13/ H. Hörber, W. Göbel, M.E. Michel-Beyerle, and E.W. Knapp: This volume.

Orientation of the Chromophores in the Reaction Center of *Rhodopseudomonas viridis*. Comparison of Low-Temperature Linear Dichroism Spectra with a Model Derived from X-Ray Crystallography

J. Breton

Service de Biophysique, Départment de Biologie C.E.N Saclay, F-91191 Gif-sur-Yvette Cedex, France

Linear dichroism (LD) and absorption (A) spectra of reaction centers from Rps. viridis included in the native chromatophores or reconstituted in planar aggregates have been recorded at 10 K. The samples were oriented in squeezed polyacrylamide gels and the primary donor P was in the reduced or (chemically) oxidized state. The LD spectra of reaction centers in these two states are in favour of a dimeric model of P in which excitonic coupling between the two non-parallel Q_Y transitions leads to a main transition at 990 nm (parallel to the membrane plane) and another one of smaller oscillator strength at 850 nm (tilted at 55-60° out of the membrane plane). These assignments are in close agreement with the ones proposed in a previous LD study at room temperature (Paillotin, G., Vermeglio, A. and Breton, J. 1979, Biochim. Biophys. Acta 545, 249-264). The main Q_X excitonic component of P has a broad absorption peaking at 620 nm and corresponds to dipoles exhibiting the same orientation as those responsible for the 850 nm transition. On the basis of the present LD study and of CD data of chemically oxidized -minus- reduced reaction centers, we propose that the minor Q_X excitonic component of P is oriented close to the membrane plane and absorbs around 660 nm.

The two monomeric bacteriochlorophylls exhibit a positive LD for both their Q_Y transitions (degenerate at 834 nm) and their Q_X transitions (resolved at 600 and 607 nm) indicating that the planes of these molecules are only slightly tilted out of the membrane plane. The two bacteriopheophytins exhibit strong negative LD with identical LD/A values for their Q_Y transitions (resolved at 790 and 805 nm) and small positive LD for their Q_X transitions (resolved at 534 and 544 nm) demonstrating that these two molecules are strongly tilted out of the membrane plane with each of the Q_Y transitions tilted at 45-50° out of that plane.

A comparison of these LD data with the qualitative model derived from X-ray crystallography (Deisenhofer, J., Epp, O., Miki, K., Huber, R. and Michel, H. 1984, J. Mol. Biol. 180, 385-398) clearly suggests that a good agreement exists between the results of the two techniques under the following conditions : (i) The C-2 symmetry axis of the reaction center runs along the membrane normal, (ii) excitonic coupling is present only in the primary donor special pair and (iii) the direction of the optical transitions of the monomeric bacteriochlorophylls and of the bacteriopheophytins is not significantly perturbed by the interactions among the pigments.

In addition, a carotenoid is detected in the isolated reaction center with an orientation rather perpendicular to the C-2 symmetry axis. Finally a comparison of these data with similar ones previously obtained on bacteriochlorophyll a - containing reaction centers points towards a geometrical arrangement of the chromophores which is indistinguishable from the one observed in the reaction center of Rps. viridis.

INTRODUCTION :

The reaction centers from photosynthetic bacteria are highly organized transmembrane pigment-protein complexes, which perform the primary charge-separation and stabilization steps with a remarkable efficiency. In the case of purple bacteria they can be isolated in a functional state, and contain four bacteriochlorophyll molecules, two bacteriopheophytins and one iron-quinone complex, which are anchored to two polypeptides called L and M. In most of the bacterial species, the pigments are of the a-type, while in Rps. viridis they are of the b-type. The molecular organization of the pigments in these reaction centers has been investigated by various spectroscopic techniques such as resonance Raman, electron-spin resonance, absorption, circular dichroism (CD) and linear dichroism (LD). This last technique allows for a description of the orientation of the transition moments of the various pigments with respect to a macroscopic axis in the reaction center, for example the normal to the plane of the photosynthetic membrane in which the reaction center is embedded. In the related technique of photodichroism, one measures the relative angles between pairs of transition moments. There have been several LD and photodichroism studies (1-9) on reaction centers of purple bacteria, which have led to specific models for the organization of the pigments. For a review of the work prior to 1981 see ref (1). In the case of the reaction center of Rps. viridis two quite different models have been proposed : The first one, described by Shuvalov and Asadov (8), is based upon CD and photodichroism of isolated reaction centers, while the second one proposed by Vermeglio, Paillotin and Breton (1,3,9,10) is based upon LD of oriented Rps. viridis cells (9) and photodichroism of isolated reaction centers (1,3). Regarding the relative geometry of the transition dipoles of the primary electron donor (P), these two models are quite mutually exclusive, so that other spectroscopic observations (11-13) have been utilized to favor one or the other of these models.

The reaction center from Rps. viridis has been crystallized (14) and a X-ray diffraction study has recently led to a 3-D model, indicating the general organization of the pigments within this chlorophyll-protein complex (15). Although only qualitatively, this model allows the orientation of the X and Y optical axes, along which the optical transitions are polarized to be observed for each of the six pigments. This has prompted us to reinvestigate the low-temperature LD spectra of reaction centers (either in the isolated state or still included in the native chromatophore membrane) with P in the reduced or oxidized states. It is thus possible to significantly extend our previous model of the reaction center of Rps. viridis (1,9), to draw analogies with the case of bacteriochlorophyll a-containing reaction centers and to critically compare the models derived from the polarized optical spectroscopy investigation versus the one obtained from the X-ray crystallography study.

MATERIALS AND METHODS :

Reaction centers were prepared according to (16, 17) from chromatophores isolated by French press treatment. In some cases further purification, as well as exchange of the lauryl dimethyl amine oxide detergent for sodium cholate, were performed by high-performance liquid chromatography (18). Reaction centers were stored at - 90°C. The reaction centers or the chromatophores (purified by centrifugation on a sucrose density gradient) were included in polyacrylamide gels containing 60 % glycerol and oriented by uniaxial squeezing of the gel (19, 20). Chemical oxidation of the reaction centers was achieved by incubating the gels with 20 mM potassium ferricyanide in 60 % glycerol at 4°C.

The samples were cooled in a cryostat operating with a temperature-regulated flow of helium gas (SMC, France). LD spectra were recorded on a previously described apparatus (21) equipped with a S1 phototube (R316, Hamamatsu, Japan) and a 850 W tungsten-halogen lamp. Absorption (A) spectra were recorded on the same instrument and with the same beam geometry as used for the LD spectra. They were obtained by recording the intensity of the transmitted light (Io, blank gel with no sample ; I, gel containing the sample) as a function of wavelength in a Tracor 1710B (Tracor Northern, USA) equipped with a photometric unit, and calculating log Io/I. The concentration of the sample was adjusted to give optical densities of 0.4-0.6 at 830 nm at low temperature. For reaction centers at 10 K the absorption and LD spectra were arbitrarily normalized to 1 at the 830 nm maximum and the LD/A spectrum was calculated. The LD set-up was optimized to measure precisely small LD and absorption signals on the same instrument and with the same beam geometry in the spectral range 450-1000 nm. This constraint introduces some technical difficulties (baseline drift, deformation of the signals) which imply that some caution should be exercized in interpreting the precise shape of the absorption (and to a lesser extent the LD) spectra above 990-1000 nm. Furthermore,while LD and absorption spectra can be obtained precisely, the exact position of the baseline in these spectra is much less accurately known. This introduces uncertainties in the LD/A values observed in the regions of low absorbance.

CD spectra were recorded at room temperature on a Mark V instrument (Jobin-Yvon, France) using 0.5 cm optical cells.

RESULTS AND DISCUSSION :

Orientation of the Reaction Center

The absorption, LD and LD/A spectra at 10 K of aggregated reaction centers from *Rps. viridis* oriented in polyacrylamide gels are presented in fig. 1. The large absorption at 990 nm clearly shows that P is in the

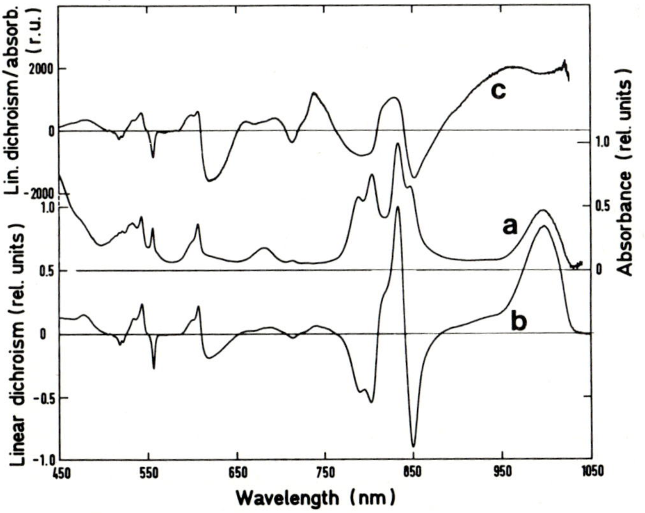

Fig. 1 - a : absorption (A), b : linear dichroism (LD) and c : LD/A spectra at 10 K of reaction centers from *Rps. viridis* oriented in polyacrylamide gel with the primary donor (P) in the reduced state. Above 1000 nm the spectra are slightly distorted. See text for details.

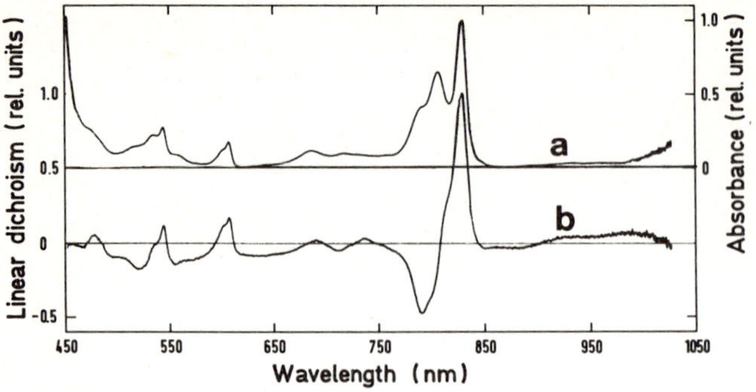

Fig. 2 - a : absorption and b : linear dichroism spectra at 10 K of reaction centers from Rps. viridis oriented in polyacrylamide gel with the primary donor chemically oxidized (state P^+).

reduced state. With our conventions a positive LD is associated with transition moments orientated close (less than 35°) to the plane of the aggregate. At such a low temperature a good resolution of the various spectral components is observed, as shown by the close correspondence between the absorption (fig. 1a) and LD (fig. 1b) peaks and also by the relative simplicity of the LD/A spectrum (fig. 1c). After incubating the gels for 24 hours with ferricyanide, the absorption spectrum (fig. 2a) shows complete oxidation of P. Due to some swelling of the gel during the incubation, the LD spectrum (fig. 2b) is of a lesser quality than the one presented in fig. 1b and comparison with several other spectra of oxidized reaction centers indicates some downward drift of the baseline. In the spectra shown in fig. 1a and 2a, a relatively large 685 nm band, which corresponds to a breakdown product of bacteriochlorophyll-b (13), is visible. Other samples with significantly less 685 nm contribution did not orient quite as well in the gel. However, the signs and relative magnitudes of the various LD bands were the same for all samples.

The antenna Q_Y transitions of Rps. viridis absorb maximally above 1000 nm and the reaction center bands around 830 nm can already be distinguished in the room-temperature LD spectra (9). This is also the case for isolated chromatophores at 10 K (fig.3), where the improved spectral resolution clearly allows the bands seen in the LD spectra of aggregated reaction centers (fig. 1b) to be recognized. The sample of oriented chromatophores used for these spectra had been incubated for six hours with ferricyanide so that a mixture of P and P^+ states was present. Unoxidized and fully oxidized samples led to absorption and LD spectra similar in the 830 nm spectral range to the ones represented in fig. 1 and 2 respectively (data not shown). The spectra of fig. 3 also show the Q_X transition of the antenna at 600 nm with its large negative LD (9) and the orientation of some reduced c-558 cytochrome (negative LD at 556 nm) as well as of the Q_X transitions of the H molecules (positive dichroism around 540 nm). The identical orientation which is observed for various transition moments with opposite dichroism when comparing the LD spectra of reaction centers included either in the chromatophore membrane (fig. 3b) or in the aggregates (fig. 1b) clearly demonstrates that in the latter case the transmembrane axis of the reaction center is perpendicular to the plane of the aggregate. Accordingly, in the following discussion we will always describe the orientation of the chromophores with respect to the membrane plane irrespective of the presence or absence of a natural membrane environment.

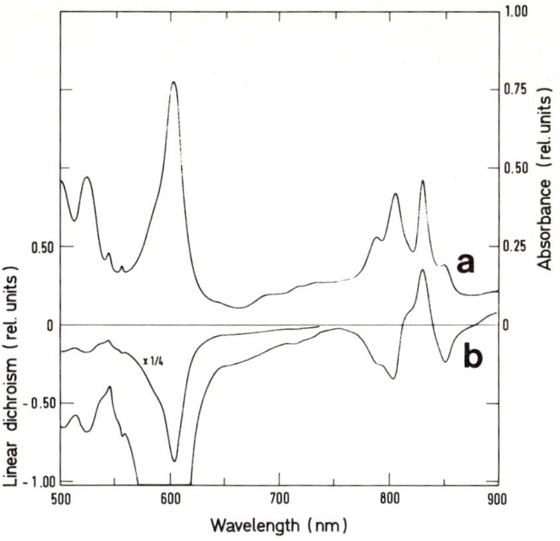

Fig. 3 - a : absorption and b : linear dichroism spectra at 10 K of chromatophores from Rps. viridis oriented in polyacrylamide gel with a fraction (about 50 %) of the reaction centers in the oxidized state.

Deisenhofer et al. (15) have proposed that the C-2 symmetry axis evident from their X-ray crystallography data runs along the membrane normal. This proposal was based upon (i) the transmembrane character of the charge separation and (ii) the orientation of the chromophores as described in (9). In the following discussion we will further stress that the orientation of the pigments as seen by LD and by X-ray spectroscopy is strongly in favor of this positioning of the symmetry axis. However, we can also add another observation in favor of this assignment, based upon the transmembrane orientation of the α-helices clearly demonstrated by both the X-ray analysis (15) and the polarized IR spectroscopy data. In the case of the reaction center from Rps. sphaeroides (22), of its LM subunit (23) and more generally of many of the hydrophobic chlorophyll-protein complexes (24), a large fraction (about 50 %) of the protein secondary structures are α -helices oriented on the average at less than 30° from the membrane normal. A transmembrane orientation of the α-helices has also been predicted from the primary structures of the L and M subunits (25). It has been recently shown by polarized IR spectroscopy that a similar geometry of the α-helices was also present in the reaction center of Rps. viridis (E. Nabedryk, G. Berger, S. Andrianambinintsoa and J. Breton, submitted) thus further strengthening the positioning of the C-2 symmetry axis perpendicular to the membrane plane.

Structure and orientation of the primary donor P

Upon oxidation of the reaction center the bleaching of the 990 nm absorption band is accompanied by a bleaching of the 850 nm band and of the broad shoulder at 610-650 nm (fig. 1a and 2a). Similar bleachings are also observed in the corresponding LD spectra (fig. 1b and 2b). The positive LD of the 990 nm band and negative LD of the 850 and 620 nm bands are in good agreement with the previous results of the LD study of the photooxidation of P in oriented cells of Rps. viridis at room temperature (9). The LD/A spectrum indicates that the 850 and 620 nm components of P are strongly tilted out of the membrane plane (at 55-60° assuming that the 990 nm transition is parallel to the membrane). By analogy with results previously obtained on bacteriochlorophyll a -containing reaction centers (26,27), we

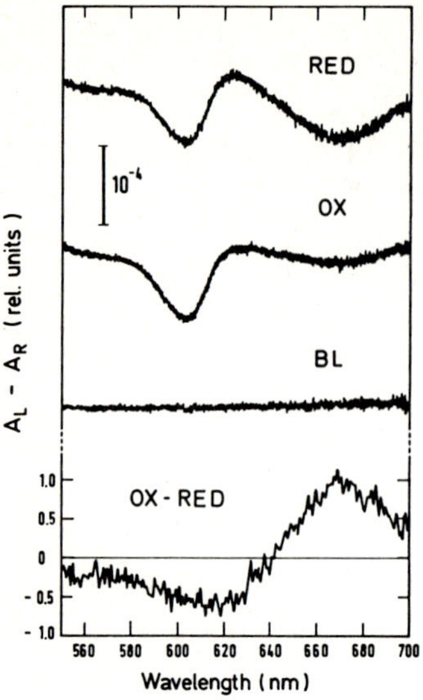

Fig. 4 Circular dichroism spectra at 20°C of reaction centers from Rps. viridis with the primary donor reduced (Red) or partially (75 %) oxidized (Ox). The reaction centers (in a 0.5 cm cuvette) had an absorbance at 830 nm of 0.9 OD units. Base line (BL) with 10mM tris buffer alone. The (OX-RED) spectrum was obtained by substracting the spectrum of the reduced reaction centers from the spectrum of the oxidized reaction centers.

have attempted to locate a lower energy exciton state associated with the Q_X transition of P. The LD/A spectrum (fig. 1c) gives an indication that a small transition with a positive LD is present around 660 nm. That this transition could be ascribed to the second exciton state of the Q_X transition of P is confirmed by the CD spectra (fig. 4) which demonstrate that at room temperature a S-shaped signal with lobes at 620 (+) and 670 (–) nm disappears upon oxidation of P.

Upon chemical oxidation of P, a 4 nm blue shift of the 834 nm band is observed, together with the bleaching of the 990 and 850 nm bands. It has been proposed that the monomer in P^+ absorbs at 805–810 nm (28) with its Q_Y transition moment oriented at about 20° with respect to the membrane plane (9). The significant reduction of the negative LD signal at 805 nm (fig. 1b) observed upon oxidation of the reaction center (fig. 2b) fully supports this model. However, the amplitude of this absorption band of the monomer is very small compared to the integrated intensity of the 990 and 850 nm transitions. This effect is even more pronounced in the Q_X region, where fig. 1b and 2b give no evidence for a monomer band appearing with a negative LD. These observations suggest that the bacteriochlorophyll molecules in P^+ are still coupled electronically.

Orientation of the bacteriopheophytin (H) and monomeric bacteriochlorophyll (B) molecules

At 10 K in reduced reaction centers the H absorption bands (fig. 1a) are resolved both in the Q_Y (at 789 and 805 nm) and in the Q_X regions (at 544 and 534 nm). These bands are only slightly affected upon oxidation of P (fig. 2a). The corresponding peaks of the LD spectra (fig. 1b) are located at 790 and 803 nm and at 544 and 534 nm. The LD/A spectrum (fig. 1a) shows a broad featureless band around 800 nm which clearly indicates that the two Q_Y transitions are making the same angle with the membrane plane and are significantly titled out of that plane (at 45–50°). A different situation is observed for the two Q_X transitions (both tilted at 25–30° out of the membrane plane) with the 544 nm transition being slightly closer to the membrane plane than the one at 534 nm. It must be further noticed that the 805 and 544 nm transitions are associated with the H molecule, which undergoes reduction upon illumination of prereduced reaction centers at low temperature (3,29,30).

Both the absorption and LD spectra (fig. 1a and 1b) show a single band around 834 nm indicating that the Q_Y transitions of the two B molecules absorb at the same energy. The LD/A spectrum (fig.1c) indicates that the average of these two transitions is oriented at a small angle (20-25°) from the membrane plane as previously reported (9). On the other hand, two bands can be observed at 607 and at approx. 600 nm in the absorption and LD spectra in the Q_X region. These two bands, although of different amplitudes and widths, correspond to two transitions which have an almost identical tilt of 25-30° with respect to the membrane.

Comparison with the X-ray data

Although only a 3-D picture of the arrangement of the pigments in the reaction center without precise atomic coordinates has been reported (15), stereoscopic inspection of the model allows for an estimation of the tilt angle of the X and Y axes of each of the six molecules with respect to a plane perpendicular to the C-2 symmetry axis (which represents the plane of the membrane). It is rather clear, for example, that the plane of each of the two monomeric B molecules is tilted at about 25-30° away from the plane of the membrane, implying that the LD of each of the individual transitions should be positive. This situation is observed for the two Q_X transitions at 600 and 607 nm as well as for the average of the two Q_Y transitions unresolved at 834 nm (fig. 1b). We note that the transmembrane orientation of the C-2 symmetry axis discussed above implies that transition moments belonging to symmetrically arranged molecules which are not in excitonic interactions should exhibit identical LD/A. This is verified for the Q_X of the two B molecules. The geometry of the two H molecules is also rather clear in the model derived from the crystallographic data.and shows that the Y optical axes of the two molecules are tilted out of the membrane plane (although not perpendicular to it) while the X axes are much closer to the membrane plane. This geometry is in excellent agreement with the LD data presented in fig. 1 which, in addition, indicate that the tilt angles of the Q_Y transition of the two H molecules are identical. The LD/A spectrum (fig. 1c) also indicates that there is a small difference in the tilt angle of the two Q_X transitions, and that accordingly either the C-2 symmetry of the X directions is slighly less well observed than for the Y directions or the two Q_X transitions are experiencing different interactions.

As deduced from the X-ray study, the planes of the two bacteriochlorophylls constituting the special pair are both rather perpendicular to the membrane plane. They are tilted at 15° from each other, and the distance between the two overlapping rings I is approx. 3A° implying a strong electronic coupling between them. The X directions are located side-by-side and rather perpendicular to the membrane plane, while one of the two Y directions is tilted slightly above and the other slightly below this plane. The angle between the two Y directions is close to 35° (H. Michel, personal communication). This geometrical arrangement is remarkably close to the one proposed in our model (9). The oscillator strength and orientation of the transitions of P responsible for the bands at 990 nm and 850 nm and at 620 and 660 nm observed in fig. 1 can thus be rationalized according to the theory of excitonic coupling (31) as the vectors sum and difference of the transition moments of the monomers seen in the model derived from the X-ray study.

Thus a qualitative inspection of the reaction center model (15) and of the LD data presented here indicates that a good correlation can be found between the geometrical arrangement of the X and Y axes of the six pigment molecules,as seen by X-ray and the orientation of the various transition moments giving rise to the different absorption bands as seen by LD under the following conditions : (i) the C-2 symmetry axis is located along the membrane normal, (ii) excitonic coupling exists within the special pair with non-parallel Q_Y transition moments and (iii) the electronic interactions which may occur among the four other molecules do not appreciably perturb the direction of their optical transitions. Although we estimate that the present agreement of the results of the two techniques is reasonably good, a more quantitative comparison between the X-ray data and the LD results should give interesting information regarding the interactions among the various pigments (32) and between the pigments and the protein (33). In addition,this comparison will constitute an unique calibration for the LD and photodichroism techniques by allowing the determination of such important parameters as the angle between the Q_X and Q_Y transitions within each molecule,which for isolated bacteriochlorophyll a in solution has been shown to be 74° (34) and not 90° as usually assumed.

Nevertheless,with the help of the qualitative model presently available,and of the LD data described here,we can tentatively assign each resolved band in the spectra to one of the twelve transition moments of the six molecules. These results are presented in Table I and they generally agree with the previous assignments made in (1,3,9,28-30,35). In addition to the tentative assignment of the 660 nm band discussed above, we have also ascribed the 600 and 607 nm bands to the Q_X transition of each of the two B molecules. This assignment is based upon the observation that only for the 607 nm transition an electrochromic blue shift is clearly observed upon 100 K photoreduction of the H molecule (30), which most probably resides on the L side (Table I),as can be deduced from picosecond spectroscopy (36). This possibility of assigning the absorption bands resolved in the 10 K spectra to all but two (B_{YL}, B_{YM}) of the twelve transition moments of the six chlorophylls will probably be useful when the molecular coordinates obtained from the X-ray study will be released. Not only will it be possible to compare the tilt angles obtained by the two techniques,but also to compare the angles between pairs of transition moments obtained by the X-ray study to the ones that can be obtained very precisely by photodichroism on the states P^+ (1,3,4,8) or H^-_L (30) upon excitation within a pure absorption band.

TABLE I :

WAVELENGTH ASSIGNMENT OF THE OPTICAL TRANSITIONS OF THE PIGMENTS IN THE REACTION CENTER OF RPS. VIRIDIS (AT 10 K). P, B and H refer to the primary donor, the monomeric bacteriochlorophylls and the bacteriopheophytins respectively, X and Y are the axes of polarization of the transitions and L and M refer to the polypeptide chains,to which are attached the pigments (the L chain represents the photosynthetic active branch of the electron pathway from P to the primary quinone Q_A)

Transition	P_{Y-}	P_{Y+}	P_{X-}	P_{X+}	B_{YL}	B_{YM}	B_{XL}	B_{XM}	H_{YL}	H_{YM}	H_{XL}	H_{XM}	Car
Wavelength (nm)	990	850	660	620	834	834	607	600	805	790	544	534	475
Sign of LD	+	−	+	−	+	+	+	+	−	−	+	+	+
Sign of CD	+	−	−	+									

Comparison with previously proposed models

Two models have been proposed to explain the LD and CD spectroscopy of Rps. viridis reaction centers. In the model of Shuvalov and Asadov (8) the directions of the Q_Y transitions of the two molecules involved in P are strictly parallel, giving only one transition at 990 nm while in our model (1,3,9,10) they are at approx. 35° from each other, giving rise to a second exciton band in addition to the 990 nm transition. This model constituted an extension of a model previously proposed by Vermeglio et al. for the reaction center of Rps. sphaeroides and derived from LD (37) and photodichroism studies (38). On the basis of its orientation at a large angle from the 990 nm transition (9), the 850 nm transition of Rps. viridis was ascribed to the higher energy exciton component of P, an assignment also proposed earlier to rationalize (i) picosecond absorption changes accompanying the formation of P^* (36) and (ii) low-temperature absorption spectra (28,35). The ratio of the areas under the 990 and 850 nm bands (approx. 10) was utilized to estimate (31) the 35° angle between the Y directions of the monomers (9). Furthermore, the orientation of the 990 and 620 nm transition dipoles at 10° and approx. 60° from the membrane plane respectively and the small tilt (25°) of the Q_Y transition dipole of the monomer band appearing in P^+, led us to propose that the two molecules constituting the special pair are both almost perpendicular to the membrane plane, with their Q_Y transitions tilted at approx +17° and −17° with respect to the membrane plane (1,9). The assignment of the 850 nm transition to P_{Y+} (Table I) has been subsequently strengthened by photoselection experiments (1,3), but also criticized (39,40), and rejected mainly on the grounds that no large bleaching was observed at this wavelength in the spectra of ^3P, the triplet state of P (11, 12). The model of Shuvalov was thus preferred by these authors. In this model (8) the 850 nm transition is ascribed to the Q_Y transition of one of the two monomeric B molecules interacting both with the Q_Y transition (at 834 nm) of the other B molecule and with the 990 nm transition of P. This type of assignment for the 850 nm band has also been later favored by Kirmayer et al. (4) and Seftor and Thornber (13).

As described in the preceding section, the model derived from the X-ray data seems to reflect rather closely the organization of the pigments proposed in (9). Relevant to the 850 nm transition, we note that its strongly negative LD (fig. 1b), indicating an orientation at approx. 60° out of the membrane, cannot be easily reconciled with the assignment propounded by Shuvalov and Asadov (8) to one of the two B molecules which, as seen by X-ray and LD, have both their planes tilted at a small angle to the membrane. On the other hand, an application of the first-order exciton coupling theory (31) to the two molecules assigned to P in the model derived from X-ray leads to four new transitions which have qualitatively the orientations and oscillator strengths observed in the present study for the 990, 850, 620 and possibly the 660 nm transitions. When a more quantitative comparison of our LD results to the X-ray data will become available, it will be possible to further refine our simple model. For example, our data show that both the 850 and the 620 nm transitions are not strictly parallel to the C-2 symmetry axis. This could be due to such effects as an asymmetric environment of P (48), some coupling between B and P, overlap of bands or non-perfect linear polarization of the transitions.

Regarding the features of the ^3P spectra which have led to the rejection (11,12) of our model, it should be stressed that the oscillator strength of the 850 nm component is only about 10 % of that of the 990 nm component. Furthermore the nature of the electronic interactions among the two bacteriochlorophylls of the special pair in the state ^3P or in P^+ can be sufficiently different, so that the amplitude and wavelength position of the first excited singlet of the remaining monomer can be very dissimilar.

Electrochromic shifts will affect more strongly the pigments in the state P^+ than in the state ^3P. Finally Scherz and Parson (32) have recently proposed that intensity borrowing could occur among all the six interacting pigments. In these conditions, a redistribution of oscillator strength among all the pigments can be expected whenever the electrostatic interactions are locally modified. Such a mechanism could lead to large differences between the P^+ - minus -P and the ^3P - minus - P spectra. All these factors make a direct comparison of the two types of spectra (depicted in fig. 11 of ref. 1 and in fig. 4 of ref. 12) rather complicated. However we suggest, as an alternative explanation to the red-shift of the 850 nm band proposed by den Blanken and Hoff (12), that the monomer in ^3P absorbs at lower energy and has a larger oscillator strength than in the state P^+. In this case the peak observed around 870 nm in the ^3P spectrum corresponds to the appearance of the long wavelength wing of the monomer absorption band, and the trough at 850 nm to the bleaching of the 850 nm component of P superimposed to the appearance of this monomer. Under these conditions the data from ^3P - minus - P spectra (11,12) can be reconciled with our model of the special pair. We further notice that the polarization values at 870 nm and 850 nm recently obtained on ^3P (41) are compatible with the interpretation proposed above.

Orientation of other components

In the LD spectra depicted in fig. 1c, 2b and 3b a small negative signal is clearly present at 715 nm. It is associated with a small absorption band at the same wavelength. Its presence in the chromatophores indicates that it is not an artefact of the reaction center isolation. This band does not bleach upon oxidation, and is thus not part of P. It could be due to a vibrational satellite of the Q_Y transitions of the H molecules.

The high-potential cytochrome c-558 absorbs at 556.3 nm at 10 K and exhibits a large negative LD signal peaking at 557.2 nm (fig. 1). This result is consistent with the LD data previously obtained on chromatophores at room temperature (42), indicating an orientation preferentially perpendicular to the membrane of this cytochrome.

At 475-480 nm a shoulder is discernible in the absorption spectra of the reaction centers used both in this study (fig. 1a) and also for preparing the crystals (5). The presence of about one carotenoid molecule per reaction center has been reported (35), and it is thus reasonable to assign the 475 nm optical transition to this species. The positive LD signal associated with it (fig. 1b) indicates that the carotenoid is oriented rather parallel to the membrane, as it has also been observed in the case of reaction centers isolated room Rps. sphaeroides G1C oriented in electric field (43) and Rps. sphaeroides 241 oriented in polyacrylamide gels (J. Breton, unpublished). This carotenoid has not been located in the X-ray study probably because, like for the secondary quinone Q_B (44), the conditions which are required for the crystallization of the reaction centers lead to a partial loss of these molecules. However, utilizing several spectroscopic observations on the carotenoid, we can already speculate on the location of this molecule in the reaction center model. On the basis of the remarkable similarity of the geometries of bacteriochlorophyll a and bacteriochlorophyll b - containing reaction centers which will be discussed in the next section, we will assume that the orientation of the carotenoid relative to the long wavelength transition of the special pair (P_{Y-}) is the same for both types of reaction centers. In the case of Rps. sphaeroides 241 this angle has been estimated to be approx. 75° by photoselection (38) and magnetophotoselection (45). Accordingly, the carotenoid molecule has its transition moment rather perpendicular to both the C-2 symmetry axis and the 990 nm transition.

Furthermore, it must be located quite close to the special pair in order to meet the requirements for the transfer of triplet character from ^3P to the carotenoid (46) which implies orbital overlap between the two partners. Finally, in order to take into account the extractability of the carotenoid (35) as well as the ease of reconstitution (47) which bear analogies with the case of the Q_B molecule, we will tentatively locate this carotenoid within the inner pocket between the L and M subunits (15). In order to maintain the remarkable C-2 symmetry of the system, it is further proposed to locate this carotenoid on the symmetry axis.

Comparison with bacteriochlorophyll a - containing reaction centers

There is much evidence pointing toward a very similar geometry of the special pair in all the reaction centers from purple bacteria. For example an arrangement similar to the one of Rps. viridis has been predicted for the reaction center of Rsp. rubrum (48,49). The oxidized-minus-reduced CD difference spectra of various reaction center preparations have similar S-shaped features with conserved sign both in the Q_Y (50, 51) and in the Q_X regions (50,51 and fig. 4). The orientation of each of these excitonic transitions relative to the membrane plane is also similar(1). Besides LD and CD data, the results of photodichroism studies on the two types of reaction centers also point towards identical values for the relative angles between pairs of transition moments (1,3,4,7,30,38,52-54). The results from these various studies (see also (55)) all point toward the view that the six chlorophylls and the carotenoid adopt quite similar geometrical orientations in the two types of reaction centers. The small differences which are observed in the relative positions of the absorption bands among different strains (7) or species (4,13) appear to us more probably related to subtle changes in the electronic interactions of the pigments with amino acid residues (33) as well as with one another leading to slight modifications of the intensity borrowing (32) and charge-transfer character (3,11,38,56,57) of the electronic transitions rather than to any large difference in the geometrical organization of the chromophores. Further work in progress on the structure of the reaction centers of Rps. sphaeroides as seen by low-temperature LD (J. Breton, unpublished) and by X-ray crystallography should allow for a test of this interpretation.

AKNOWLEDGEMENTS :

We enjoyed stimulating discussions with N.E. Geacintov, A.J. Hoff, B. Honig, W. Mantele, E. Nabedryk, W.W. Parson and A.W. Rutherford. We also thank D.M. Tiede and Y. Choquet for their help in the initial phase of the project, J.P. Thornber, C. Peterson, A. Vermeglio and S. Andrianambinintsoa for samples of reaction centers and A.M. Bardin for recording the CD spectra. Part of this project was funded by AFME grant 3.320 1402

REFERENCES :

1 BRETON, J. and VERMEGLIO, A. (1982) in Photosynthesis : Energy Conversion by Plants and Bacteria vol. 1 (Govindjee, ed) Acad. Press, New York pp 153-194

2 EROKHIN, Y.E., ABDOURAKHMANOV, I.A. and PROKHORENKO, I.R. (1981) in Photosynthesis III. Structure and Molecular Organisation of the Photosynthetic Apparatus (Akoyunoglou, G. ed) pp 959-968

3 VERMEGLIO, A. and PAILLOTIN, G. (1982) Biochim. Biophys. Acta 681, 32-40

4 KIRMAYER, C., HOLTEN, D. and PARSON, W.W. (1983) Biochim. Biophys. Acta 725, 190-202

5 ZINTH, W., KAISER, W. and MICHEL, H. (1983) Biochim. Biophys. Acta 723, 128-131
6 ALLEN, J.P. and FEHER, G. Proc. Natl. Acad. Sci. USA (1984) 81, 4795-4799
7 MAR, T. and GINGRAS, G. (1984) Biochim. Biophys. Acta 765, 125-132
8 SHUVALOV, V.A. and ASADOV, A.A. (1979) Biochim. Biophys. Acta 545, 296-308
9 PAILLOTIN, G., VERMEGLIO, A. and BRETON, J. (1979) Biochim. Biophys. Acta 545, 249-264
10 PAILLOTIN, G. and BRETON, J. (1983) in Organic Molecular Aggregates (Reineker, P., Haken, H., Wolf, H.C. eds.) Springer-Verlag Berlin pp 272-283
11 SHUVALOV, V.A. and PARSON, W.W. (1981) Biochim. Biophys. Acta 638, 50-59
12 DEN BLANKEN, H.J. and HOFF, A.J. (1982) Biochim. Biophys. Acta 681, 365-374
13 SEFTOR, R.E.B. and THORNBER, J.P. (1984) Biochim. Biophys. Acta 764, 148-159
14 MICHEL, H. (1982) J. Mol. Biol. 158, 567-572
15 DEISENHOFER, J., EPP, O., MIKI, K., HUBER, R. and MICHEL, H. (1984) J. Mol. Biol. 180, 385-398
16 PRINCE, R.C., TIEDE, D.M., THORNBER, J.P. and DUTTON, P.L. (1977) Biochim. Biophys. Acta 462, 467-490
17 CLAYTON, R.K. and CLAYTON, B.J. (1978) Biochim. Biophys. Acta 501, 478-487
18 BERGER, G., TIEDE, D.M. and BRETON, J. (1984) Biochem. Biophys. Res. Comm. vol. 121, 47-54
19 ABDOURAKHMANOV, I.A., GANAGO, A.O., EROKHIN, Y.E., SOLOV'EV, A.A. and CHUGUNOV, V.A. (1979) Biochim. Biophys. Acta 546, 183-186
20 HAWORTH, P., ARNTZEN, C.J., TAPIE, P. and BRETON, J. (1982) Biochim. Biophys. Acta 679, 428-435
21 BRETON, J. (1977) Thèse Doctorat d'Etat, Univ. Paris-sud
22 NABEDRYK, E., TIEDE, D.M., DUTTON, P.L. and BRETON, J. (1982) Biochim. Biophys. Acta 682, 273-280
23 NABEDRYK, E., TIEDE, D.M., DUTTON, P.L., and BRETON, J. (1984) in Advances in Photosynthesis Research vol. II (Sybesma, C. ed.) Martinus Nijhoff/Dr W. Junk Publishers, The Hague pp 177-180
24 BRETON, J. and NABEDRYK, E. (1984) FEBS Lett. 176, 355-359
25 WILLIAMS, J.C., STEINER, L.A., FEHER, G. and SIMON, M.I. (1984) Proc. Natl. Acad. Sci. USA 81, 7303-7307
26 RAFFERTY, C.N. and CLAYTON, R.K. (1979) Biochim. Biophys. Acta 546, 189-206
27 MAR, T. and GINGRAS, G. (1984) Biochim. Biophys. Acta 764, 86-92
28 THORNBER, J.P., DUTTON, P.L., FAJER, J., FORMAN, A., HOLTEN, D., OLSON, J.M., PARSON, W.W., PRINCE, R.C., TIEDE, D.M. and WINDSOR, M.W. (1978) Proceedings of the Fourth International Congress on Photosynthesis (Hall, D.D., Coombs, J. and Goodwin, T.W. eds.) pp 55- 70 The Biochem. Soc., London
29 THORNBER, J.P., SEFTOR, R.E.B. and COGDELL, R.J. (1981) FEBS Lett. 134, 235-239
30 TIEDE, D.M., CHOQUET, Y. and BRETON, J. (1985) Biophys. J. vol. 47, 443-447
31 KASHA, M., RAWLS, H.R. and EL-BAYOUMI, M.A. (1965) Pure Appl. Chem. 11, 371-392
32 SCHERZ, A. and PARSON, W.W. (1984) Biochim. Biophys. Acta 766, 666-678
33 ECCLES, J. and HONIG, B. (1983) Proc. Natl. Acad. Sci. USA 80, 4959-4962
34 BRETON, J., VERMEGLIO, A., GARRIGOS, M. and PAILLOTIN, G. (1981) in Photosynthesis III Structure and Molecular Organisation of the Photosynthetic Apparatus (Akoyunoglou, G. ed.) pp 445-459
35 THORNBER, J.P., COGDELL, R.J., SEFTOR, R.E.B. ans WEBSTER, G.D. (1980) Biochim. Biophys. Acta 593, 60-75
36 HOLTEN, D., WINDSOR, M.W., PARSON, W.W. and THORNBER, J.P. (1978) Biochim. Biophys. Acta 501, 112-126

37 VERMEGLIO, A. and CLAYTON, R.K. (1976) Biochim. Biophys. Acta 449, 500-515
38 VERMEGLIO, A., BRETON, J., PAILLOTIN, G. and COGDELL, R. (1978) Biochim. Biophys. Acta 501, 514-530
39 PEARLSTEIN, R.M. (1982) in Photosynthesis : Energy-Conversion by Plants and bacteria (Govindjee, ed.) vol. I Acad. Press. New York pp 293-330
40 PARSON, W.W (1982) Ann. Rev. Biophys. Bioenerg. 11, 57-80
41 DEN BLANKEN, H.J., MEIBURG, R.F. and HOFF, A.J. (1984) Chem. Phys. Lett. 105, 336-342
42 VERMEGLIO, A., BRETON, J., BAROUCH, Y. and CLAYTON, R.K. (1980) Biochim. Biophys. Acta 593, 299-311
43 GAGLIANO, A., BRETON, J. and GEACINTOV, N.E. (1985) Photobiochem. Photobiophys. in press
44 GAST, P., MICHALSKI, T.J., HUNT, J.E. and NORRIS, J.R. (1985) FEBS Lett. 179, 325-328
45 McGANN, W.J. and FRANK, H.A. (1985) Biochim. Biophys. Acta 000, 000-000
46 PARSON, W.W. and MONGER, T.G. (1977) Brookhaven Symp. Biol. 28, 195-212
47 AGALIDIS, I., LUTZ, M. and REISS-HUSSON, F. (1980) Biochim. Biophys. Acta 589, 264-274
48 MAR, T. and GINGRAS, G. (1984) Biochim. Biophys. Acta 764, 283-294
49 LUBITZ, W., ISAACSON, R.A., ABRESCH, E.C. and FEHER, G. (1984) Proc. Natl. Acad. Sci. USA 81, 7792-7796
50 PHILIPSON, K.D. and SAUER, K. (1973) Biochemistry 12, 535-539
51 REED, D.W. and KE, B. (1973) J. Biol. Chem. 248, 3041-3045
52 FRANK, H.A., BOLT, J., FRIESNER, R. and SAUER, K. (1979) Biochim. Biophys. Acta 547, 502-511
53 TROSPER, T.L., FRANK, H.A., NORRIS, J.R. and THURNAUER, M.C. (1982) Biochim. Biophys. Acta 679, 44-50
54 BOXER, S.G. and ROELOFS, M.G. (1979) Proc. Natl. Acad. Sci. USA 76, 5636-5640
55 TIEDE, D.M. (1985) in Photosynthetic Membranes (Staehelin, L.A. and Arntzen, C.J., eds.) Encyclopedia of Plant Physiol. (New Series) Springer-Verlag. In press
56 WARSHEL, A. (1980) Proc. Natl. Acad. Sci. USA 77, 3105-3109
57 MASLOV, V.G., KLEVANIK, A.V., ISMAILOV, M.A. and SHUVALOV, V.A. (1983) Dokl. Akad. Nauk SSSR (Engl. transl.) 269, 87-90

Calculations of Spectroscopic Properties of Bacterial Reaction Centers

W.W. Parson[1], *A. Scherz*[2], *and A. Warshel*[3]

[1]Dept. of Biochem., Univ. of Washington, Seattle, WA 98195, USA
[2]Dept. of Biochem. Weizmann Inst., 76100 Rehovot, Israel
[3]Dept. of Chem., Univ. Southern Calif., Los Angeles, CA 90007, USA

1. Introduction

The recent crystallographic work of DEISENHOFER et al. [1] has provided a beautifully detailed map of the structure of a photosynthetic bacterial reaction center (RC). With such a map in hand, it now is possible to explore the connection between the structure and the RC's complicated spectroscopic properties. An understanding of this connection is critical, if we are to have an accurate picture of how the RC captures the energy of light.

The RC of *Rhodopseudomonas viridis* is built around a pair of strongly overlapping molecules of bacteriochlorophyll-b (BChl-b). This pair (P) transfers an electron to a molecule of bacteriopheophytin (BPh) when the RC is excited with light [2,3]. Two additional BChls and a second BPh are located nearby, but their roles in the electron-transfer reactions are still unclear. We have tried to calculate how the absorption and circular dichroism (CD) spectra of the RC (Fig. 1) depend on the interactions between the two BChl molecules of P, and on the interactions of P with the neighboring BChls. The contributions that charge-transfer (CT) states make to the spectra are of particular interest, because of the possible participation of these states in electron-transfer.

2. Methods

Monomeric BChl has four main optical absorption bands, called Q_y, Q_x, B_x and B_y (Fig. 1). The four bands represent $\pi\pi^*$ excitations from the top two filled molecular orbitals (MOs) to the two lowest unfilled [4]. We shall refer to these MOs as Φ_1 to Φ_4 in order of increasing energy, so that the highest filled MO in the ground-state is Φ_2. The MOs can be expressed as linear combinations of atomic p orbitals: $\Phi_k = \Sigma v_{kt}\chi_t$, where χ_t is a p_z orbital on atom t and the coefficients v_{kt} are found by solving a self-consistent field equation (see, e.g. [5]). Wavefunctions for the excited states then can be written as $\Psi_n = \Sigma c_{nk}\Delta_k$, where Δ_k represents the singlet wavefunction corresponding to excitation from Φ_{k1} to Φ_{k2}. The mixing coefficients c_{nk} are found by diagonalizing the configuration interaction matrix [5], in which the elements depend on electron-electron repulsion integrals. The Q_y transition consists mainly of an excitation from Φ_2 to Φ_3, and to a lesser extent, of Φ_1 to Φ_4. B_y is mainly Φ_1 to Φ_4, with a smaller amount of Φ_2 to Φ_3. Q_x and B_x consist mainly of mixtures of Φ_2 to Φ_4 and Φ_1 to Φ_3. The Q_y and B_y transition dipoles are polarized approximately along the axis connecting rings I and III of the BChl; the Q_x and B_x, along the axis from ring II to ring IV [4].

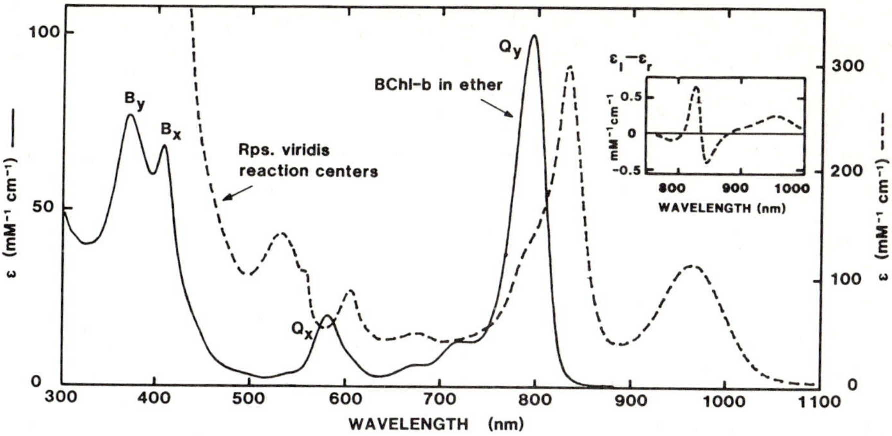

Fig. 1 Absorption spectra at 295 K of monomeric BChl-b in ether, purified from *Rps. viridis* essentially as described for BChl-a [6] (——), and of *Rps. viridis* RCs, purified as described [7] (---). To calculate the BChl extinction coefficient (ε), Mg was measured by atomic absorption; the uncertainty in ε is approx. ±10%. Inset: CD spectrum of RCs, from [8] with ε scaled as in [7].

To treat a BChl dimer, one could consider the two BChls together as a supermolecule, and solve for the MOs of the entire system. Although this approach has been fruitful in some cases [9-12], it produces the final states directly, without providing a convenient description of the interaction between exciton and CT states. Here we take the alternative approach of constructing an interaction matrix of the following form:

$$\mathbf{U} = \left|\begin{array}{cc|cc} E_a & U_{ex} & \multicolumn{2}{c}{U_{ex,ct}} \\ U_{ex} & E_b & & \\ \hline \multicolumn{2}{c|}{U_{ex,ct}} & E_{ab} & U_{ct} \\ & & U_{ct} & E_{ba} \end{array}\right|$$

The diagonal matrices E_a and E_b contain the excitation energies of local $\pi\pi^*$ transitions of monomers a and b (Q_y, Q_x, etc.). Matrix $\mathbf{U}_{ex}$ describes the exciton interactions among these states. E_{ab} represents the four singlet CT states in which an electron goes from Φ_1 or Φ_2 of molecule a to Φ_3 or Φ_4 of molecule b; E_{ba} represents transfer in the opposite direction. Matrices U_{ct} and $\mathbf{U}_{ex,ct}$ describe the interactions of the eight CT states with each other and with the eight $\pi\pi^*$ exciton states. $\mathbf{U}$ can be enlarged straightforwardly to include additional BChls.

A more formal discussion of $\mathbf{U}$ will be presented elsewhere, along with convenient prescriptions for calculating the matrix elements. In brief, we proceeded as follows. The excitation energies E_a and E_b were obtained empirically from the absorption spectrum of monomeric BChl-b (Fig.

1), except that the Q_x and Q_y energies were decreased slightly to reflect what appear to be perturbations caused by the protein (see below). For simplicity, the same set of energies was used for all four BChls of the RC. The BChl-b absorption spectrum also provided the transition dipole moments, which were used to contruct $\mathbf{U}_{ex}$ by a point-dipole treatment [13]. Diagonalization of a matrix that includes only $\mathbf{E}_a$, $\mathbf{E}_b$ and $\mathbf{U}_{ex}$ is the familiar exciton problem. The importance of including all four $\pi\pi^*$ states of each molecule in the calculation has been discussed previously [13]. To explore the relationship between the RC's structure and its spectroscopic properties, the geometrical parameters used in the calculation of $\mathbf{U}_{ex}$ and $\mathbf{U}_{ex,ct}$ were varied about the values seen in the x-ray structure [1]. Since the structure is still undergoing refinement, the actual geometry is subject to some uncertainty. We took advantage of the approximate C_2 symmetry in the structure by using a modified set of basis states that are either symmetric or antisymmetric with respect to the C_2 axis. These are simply the zero-order exciton states, involving identical $\pi\pi^*$ transitions in the two corresponding BChls on opposite sides of the axis, and analogous combinations of the CT states, in which an electron goes from molecule a to b or from b to a. This basis facilitates the diagonalization of $\mathbf{U}$, because the antisymmetric and symmetric states do not mix.

The CT energies $\mathbf{E}_{ab}$ were calculated by using a QCFF/PI supermolecule treatment [5] of a BChl-b dimer with C_2 symmetry, in which the two macrocyclic rings were parallel and positioned to overlap in ring 1 essentially as seen in the RC x-ray structure. The separation between the planes of the two molecules was taken to be r_z = 3.6 Å. The calculated energies were 14,580 cm^{-1} (for transfer from Φ_2 of molecule a to Φ_3 of molecule b), 17,800 (Φ_1 to Φ_3), 27,060 (Φ_2 to Φ_4), and 33,000 cm^{-1} (Φ_1 to Φ_4). This calculation was not repeated when the structure was varied. Instead, the energies of the CT states were assumed to depend on r_z in the manner found previously for the lowest CT state of a similar chlorophyll-a dimer [11]. Decreasing r_z to 3.3 Å lowers the energies by 140 cm^{-1}. The CT energies cannot be calculated rigorously for the RC, partly because we have no information on the disposition of charged or polar amino acids near the BChls. We therefore varied the energies about the values just listed, and considered how this affected the spectra.

The interaction matrix elements of $\mathbf{U}_{ct}$ and $\mathbf{U}_{ex,ct}$ were obtained by using the same π-electron parameters as in [10,11], except that the MO calculations were constrained to the individual BChls and intermolecular mixing was considered only at the level of configuration interaction. These calculations involve summing parametrized overlap integrals or repulsion integrals for the atomic p_z orbitals of the two different BChls, with weighting by the coefficients v_{kt} and c_{nk} for the MOs involved in the two transitions that the matrix element mixes.

After diagonalizing $\mathbf{U}$, one obtains a matrix of coefficients that express the contributions of the various $\pi\pi^*$ or CT basis states to the excited states of the complex. Absorption and CD spectra then can be calculated by vector addition [13]. The CT absorption bands are assumed to have no intrinsic dipole strength, but become allowed by mixing with $\pi\pi^*$ transitions.

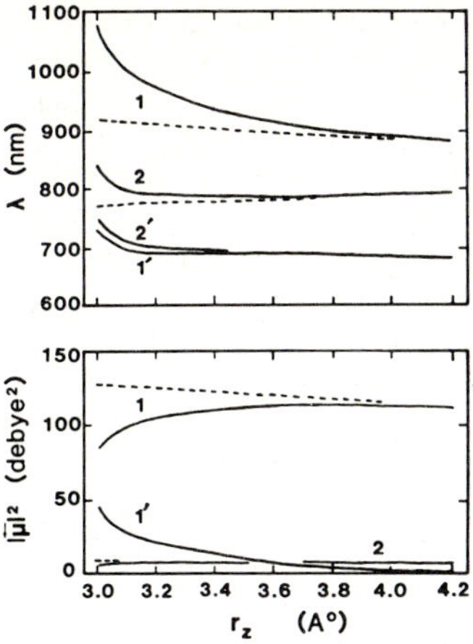

Fig.2 Calculated wavelengths and dipole strengths of a BChl-b dimer. BChl 1 is at the origin of a Cartesian coordinate system, with the vectors from ring I to III and from ring II to IV on the y and x axes [13]. BChl 2 is centered at $(-2,-6,r_z)$ Å. The direction angles for the Q_y axis of BChl 2 are $\alpha_y \approx 126°$, $\beta_y \approx 144°$, and $\gamma_y = 90.5°$; those for Q_x are $\alpha_x \approx 36°$, $\beta_x \approx 126°$ and $\gamma_x \approx 88°$. The dimer has C_2 symmetry. As r_z is varied, γ_y is held constant and the other angles are adjusted to maintain the symmetry. Wavelengths and dipole strengths used for the monomeric BChl-b are: Q_y, 830 nm and 50.0 debye^2; Q_x, 610 and 8.5; B_x, 410 and 47.1; B_y, 370 and 74.5. The energies of the CT basis states are adjusted with r_z.

3. Results

Figure 2 shows the calculated wavelengths and dipole strengths for some of the absorption bands of a BChl-b dimer, as functions of r_z, the distance between the planes of the two macrocyclic rings. The molecular planes are held nearly parallel, with the two BChls overlapping in ring I as they do in the Rps. viridis x-ray structure. The curves labeled 1 refer to the lowest-energy absorption band. When CT states are omitted from the calculation (dashed curves), this absorption band results mainly from an antisymmetric combination of the two Q_y $\pi\pi^*$ states. When CT states are added (solid curves), they contribute to the excited state and lower its energy. The extend of this mixing grows rapidly as r_z is decreased below about 3.4 Å. The curves labeled 1' represent the lowest antisymmetric CT absorption band. This band increases in dipole strength at the expense of band 1 as the BChls are moved together. At distances that are appropriate for the RC ($r_z \approx 3.3$ Å), band 1' is calculated to be about 1/5 as strong as band 1. The extinction coefficient of band 1' probably would be less than 1/5 that of band 1, because CT bands usually are very broad. Bands 2 and 2' refer to analogous symmetric transitions, which are both relatively weak.

According to the x-ray structure, the planes of the two BChls of P are not exactly parallel, but are rotated so that the angle between their normals is about 165°. For the calculations described below, the BChls were rotated into approximately this geometry. The rotation moves band 1 to slightly shorter wavelengths, and slightly increases its dipole strength.

The two BChl molecules that are near neighbors of P in the RC interact with each other only weakly. If one considers these two BChls alone, the calculated Q_y exciton bands are split by only about 20 cm^{-1}. The symmetric transition occurs at lower energy than the the antisymmetric, and is the weaker of the two. When the BChls of P (molecules 1 and 2) are considered along with neighboring BChls (molecules 3 and 4), there is substantial mixing of their excited states. The effects of this mixing are illustrated in Fig. 3. The panels on the right show the dipole and

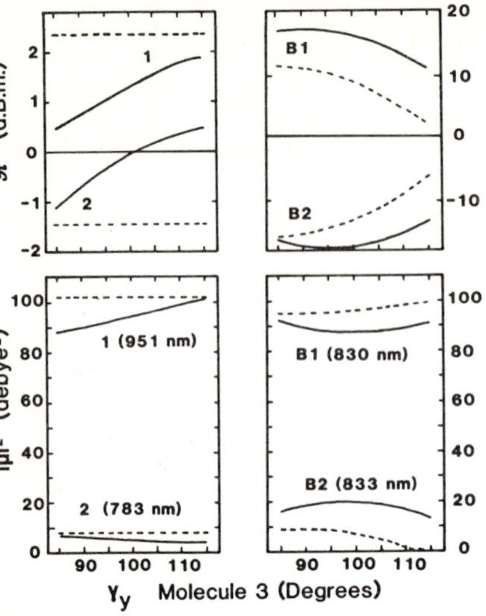

Fig. 3 Effects of interactions between P (BChls 1 and 2) and the nearby BChls (3 and 4). Dashed curves are calculations for BChls 1 and 2 alone (left panels) or for BChls 3 and 4 alone (right panels); solid curves, for all four BChls. BChls 2, 3, and 4 are centered at (-2,-6,3.3), (1.8,-2,12.6), and (2.01,-5.86, -8.61) Å. Angles for BChl 2 are α_y = 126°, β_y = 149.2°, γ_y = 93°, α_x =33°, β_x = 120.6°, and γ_x = 79.1°. The mean distance of the atoms in ring I of BChl 2 to the plane of BChl 1 is 3.5 Å. Fixed angles for BChl 3 are β_y = 140° and γ_x = 30°. BChl 4 is rotated to maintain maintain C_2 symmetry as γ_y for BChl 3 is changed. Wavelengths indicated are average values obtained with all four BChls; they change by ± 2 nm as γ_y is varied over this range. Bands B2 and B2 are slightly closer together in energy if BChls 1 and 2 are omitted.

rotational strengths of some of the absorption bands that originate mainly in BChls 3 and 4. Dashed curves represent calculations for these two BChls alone; solid curves, calculations for all four BChls. The left panels pertain to bands that originate mainly in P. Here, dashed curves are for P alone, and solid curves again for all four BChls. The molecular centers are fixed, and BChls 3 and 4 are rotated together, in a manner preserving C_2 symmetry. The abscissa gives γ_y for BChl 3, the angle between the Q_y vector of BChl 3 and the normal to BChl 1. (BChl 4 is oriented similarly with respect to BChl 2.)

With all of the geometries considered in Fig. 3, there is a transfer of dipole strength from P to the symmetric Q_y exciton band of BChls 3 and 4 (band 2). Band B1, the antisymmetric Q_y band of BChls 3 and 4, is affected oppositely. There are simultaneous transfers of rotational strength, which increase the rotational strengths of bands B1 and B2, and decrease those of bands 1 and 2. These effects are particularly dramatic when γ_y (mol. 3) ≈ 100°, which is approximately the angle seen in the x-ray structure. Here the rotational strength of band 2 approaches zero.

Figure 4 shows the effects of changing the energies of the CT basis states. The positions and orientations of the four BChl molecules are held constant at the values used for Fig. 3, with γ_y (mol. 3) ≈ 100°. In this geometry, the lowest CT basis state of BChls 1 and 2 is calculated to lie at 14,440 cm^{-1}. If the CT energies are decreased below the calculated values, the mixing of the lowest CT state and the antisymmetric Q_y exciton state increases. The CT absorption band (1') grows in strength, and band 1 gets weaker (Fig. 4, left). As the lowest CT state falls below 11,400 cm^{-1}, band 1 becomes mainly CT in character, and band 1' becomes mainly Q_y. A similar change of character occurs in the symmetric bands 2 and 2' (Fig. 4, right). Putting the CT energies much below the calculated

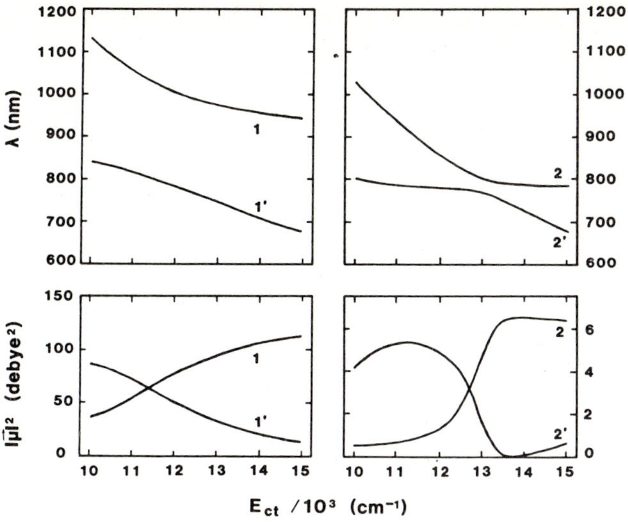

Fig. 4 Effects of the energies of the CT basis states. Four BChls were included, in the geometry used for Fig. 3, with γ_y (mol. 3) = 100°. Specifying γ_y fixes α_y = 128.2°, α_x = 74.1°, and β_x = 114.7° for BChl 3. The abcissa gives the energy of the lowest CT basis state for electron transfer between BChls 1 and 2. (The symmetric and antisymmetric CT basis states are degenerate.) The higher CT states are adjusted in energy to maintain the spacing indicated in the text.

values gives results that seem inconsistent with the near-IR absorption spectrum of P, which has a single, strong band near 960 nm (Fig. 1). Much higher energies also seem unlikely to be correct, because they cause a greater mixing of the lowest CT states with the Q_x states, modifying the spectrum considerably in the 600-nm region (not shown).

The CT states of BChls 1 and 2 thus are calculated to influence the absorption and CD spectra of the RC significantly. CT states involving BChls 3 and 4, on the other hand, probably make only negligible contributions to the spectra. The orbital overlap of these BChls with each other and with BChls 1 and 2 is much smaller than the overlap between BChls 1 and 2. It has been suggested [14,15] that a CT state involving electron transfer from BChls 1 and 2 to BChl 3 or 4 accounts for a pronounced shoulder that is seen on the long-wavelength side of the 960-nm band at low temperature. However, we calculated the dipole strength for such a CT absorption band to be only about 0.03 debye^2, compared to the dipole strength of about 100 debye^2 calculated for the main absorption band. (For this calculation, we arbitrarily lowered the energy of the CT basis states involving electron-transfer to BChls 3 and 4, to put the lowest CT absorption band in the suggested position. This required a decrease of about 4100 cm^{-1}.)

For illustration, it is helpful to dress the calculated stick spectra with gaussians having widths similar to those of BChl-b in vitro.

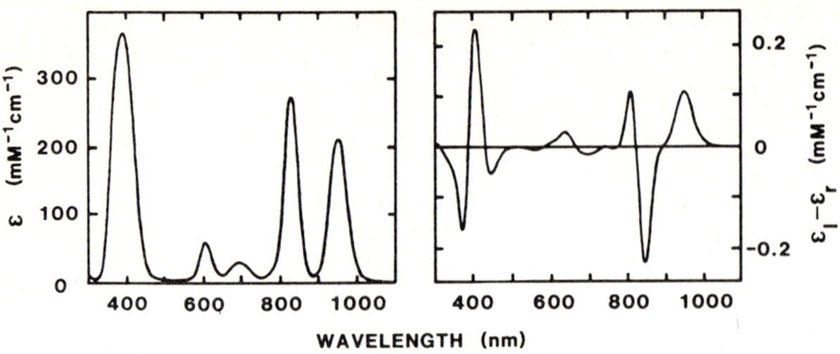

Fig. 5 Absorption and CD spectra calculated with the parameters specified in Fig. 4. The energies of the CT basis states are fixed at the calculated values, with adjustments for r_z. Absorption bands are assumed to be gaussian, with widths similar to those of monomeric BChl-b in vitro, except that the width of band 1' is made twice that of band 1.

Absorption and CD spectra drawn in this way (Fig. 5) are similar in many respects to the spectra of Rps. viridis reaction centers (Fig. 1). However, we have not tried to fit the spectra closely by adjusting the widths and shapes of the various bands. The CD spectrum is particulary sensitive to the widths of the CT bands, which can only be assigned arbitrarily. In addition, we have not considered the two BPh molecules of the RC, which contribute absorption bands near 530 and 795 nm in Fig. 1. Also, no attempt was made to optimize the fit by using different energies or dipole strengths for the four different BChls. The Q_y wavelengths of all four BChls were taken to be 830 nm, to reflect the fact that BChls 3 and 4 of the RC evidently absorb in this region (Fig. 1). The shift of the Q_y bands from 795 to 830 nm cannot be due to exciton interactions, because the overall effect of these is only a movement by about 1 nm in the opposite direction (Fig. 3). Excluding chemical modification of the BChl, for which there is no evidence, interactions with charged or polar groups on the protein remain as the most plausible explanation of the shift [11,12,16,17]. Interactions of this type presumably have a similar effect on BChls 1 and 2, but we have no way of measuring this. Calculations on this point should be possible when the amino acid sequence of the protein has been fit to the x-ray map. Agreement between the observed and calculated spectra could be improved by assuming that the Q_y wavelengths for BChls 1 and 2 are somewhat longer than those for 3 and 4, and that the latter two wavelengths are not exactly the same.

4. Discussion

Several features of the results are of particular interest. First, the mixing with CT states appears to play a significant role in determining the energy of the lowest excited state of the RC (Figs. 2 and 4). This mixing can account for about 1/3 of the movement of the long-wavelength absorption band from 795 nm, where monomeric BChl-b absorbs in vitro, to 960 nm.

Second, although the absorption bands near 830 nm originate in BChls 3 and 4, their dipole and rotational strengths depend strongly on the interactions of these BChls with P (Fig. 3). Thus, there should be absorbance changes in this region if P is raised to an excited singlet or triplet state, or oxidized to P^+. One should not attribute such absorbance changes to the higher-energy Q_y exciton band of P, which is calculated to be near 790 nm and to have very little dipole or rotational strength (Figs. 2-4). These considerations are important for an understanding of the transient absorbance changes that accompany electron transfer in the RC [2,3].

The third point concerns the relatively weak CT absorption band that we calculate to be near 670 nm. RCs do absorb in this region (Fig. 1), but the absorbance is difficult to evaluate, because it could be due to a chlorin impurity. However, photooxidation of P to P^+ causes an absorbance decrease near 670 nm that has the predicted linear dichroism parallel to the 960-nm transition [18]. RCs also lose a negative CD band in this region when P is oxidized [19]. As pointed out above, the observed spectra would be hard to reconcile with a much higher or lower energy for the CT state. We therefore conclude tentatively that the CT absorption band-reflecting electron-transfer between BChls 1 and 2 is near 670 nm, i.e. about 4000 cm^{-1} above the lowest excited state of the RC.

The experimental spectra give no such information on the energy of charge-transfer from P to BChls 3 or 4. This CT band is calculated to be too weak to observe. It probably is not responsible for the shoulder on the 960-nm absorption band, in agreement with recent work [20] suggesting that the shoulder reflects an inhomogeneous broadening of the band. However, the distance from P to BChl 3 or 4 (11 to 13 Å between molecular centers) is greater than the distance between the two BChls of P (7 Å). This factor would tend to raise the energy of the P^+BChl^- state even higher, relative to the energy for internal CT within P. Thus it is not obvious how P^+BChl^- could be an intermediate in photochemical electron transfer from P to the nearby BPh. Estimates of the energy of P^+BChl^- must be viewed cautiously, however, because the energy could be decreased by interactions with charged amino acids, or by H-bonding of the carbonyl groups of the BChl. The x-ray structure suggests that BChl 3 plays some role in electron-transfer from P to the BPh, because the BChl is located almost in between P and the BPh.

A possible role for BChl 3 is suggested by the consideration the Q_y states of BChls 3 and 4 mix into the lowest excited state of P, just as P contributes to the absorption bands near 830 nm. Although this mixing has essentially no effect on the wavelength of the 960-nm absorption band, it could be important because it expands the region of space that is occupied by the excited state wavefunction of P. It seems possible that mixing of the additional BChls into the excited state allows electron-transfer to occur directly from the excited state to the BPh, without the formation of P^+BChl^- as a distinct intermediate.

5. Acknowledgements

This work was supported by National Science Foundation Grants PCM-8303385 and PCM-8316161. We thank J. Breton, M. Gouterman, D. Holten, C. Kirmaier, D. Middendorf, N. Woodbury and W. Zinth for helpful discussion, P. Linnemeyer for purifying the RCs, and C. Kirmaier for measuring their absorption spectrum.

6. References

1. J. Deisenhofer, O. Epp, K. Miki, R. Huber, H. Michel: J. Mol. Biol. 180, 385 (1984)
2. D. Holten, M.W. Windsor, W.W. Parson, J.P. Thornber: Biochim. Biophys. Acta 501, 112 (1978)
3. C. Kirmaier, D. Holten, W.W. Parson: Biochim. Biophys. Acta 725, 190 (1983)
4. Gouterman: J. Mol. Spectrosc. 6, 138 (1961)
5. A. Warshel: in Modern Theoretical Chemistry, ed. G. Segal (Plenum, New York, 1977), Vol 7 Part A, pp. 133-172
6. A. Scherz, W.W. Parson: Biochim. Biophys. Acta 766, 653 (1984)
7. R.K. Clayton, B.J. Clayton: Biochim. Biophys. Acta 501, 478 (1978)
8. K.D. Philipson, K. Sauer: Biochem. 12, 535 (1973)
9. L. Salem, J. Am. Chem. Soc. 90, 543 (1968)
10. A. Warshel, E. Huler: Chem. Phys. 6, 463 (1974)
11. A. Warshel: J. Am. Chem. Soc. 101, 744 (1979)
12. A. Warshel: Proc. Natl. Acad. Sci. U.S. 77, 3105 (1980)
13. A. Scherz, W.W. Parson: Biochim. Biophys. Acta 766, 666 (1984)
14. Vermeglio, G. Paillotin: Biochim. Biophys. Acta 681, 32 (1982)
15. V.A. Shuvalov, A.V. Klevanik: FEBS Lett. 160, 51 (1983)
16. Warshel, S.T. Russel: Quart. Rev. Biophys. 17, 283 (1984)
17. J. Eccles, B. Honig: Proc. Natl. Acad. Sci. U.S. 80, 4959 (1983)
18. W. Zinth, M. Sander, J. Dobler, W. Kaiser, H. Michel: this volume
19. J. Breton: this volume
20. H.J. den Blanken, A.J. Hoff: Chem. Phys. Lett. 98, 255 (1983)

On the Temperature-Dependence of the Long Wavelength Fluorescence and Absorption of *Rhodopseudomonas viridis* Reaction Centers

P.O.J. Scherer, S.F. Fischer, J.K.H. Hörber, and M.E. Michel-Beyerle

Physik-Department und Institut für Physikalische und Theoretische Chemie, Technische Universität München, D-8046 Garching, F. R. G.

H. Michel

Max-Planck-Institut für Biochemie, D-8033 Martinsried, F. R. G.

1. Introduction

Recent measurements of absorbance and fluorescence spectra of reaction centers of photosynthetic bacteria indicate that lowering the temperature induces a significant narrowing and red shift of the long wavelength dimer band, whereas the positions of all the other absorption bands are unaffected. This phenomenon has been observed in absorbance /1,2/ and emission /2/ spectra (300 K, 77 K) of reaction centers of Rp. sphaeroides. Reaction centers of Rp. viridis behave similarly when comparing absorbance spectra at various temperatures between 300 K and 2 K /3-6/, magnetooptical absorbance difference spectra at 285 K and 115 K /7/ and fluorescence spectra at 77 K /6/ and 2 K /6,8/.

In this paper, the changes in width and position of the dimer band upon cooling are interpreted in terms of thermal contraction and other changes in the structure of the protein matrix. In order to put such an interpretation to test, the volume of experimental data has been extended by temperature - dependent measurements of the fluorescence spectrum of reaction centers of Rp. viridis in solution and crystal.

2. Experimental Results

The fluorescence spectra depicted in Fig. 1 exhibit three main bands which are attributed as follows: The band around 700 nm to the emission of oxidation products /3/; the band at 840 to the emission of disintegrated bacteriochlorophyll monomer and -pheophytins as concluded from the excitation spectra /9/; the long wavelength emission is the fluorescence of the bacteriochlorophyll dimer. Upon lowering the temperature, the maximum of the dimer band is shifted to the red, whereas the positions of the other bands remain unaffected. Interestingly, a shoulder in the emission band attributed to a charge transfer state /6/ was not observed, neither at 77 K nor at 1.7 K.

The quantum yield of the dimer emission is as low as ~10^{-4} /10/ for reaction centers containing only one menaquinone, which is photochemically reduced as in Fig. 1. This low quantum yield has to be considered when comparing the intensity of the dimer fluorescence to that of the other emission bands.

The dimer fluorescence spectrum is identical for reaction centers in solution and crystal at all temperatures. This is shown in Fig. 2 for room

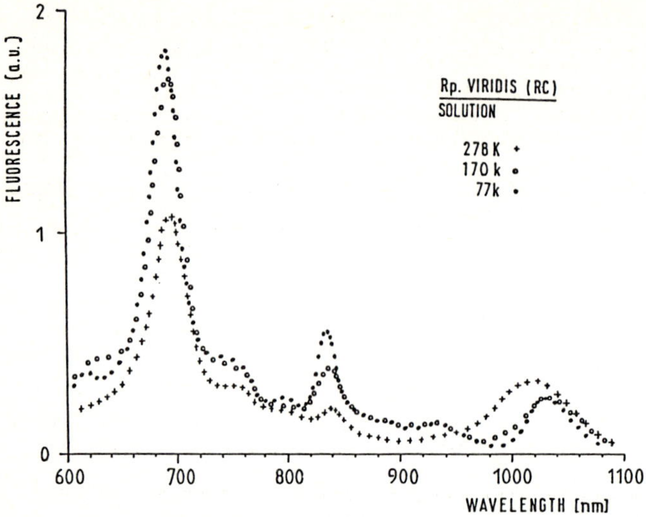

Fig. 1: Fluorescence spectra of reaction centers of Rp.viridis in solution; preparation of reaction centers: 11 M in 70 % succrose frozen under illumination in presence of 10 mM ascorbate /11/; excitation wavelength and intensity: 590 nm with 10 photons/sRC; fluorescence detection: single photon-counting using a S1-photomultiplier (Hamamatsu 63201) operating at low dark current (5 c/s) /9/.

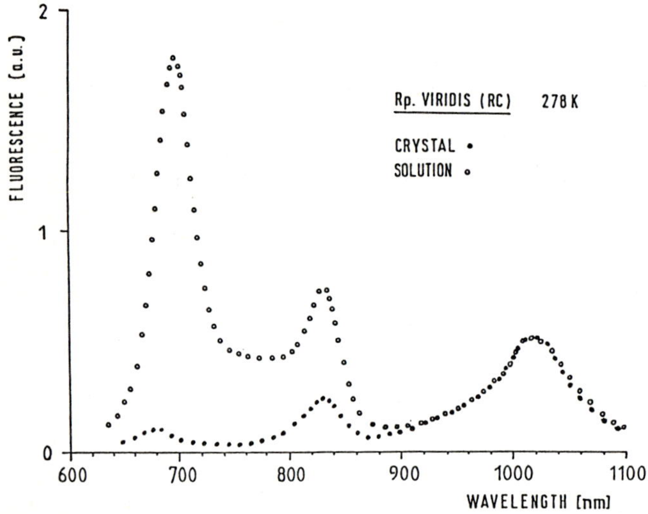

Fig. 2: Fluorescence spectra of reaction center crystal and solution at 278 K; the crystal spectrum has been corrected for reabsorption, experimental conditions otherwise as in Fig. 1.

temperature. Interestingly, the low intensity of the 700 nm and 840 nm bands reflects the minimization of oxidation products and disintegrated pigments in crystals. For reaction centers in solution the intensity of these two bands is extremely sensitive to the isolation technique.
In all preparations investigated, a simultaneous red shift of the dimer emission and absorption band is only observed when the temperature is lowered. This suggests to us that thermal changes of the protein environment affect the spectroscopic properties of the dimer. Such a view is also supported by room-temperature absorbance measurements on chromatophors /12/ showing the peak of the dimer band at 995 nm rather than at 965 nm as observed for reaction centers after isolation.

3. Description of the Model

In this section we discuss a simple model for the dimer absorption and emission band /13/ which allows us to understand thermal shift and broadening of the experimental spectra.
As thermal changes are only observed for the dimer band, they are unlikely to reflect merely the temperature dependence of dielectric constant and index of refraction.
The two monomers forming the special dimer are subject to strong excitonic coupling of 905 cm^{-1} /14,15/. The coupling to the other monomers in the RC is much weaker (at least by a factor 4.5). This encourages us to treat the dimer separately and consider the rest of the RC as part of the dimer environment.
Let q be the normal coordinates of the dimer, which in the electronic groundstate oscillate around their equilibrium values q_i^o. These equilibrium coordinates will be temperature-dependent due to thermal expansion and thermal structural changes. For simplicity we take the oscillations of q_i as harmonic. The groundstate hamiltonian of our model then is a sum of harmonic oscillators.

$$H_o = \sum_i m\omega_i^2 (q_i - q_i^o)^2/2 + p_i^2/2m \tag{1}$$

In the excited state, the two monomers are coupled by the electronic interaction V which depends on distance and orientation of the two monomers. The following model hamiltonian for a symmetric dimer in the electronic excited state

$$H = H_0 + \begin{bmatrix} W_1(q_i) & V(q_i) \\ V(q_i) & W_2(q_i) \end{bmatrix} \tag{2}$$

contains a matrix of dimension 2 as two electronic excited states are coupled. The diagonal energies $W_{1,2}$ depend on the surrounding of the dimer. Structural changes, e.g. reorientation of neighbouring charges may lead to an explicit temperature dependence.

For a symmetric dimer which is a reasonable approximation as long as the coupling V is large compared with the energy difference $W_1 - W_2$, it is convenient to consider delocalized vibrations $q_\pm$ which are symmetric and antisymmetric combinations of the local vibrations q_i. In the limit of strong exciton interaction, the antisymmetric modes are essentially narrowed /16/ and the spectral properties may be described with the simplified hamiltonian

$$H = H_0 \pm V(d^0) + W(q_{s-}^0, q_{s+}) \tag{3}$$

The interaction V mainly depends on the distance d of the two monomers which is a symmetric normal coordinate. The dependence on other normal coordinates is neglected. Thermal expansion of the system leads to a temperature-dependence of the exciton splitting 2V whereas the temperature dependence of the diagonal W equally shifts both dimer states. From a detailed inspection of temperature-dependent CD spectra, it should be possible to determine the relative importance of these two different frequency-shift mechanisms, since the CD spectra are sensitive to the position of the upper exciton band of the dimer /15/.

4. Temperature-Dependence of the Dimer Absorption and Fluorescence

For our model calculations we expand the diagonal energy $W(q_+)$ up to linear and quadratic terms, i.e. we consider the vibration of the excited dimer to be harmonic around a new equilibrium position and with a new frequency. Introducing creation and destruction operators b_s^* and b_s we have

$$H_0 = \sum_s \hbar\omega_s b_s^* b_s \tag{4}$$

$$W = W_0(T) + \sum_s \lambda_s \hbar\omega_s (b_s^* + b_s) + \sum_s \frac{g_s}{2} \hbar\omega_s (b_s^* + b_s)^2 \tag{5}$$

We neglect a mixing of the different modes in the excited state as well as a possible temperature-dependence of the parameters λ_s and g_s as higher order effects. Then each mode is treated separately. The first two moments of the absorption spectrum are evaluated straightforwardly as the thermal average over according powers of the operator $(H-H_0)$

$$\begin{aligned} E_{abs} &= \langle\langle H - H_0 \rangle\rangle \\ &= V(d^0(T)) + W_0(T) + \sum_s \tfrac{1}{2} g_s \hbar\omega_s \coth(\hbar\omega_s/2kT) \end{aligned} \tag{6}$$

$$\Delta_{abs}^2 = 8\ln 2 \left[\langle\langle (H-H_0)^2 \rangle\rangle - \langle\langle H-H_0 \rangle\rangle^2 \right]$$

$$= 8\ln 2 \sum_s \left[(\lambda_s \hbar\omega_s)^2 \coth(\hbar\omega_s/2kT) + \tfrac{1}{2} (\hbar\omega_s g_s)^2 \coth^2(\hbar\omega_s/2kT) \right] \tag{7}$$

For the emission spectrum the thermal average has to be taken with the excited state wavefunctions. To this end a canonical transform is useful which brings $H_0 + W$ into the form of harmonic oscillators with shifted equilibrium positions and new frequences

$$\omega_s' = \omega_s (1+2g_s)^{1/2} \tag{8}$$

The resulting first two moments of the emission spectrum are

$$\begin{aligned} E_{em} = V(d^0(T)) + W_0(T) + \sum_s \Bigg[& \tfrac{1}{2} \hbar\omega_s g_s \frac{\omega_s}{\omega_s'} \coth(\hbar\omega_s'/2kT) \\ & - 2\lambda_s^2 \hbar\omega_s (g_s+1)(2g_s+1)^{-2} \Bigg] \end{aligned} \tag{9}$$

$$\begin{aligned} \Delta_{em}^2 = 8\ln 2 \sum_s \Bigg[& (\lambda_s \hbar\omega_s)^2 \frac{\omega_s^5}{\omega_s'^5} \coth(\hbar\omega_s'/2kT) \\ & + \tfrac{1}{2} (\hbar\omega_s g_s)^2 \frac{\omega_s^2}{\omega_s'^2} \coth^2(\hbar\omega_s'/2kT) \Bigg] \end{aligned} \tag{10}$$

5. Comparison with the Experimental Data

Table 1 shows experimental data on the position of absorption and fluorescence maxima as a function of the temperature. As the temperature increases from 2 K to 160 K both peaks are shifted some 30 cm^{-1} to higher frequencies, whereas between 160 K and 300 K a larger shift of additional 130 cm^{-1} is observed.

Table 1:

T (K)	(cm^{-1}) in absorption	Reference		(cm^{-1}) in emission	Reference	
2	10130	□	/6/	9635	■	/6,8,17/
77	10160	□ △	/6,4/	9660	● ■	/6/
100	10165	△	/4/	9660	↑	this paper
130	-		-	9660		
160	10169	▽	/5/	9665		
200	-		-	9710	↓	
300	10309	○	/3/	9785	●	

In contrast to this behaviour the widths of the bands increase more smoothly with temperature. Fig. 3 shows the squared width Δ^2 of emission and absorption (FWHM). The uncertainty in determining the width is large since the bands are asymmetric and overlapping with neighbouring ones. Within this uncertainty we observe a nearly linear increase of Δ^2 with temperature between 2 K and 300 K.

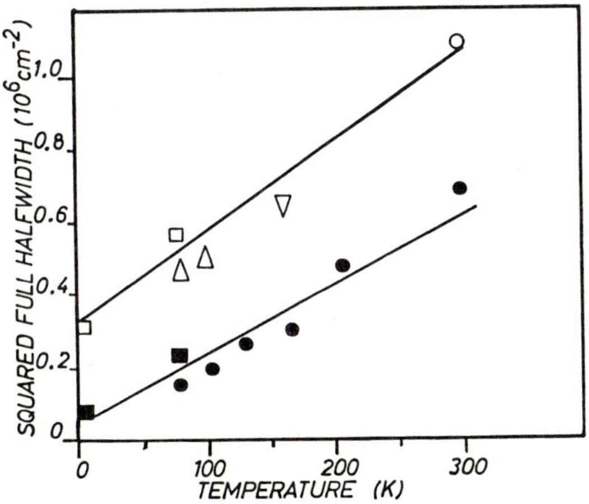

Fig.3: Temperature dependence of the dimer bandwidth in absorption (open symbols) and emission (closed symbols), References see table 1; solid line: model calculation for a linear mode with $\hbar\omega\lambda^2 = 290\ cm^{-1}$, $\hbar\omega = 30\ cm^{-1}$ and $\omega'/\omega = 1.05$. A temperature - independent contribution due to higher frequency modes has been added.

For the following discussion we distinguish between two types of vibrations: Linear modes with large Stokes shift $\hbar\omega\lambda^2$ and frequencies below 50 cm^{-1} provide an efficient line-broadening mechanism in molecular physics. Quadratic modes with negligible Stokes shift and large quadratic coupling $\hbar\omega g_s$ lead to a temperature-dependent shift of absorption and fluorescence maxima /18/.
For linear modes in the high-temperature limit $kT > \hbar\omega$ we have approximately $\coth(\hbar\omega/2kT) = 2kT/\hbar\omega$. If $\lambda^2 \doteq \Sigma\lambda_s^2$ is large and the frequency changes are small, then Δ^2 varies linearly with the temperature according to

$$\Delta^2_{abs} = 16\ln 2\ \lambda^2\hbar\omega kT \qquad \Delta^2_{em} = 16\ln 2\ \lambda^2\hbar\omega\ (\omega/\omega')^6\ kT \qquad (11)$$

From comparison with the experiment (cf.Fig.3) we determine a value of $\lambda^2\hbar\omega = 290\ cm^{-1}$ and of $\omega'/\omega = 1.05$. From this we calculate for the difference between absorption and emission energy

$$E_{em} - E_{abs} = -2\lambda^2\hbar\omega(g_s+1)(2g_s+1)^{-2} \qquad (12)$$

a value of 503 cm^{-1} which agrees well with the experimental Stokes shift of 500 cm^{-1}.

We conclude from this analysis that we are dealing with a strong coupling of a low frequence mode. As a consequence, the large width of the spectrum can be understood as being due to homogeneous broadening.

A possible interpretation of this low frequency mode could involve a motion of the two dimer components which changes their mutual distances. An alternative explanation could be based on broadening induced by the protein environment. However, the temperature dependence and the large Stokes shift put considerable constraints on the nature of the protein coupling. It must be sensitive to the excitation of the dimer. The different widths of absorption and emission spectra at low temperatures can be accounted for by quadratic coupling to a higher frequence mode. Actually we obtain a reasonable fit to the experimental data for a frequency of 350 cm^{-1} and a frequency change of 250 cm^{-1} to 600 cm^{-1} in the excited state. The calculated curves in this limit show a positive curvature instead of the straight line plotted in Fig. 3 and merge into the limiting final width for low temperatures. Such a quadratic mode also contributes to the blue shift of the absorption spectrum for increasing temperature and a smaller shift of the fluorescence. In any case this mechanism cannot fully explain the experimentally observed shift which is almost equally large for absorption and emission and which is quite sensitive to the phase transition from the solid to the liquid state of the samples. That is why we like to discuss other contributions to the shift of the dimer absorption and fluorescence. Roughly speaking the thermal shift reflects changes of intermolecular distances and orientations and of the mobility of vibrating and rotating groups of atoms. The protein environment may strongly affect such changes. As an example, conformational changes of the protein may cause large anharmonicities of the nuclear motion or even turn a torsional vibration into a free rotation.

Within our model three different contributions to the thermal shift can be considered (cf. Eqs. 6 and 9):

(i) The variation of the interaction V with the equilibrium distance d^o certainly is of importance. It is also discussed in the context of electron transfer and should depend exponentially on the thermal expansion coefficient. As mentioned above the dimer coupling V determines the location of the upper excitonic dimer state. Detailed studies of circular dichroism spectra (Knapp and Fischer in this book) show that the position of the upper dimer band must be very similar at high and at low temperatures. So we conclude that the temperature dependence of V is not the dominating contribution.

(ii) The energy $W_o(T)$ of a monomer from the special pair describes the temperature dependent environmental shift. A similar effect might be expected for the monomer bands unless dimer and monomer have specifically different environments. Indeed there are protein groups in the very vicinity of the dimer (within 3-4 Å) which are absent for the accessory monomer. In addition we remind that for the dimer the polarizability can be much more different in electronic ground and excited state than for the monomer. This polarizability difference enters into the solvent shift.

(iii) The thermal shift due to frequency changes points in the right direction but within the harmonic approximation the experimental temperature dependence of the shift cannot be fully accounted for, at least at temperatures above 200K. It might be necessary to consider strongly anharmonic vibrations and hindered rotations.

In conclusion we might state that the temperature dependence of the widths can be explained quite satisfactorily with a harmonic model. We point to the agreement with the observed energy distance of absorption and emission maximum. For a full understanding of the thermal shift more detailed investigations on anharmonic models may be necessary.

Acknowledgement

We like to thank Wilfried Göbel for helpful assistance performing the fluorescence measurements. This work was supported by the Deutsche Forschungsgemeinschaft, SFB 143.

References

/1/ C.N. Rafferty, R.K. Clayton: Biochim. Biophys. Acta 502, 51 (1978)
/2/ N.W.T. Woodbury, W.W. Parson: Biochim. Biophys. Acta 767, 345 (1984)
/3/ T.L. Trosper, D.L. Benson, J.P. Thornber: Biochim. Biophys. Acta 460, 318 (1977)
/4/ V.A. Shuvalov, A.A. Asadov: Biochim. Biophys. Acta 545, 296 (1979)
/5/ A. Vermeglio, G. Paillotin: Biochim. Biophys. Acta 681, 32, (1982
/7/ A.J. Hoff, E.J. Lous, K.W. Moehl, J.A. Dijkman: Chem. Phys. Lett., Vol. 114, No. 1, 39 (1985)
/6/ V.G. Maslov, A.V. Klevanik, M.A. Ismailov, V.A. Shuvalov: UDC 577 Doklady Akademiie Nauk SSSR, Vol. 269, No. 5, 1217 (1983)
/8/ A. Angerhofer, J.U. von Schütz, H.C. Wolf: Z. Naturforsch. 39c, 1085 (1984)
/9/ J.K.H. Hörber, W. Göbel, R. Feick, and M.E. Michel-Beyerle: to be published
/10/ J.K.H. Hörber, W. Göbel, M.E. Michel-Beyerle, and E.W. Knapp: this book
/11/ H. Michel: J. Mol. Biol. 158, 567 (1982)
/12/ A.S. Holt, R.K. Clayton: Photochem. Photobiol. 4, 829 (1965)
/13/ M.Z. Zgierski: J. Chem. Phys. 59, 1052 (1973)
/14/ W. Zinth, E.W. Knapp, S.F. Fischer, W. Kaiser, J. Deisenhofer, H. Michel: Chem. Phys. Lett. 119, 1 (1985)
/15/ E.W. Knapp, S.F. Fischer, W. Zinth, M. Sander, W. Kaiser: Proc. Natl. Acad. Sci. USA, in press
/16/ E.W. Knapp, S.F. Fischer: Chem. Phys. Lett. Vol.103, No.6, 479 (1984)
/17/ W. Göbel, A. Ogrodnik, C. Winscom: private communication
/18/ T.H. Keil: Phys. Rev. 2A 140, 601 (1965)

Local Environments of Pigments in Reaction Centers of Photosynthetic Bacteria from Resonance Raman Data

M. Lutz and B. Robert

Service de Biophysique, Département de Biologie, C.E.N Sacley, F-91191 Gif-sur-Yvette Cedex, France

Reaction centers from purple photosynthetic bacteria (Rhodospirillales) appear to generally contain two identical sets of pigments. One of these sets is predominantly (but most probably not exclusively, ref. 1) involved in the primary electron-transfers. Recent X-ray crystallographic data on reaction centers (RCs) from Rhodopseudomonas viridis show that these two sets of pigments are symmetrically arranged within the proteic core along a C_2 axis (2). The relative orientations and distances of molecules are very much the same in the two sets. Spectroscopic data indicate an identical situation in RCs from Rps. sphaeroides (3). Hence the very different electron-transfer properties of the two sets must largely result from differences in local environments, which may, in particular, result in different redox potentials of molecules in the two sets (4). These local environments still cannot be determined in a detailed fashion from X-ray data.

Resonance Raman (RR) spectroscopy offers this possibility for the bacterio-chlorophylls and -pheophytins of the center : previous work has shown that information can be obtained on the conformations and intermolecular binding states of their dihydrophorbin rings and their conjugated substituents, as well as on the local permittivities around them (5,9). The highest information content of the RR spectra of BChl and of Bpheo is obtained when resonance is with the Soret electronic transition (7). In particular, RR spectra of either of these molecules excited in these conditions inform on the intermolecular binding states of the conjugated ketone and acetyl carbonyls, and of the $Mg-N_4$(pyrrole)LL' group when present (6,7). In these conditions however, a limited selectivity only is obtained amid the contributions of the six bacteriochlorin pigments of the RC. Resonance Raman spectra of chemically oxidized or reduced RCs from e.g. Rps. sphaeroides excited at 363.8 nm contain 10–11 distinct spectral components in the carbonyl stretching region, where 12 unequivalent carbonyl vibrators are expected (fig. 1). These components are listed in Table 1 for dithionite-treated RCs of Rps. sphaeroides 2.4.1.

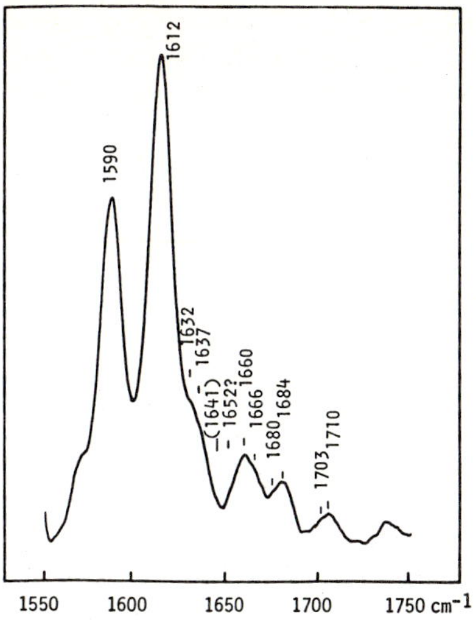

Figure 1 :
Resonance Raman spectrum (carbonyl stetching region) of reaction centers from Rps. sphaeroides 2.4.1 at 30K. Excitation wavelength : 363.8 nm. Spectral resolution : 8 cm^{-1}.

Table 1. Average wavenumbers (cm^{-1}) observed in the carbonyl stretching regions of RR spectra of BChl and BPheo in reduced reaction centers from Rps. sphaeroides, wild type at 30°K.

SORET	1632	1637	1641	1652?	1660	1666	1680	1684	1704	1710
Q_x	1627 (535)	1635 (545)			1660? (535)		1678 (545)		1703 (535)	
ASSIGN^T	H_B	H_A,P	B	B?	P,H_B?	P?	H_A	P	H_B	B

Soret : spectra excited at 363.8 nm ; contributions from the four BChls and the two BPheos. Q_x : spectra excited in the Q_x bands of the BPheos, contributions from these molecules only. (535), (545) : frequencies assigned to the BPheo molecules absorbing at 535 or at 545 nm at 30°K, respectively. Assign^t : tentative assignments of the reaction center frequencies to the primary donor BChl molecules (P), to the other two BChl molecules (B), to the primary acceptor BPheo molecule (H_A), and to the second BPheo molecule (H_B).

1 – Bacterio pheophytins

Comparing the environmental interactions assumed by the individual pigments requires that each of these carbonyl stretching frequencies is attributed to an individual molecule. The frequencies of the conjugated carbonyls of the two BPheos can be determined by inspection of RR spectra obtained at 20°K and excited in the Q_x bands of these molecules (fig. 2). Although the selectivity in observing either of the two individual molecules is not

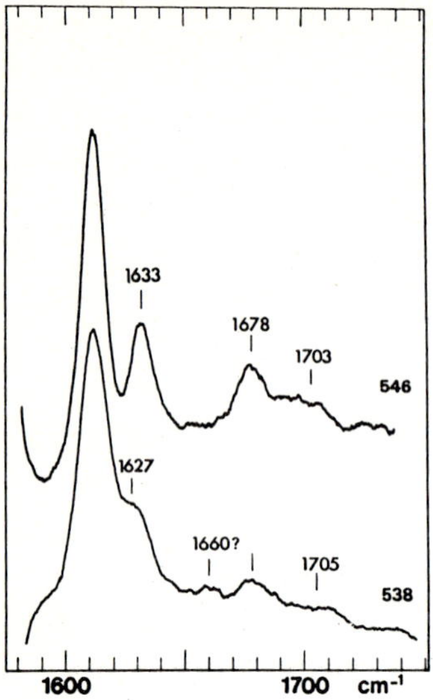

Figure 2 :
Carbonyl stetching regions of RR spectra of the bacteriopheophytins present in reaction centers from wild-type *Rps. sphaeroides* at 20K. 538, 546 : excitation wavelengths (nm). Spectral resolution : 5 cm^{-1}.

absolute (7,8), spectra excited at 545 nm primarily arise from the normal primary acceptor molecule H_A, while those excited near to 535 nm arise from the second BPheo molecule, H_B. It thus appears likely that H_A has a bound acetyl C=0, vibrating at 1635 cm^{-1} and a bound ketonic C=0 vibrating at 1678 cm^{-1}. H_B has a bound acetyl C=O vibrating at 1627 cm^{-1} and a free keto group vibrating at 1703 cm^{-1}. This latter group might also alternatively vibrate at 1660 cm^{-1}, although less likely, and then be intermolecularly bound. All of these frequencies match, within experimental uncertainty, those of the ν C=0 cluster observed in Soret resonance conditions (Table 1). Further inequivalence in the local environments of the BPheos is demonstrated by differences in frequencies and relative intensities of several of their RR-active modes arising from the dihydrophorbin rings (7,8).

2 - Primary donor

Distinguishing the carbonyl vibrations of the four BChl molecules in the Soret-excited spectra is more difficult, especially because these stretching modes are not significantly resonant with the Q_x transition near 600 nm (5). The simple-minded idea that RR spectra of the primary donor BChls, P_A and P_B, should be obtained by difference between spectra of RCs poised at high and low redox potentials actually meets two difficulties. The first one arises from the actinic effect of the laser excitation which results, in reduced RCs, in the building up of P^R (a triplet state involving P870), at the expense of the neutral species, in variable amounts depending on temperature, on the excitation conditions and on the presence and chemical nature of a carotenoid in the RC (10,11). Another difficulty arises from the still open possibility that RR spectra of oxidized $P870^+$ may involve contributions from $P_A P_B^+$ and $P_A^+ P_B$ species, i.e. may involve contributions from neutral P_A and P_B molecules with their ground-states hardly perturbed by their interactions with the adjacent cation (12).

We circumvented these difficulties (B. ROBERT and M. LUTZ, manuscript in preparation) by making use of the fact that, in RCs of the wild type of Rps. sphaeroides at low redox potential, the energy-transfer from P^R to the carotenoid (spheroidene) is strongly temperature-dependent. This transfer has been shown to occur with high yield from 300 to ca 70K and to decrease at lower temperatures (10,11). In steady-state excitation conditions, and at e.g. 80K, the actinic effect of the laser beam (e.g. 363.8 nm) used to selectively observe neutral BChl molecules thus results in the building up of sizable amounts of ^{3}Car (13), and hence in sizable amounts of neutral P870 remaining present, while at e.g. 20K, P^R builds up at the expense of neutral P870. A result of such a pump-probe pair of experiments is shown on fig. 3 : the two spectra, recorded at 20 and 75K, were normalized at 1590 cm^{-1}, a band specifically contributed by the BPheo molecules. It is seen that, as expected, the νC=C band of ground-state spheroidene at 1540 cm^{-1} is weaker at 20K than at 75K (the triplet species does not significantly resonate at this excitation wavelength). On the opposite, the 1615 cm^{-1} band arising from stretching of the methine bridges of both the BChl and BPheo molecules is enhanced, indicating that more BChl molecules are contributing at 20K than at 75K.

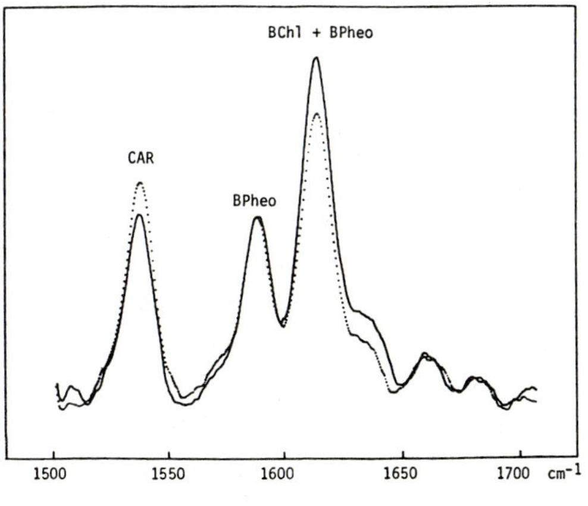

Figure 3 : Resonance Raman spectra (1500–1700 cm^{-1} region) of reaction centers from Rps. sphaeroides 2.4.1 treated with excess dithionite – Excitation : 363.8 nm. Temperature : 30K (solid line) and 75K (dotted line).

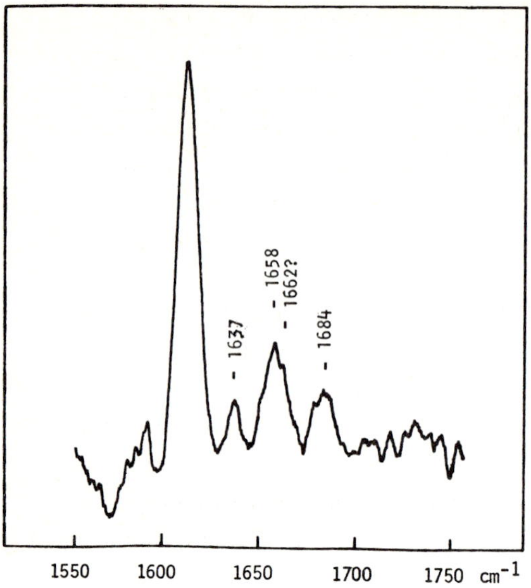

Figure 4 : Difference resonance Raman spectrum (carbonyl stetching region) resulting from subtraction of spectra of reduced reaction centers from Rps. Sphaeroides 2.4.1 obtained at 30K and at 80K, respectively. Excitation wavelength : 363.8 nm. The spectra (summation of 25 scans each) were normalized on the 1590 cm^{-1} band intensity (cf fig. 3).

A difference spectrum obtained in these conditions is shown on fig. 4. It clearly is constituted by positive contributions from BChl molecules, and no negative contributions which might have arisen from P^R. Taking into account the relatively low signal to noise ratio of these difference spectra, the minimal number of carbonyl bands observed in the 1630-1700 cm^{-1} region is three, indicating that at least two BChl molecules contribute. There is actually no need of any additional molecule to account for the observed spectrum, confirming that, at 20K, the P^R state involves the primary donor alone (14). It thus appears reasonable to ascribe the RR spectrum of fig. 4 to P870 in its neutral state.

In the present state of interpretation of these spectra, the following preliminary conclusions can be drawn about environmental interactions assumed, in the ground-state, by the P_A and P_B BChl molecules.

Magnesium liganding. The high-frequency value (1612 cm^{-1}) and the small halfbandwidth (13 cm^{-1}) of the νC_aC_m mode indicate that, in neutral P870, the magnesium atoms of both P_A and P_B should be five-coordinated, confirming conclusions previously drawn from inspection of RR spectra of the whole set of RC pigments (6). Indeed, in vitro work has established that the νC_aC_m frequency occurs around 1605 cm^{-1} for BChls with a six-coordinated Mg, and at 1615 cm^{-1} when this atom assumes five coordination (15,16).

Carbonyl stretching frequencies. Three different interpretations may be proposed at this stage for this spectral region. In the first two, which appear the more likely ones, the two acetyl carbonyl vibrators, which should contribute in the RR spectra of neutral P870, are assigned the 1637 and 1658 cm^{-1} bands, on the basis of their frequency range observed in several BChl *a*-containing systems (8). The 9-keto carbonyls should then either both vibrate at *ca* 1683 cm^{-1}, or, taking into account a persistent shoulder present at 1662 cm^{-1} in independent experiments, one of them might vibrate at this latter frequency, the other one then contributing alone to the 1683 cm^{-1} band (fig. 4). In these interpretations, the P_A and P_B BChl molecules should assume markedly different intermolecular bonding, at least through their 3-acetyl carbonyls. One of these groups indeed should be strongly bound, assuming a frequency close to that observed e.g. in $BChl_n$ oligomers in vitro (8), and the second one should be only very weakly interacting, if at all, its stretching frequency corresponding to the lower edge of the range observed for the free vibrator (8). It may be worth noting that, in this interpretation, the weakly bound acetyl C=O yields a comparatively strong band (taking the 1612 cm^{-1} band intensity as a standard.) This indicates that this group assumes a highly conjugated, in-plane conformation. The 1637 cm^{-1} band is weaker, suggesting that the corresponding acetyl carbonyl is drawn out of optimal conjugation by its intermolecular bond.

In a third, less likely interpretation of this spectral region, the 1637 cm^{-1} band might be assigned to a strongly interacting keto carbonyl, while the 1658-1662 cm^{-1} band would arise from the two acetyl groups. The 9-C=O group indeed has been observed to vibrate at frequencies as low as 1640 cm^{-1} in hydrated polymers of chlorophylls (17). However, such a frequency never has been characterised up to now for any chlorophyll in a biological environment.

The C=O stretching frequencies observed for the P870 BChls have been indicated on Table 1. It should be noted that these frequencies not only occur in RR spectra of chemically reduced RCs, but also match with observable features of chemically oxidized RCs (6,9), again suggesting that RR spectra of the latter may well contain contributions of P_A and P_B in their neutral ground states.

3 - Other bacteriochlorophylls

Table 1 shows that taking into account the assignments proposed in the preceding sections for the bacteriopheophytins and for P870, the 1641, 1652, and 1710 components of reduced RCs spectra remain available as stretching frequencies of the conjugated carbonyls of the secondary BChls. In the frame of the tentative and partial assignment proposed in Table 1, the bounding interactions assumed by the acetyl carbonyls of these BChls should be moderately unequivalent only in as much as their stretching frequencies might be 1641 and 1652 cm^{-1}. Also, the secondary BChls both should have five-coordinated magnesium atoms (6).

4 - Conclusion

Taken together, the present RR data indicate that, in reaction centers from Rps. sphaeroides, marked differences occur in the intermolecular bonding of the two sets of pigments. Much of this asymmetry is somewhat beyond the present spatial resolution of X-ray crystallographic data on RCs of Rps. viridis (2). However, some inconsistency appears between RR and X-ray data on the primary donor, in as much as RCs from these two species should have closely similar structures (3). Indeed, X-ray data indicate that the magnesiums of both P_A and P_B should be six-coordinated, simultaneously being interacting with the protein and being in close contact with the acetyl $COCH_3$ grouping of the adjacent molecule (2). The known electron donor properties of the acetyl C=O group strongly suggest that, if not prevented by steric constraints, bonding will be established in these conditions between the Mg atoms and the $_3C = O$ groups. True, the frequency of the νC_aC_m mode is indirectly sensitive to the number of ligands on the magnesium, probably through the conformation of the MgN_4 grouping : a 1612 cm^{-1} frequency conceivably might arise from a markedly unsymmetrical liganding on the magnesium, allowing an out-of-plane position for this atom close to that observed for genuine five-coordinated models, i.e. ca 4A (18,19). This hypothesis might be tested when the lower frequency regions of the RR spectra of P870 are obtained, because of their involving modes of the $MgN_4L(L')$ grouping (6,8).

However, as discussed in a preceding section, a symmetrical structure for P870, involving two equivalent $_3C{=}O{-}Mg$ bonds.is probably inconsistent with RR data. This situation conceivably may result from a non-unicity of the ground-state structure of P870 in a macroscopic sample, an average picture being given by X-ray data. Further work clearly is needed to solve this interesting problem.

References

1 B. ROBERT, M. LUTZ and D. TIEDE : FEBS Letters 183, 326 (1985)
2 D. DEISENHOFER, O. EPP, K. MIKI, R. HUBER and H. MICHEL : J. Mol. Biol. 180, 385 (1984)
3 J. BRETON : These Proceedings (1985)

4 A. DESBOIS, G. MAZZA, F. STETZKOWSKI and M. LUTZ : Biochim. Biophys. Acta 785, 161 (1984)
5 M. LUTZ, J. KLEO and F. REISS-HUSSON : Biochem. Biophys. Res. Comm. 69, 711 (1976)
6 M. LUTZ : in Photosynthesis (G. Akoyunoglou, ed.) Balaban ISS, Philadelphia, vol 3, p 461 (1981)
7 M. LUTZ : in Proc. 7th Internat. Conf. on Raman Spectroscopy (W.F. MURPHY, ed.) North-Holland, Amsterdam, p 520 (1980)
8 M. LUTZ : in Adv. in Infrared and Raman spectroscopy (R.J.H. Clark and R.E. Hester, eds) Wiley-Heyden, Chichester, vol 11, ch. 5 (1984)
9 M. LUTZ, A.J. HOFF and L. BREHAMET : Biochim. Biophys. Acta 679, 331 (1982)
10 W.W. PARSON and T.G. MONGER : Brookhaven Symp. Biol. 28, 196 (1976)
11 C.C. SCHENCK, P. MATHIS and M. LUTZ : Photochem. Photobiol. 39, 407 (1984)
12 M. LUTZ and J. KLEO : Biochim. Biophys. Acta 546, 365 (1979)
13 M. LUTZ, L. CHINSKY and P.Y. TURPIN : Photochem. Photobiol. 36, 503 (1982)
14 V.A. SHUVALOV and W.W. PARSON : in Photosynthesis (G. Akoyunoglou, ed.) Balaban ISS, Philadelphia, vol 3, p 949 (1981)
15 T.M. COTTON and R.P. VAN DUYNE : J. Am. Chem. Soc. 103, 6020 (1981)
16 B. ROBERT and M. LUTZ : Biochim. Biophys. Acta 807, 10 (1985)
17 M. LUTZ : Biochim. Biophys. Acta 460, 408 (1977)
18 H.C. CHOW, R. SERLIN and C.E. STROUSE : J. Am. Chem. Soc. 97, 7230 (1975)
19 B.W. MATTHEWS, R.E. FENNA, M.C. BOLOGNESI, M.F. SCHMID and J.M. OLSON : J. Mol. Biol. 131, 259 (1979)

The Spin-Polarization Pattern of the $\Delta m = 1$ Triplet EPR Spectrum of *Rps. viridis* Reaction Centers

F.G.H. van Wijk, P. Gast, and T.J. Schaafsma

Agricultural University, Department of Molecular Physics de Dreijen 11, NL-6703 BC Wageningen, The Netherlands

The $\Delta m = 1$ triplet EPR spectra of bacterial RC's exhibit a unique AEEAAE electron spin polarization (ESP) pattern (A = absorptive, E = emissive), which can be explained by invoking an intermediary radical pair state [1,2]. We have observed that this pattern changes into AEAEAE at ∿20K for reduced Rps. viridis RC's.

Since the spectrum remains strongly polarized at T > 20K, spin-lattice relaxation in the $^3P_{960}(P_R)$ state cannot cause the polarization-transition. Furthermore, normal intersystem crossing $^1P \rightarrow {}^3P$ would not give rise to the large polarization observed. Exchange coupling of the spin of the bacteriopheophytin-b anion (I^-) in the RP to the $Fe^{2+}Q^-$ paramagnetic complex [3] can explain the experimental results, since the spin-lattice relaxation time of high-spin Fe^{2+} is known to be strongly dependent on temperature, and may reach a value comparable to the RP-lifetime around 20K. This is confirmed by the observation that the polarization-transition does not occur when chromatophores are treated with SDS, known to affect the $Fe^{2+}Q^-$ and thus the Q^-I^- interactions [4].

References:
1. M.C. Thurnauer, J.J. Katz and J.R. Norris: Proc. Natl. Acad. Sci. 72, 3270 (1975)
2. R.E. Blankenship, T.J. Schaafsma and W.W. Parson: Biochim. Biophys. Acta 461, 297 (1977)
3. A.J. Hoff and P.J. Hore: Chem. Phys. Lett. 108, 104 (1984)
4. R.C. Prince, D.M. Tiede, J.P. Thornber and P.L. Dutton: Biochim. Biophys. Acta 462, 467 (1977).

Triplet State Investigation of Charge Separation and Symmetry in Single Crystals of *R. viridis* Reaction Centers

J.R. Norris, D.E. Budil, H.L. Crespi, M.K. Bowman, P. Gast, C.P. Lin, C.H. Chang[†]*, and M. Schiffer*[†]

Chemistry Division and [†]Division of Biological and Medical Research
Argonne National Laboratory, Argonne, IL 60439, USA

1. Introduction

The special pair model for the primary donor of bacterial photosynthesis which was originally proposed on the basis of EPR spectroscopy [1] has now been confirmed by x-ray structural data [2]. The initial interpretation of the EPR linewidth of the donor cation was based on a symmetrical sharing of the unpaired electron on a dimer of chlorophyll or bacteriochlorophyll. The ground state of the special pair as presently revealed from the x-ray study is indeed a highly symmetrical structure with approximate C_2 symmetry.

A rigorous test of the actual working symmetry of the donor is simple and straightforward. By "working symmetry", we mean the symmetry exhibited in states of the donor other than the ground state which participate in the kinetic scheme of photosynthetic charge-separation. These states include the lowest excited singlet and triplet states of the dimer and its cation. We will explore experimental evidence relevant to the lowest excited triplet state; specifically, this paper discusses EPR evidence [3] that the triplet state of the special pair and its immediate local protein environment deviates substantially from C_2 symmetry.

2. Experimental Method

The initial x-ray structure possesses a local, approximate C_2 symmetry for the ground state of the donor in photosynthetic reaction centers from R. viridis [2]. If the triplet state resides symmetrically on a pair of bacteriochlorophylls with C_2 symmetry, then one of the three axes of the triplet dipolar tensor, $\hat{x}$, $\hat{y}$, or $\hat{z}$, must lie along, or approximately along, the 2-fold rotation axis defined by the C_2 symmetry. It then follows that the two remaining triplet axes must be perpendicular to the C_2 axis. The method used to probe the working symmetry of the donor triplet will be to compare the principal directions for the zero-field splitting tensor of the triplet state in a single reaction center crystal with the direction of the C_2 axis derived from x-ray diffraction.

Deisenhofer et. al. [2] have determined a matrix which performs the C_2 rotation for the chlorophylls in the R. viridis reaction center crystal. The unit vector in the direction of the C_2 axis, $\hat{c}_2$, was determined by diagonalizing their matrix. We compare $\hat{c}_2$ to the molecular axes of the triplet reported by Gast et. al. [3] by calculating the directional angles $\theta_i = \cos^{-1} \hat{c}_2 \cdot \hat{\imath}$ for i = x,y,z. Since the reported triplet axes define an arbitrary one of the eight possible triplet orientations in the unit cell, we calculate the θ_i for the other seven sites which are obtainable via the symmetry operations of the $P_{4_1 2_1 2}$ space group of the crystal. Table 1 shows the directional angles calculated for each of the eight possible orientations of the donor triplet in the unit cell.

Table 1: Directional angles of the C_2 axis in the molecular coordinates for the eight triplet orientations in the unit cell.

Site	θ_x	θ_y	θ_z
1	27	116	95
2	138	48	89
3	100	148	60
4	142	128	94
5	118	68	37
6	71	127	137
7	54	39	78
8	69	47	129

3. Results and Conclusions

Since the x-ray crystal structure has not been refined, the C_2 symmetry and the direction of the axis is not expected to be perfect; moreover, the rotation matrix of Deisenhofer et. al. is derived from a least-squares fit to all six chromophores and not just the special pair [2]. Nevertheless, as is evident from Table 1, no zero-field triplet axis is even approximately parallel (or antiparallel) to the x-ray-determined C_2 axis. The closest approach is about 27°, much larger than possible errors in the determination of $\hat{c}_2$ or the zero-field axes. Although there are three orientations where a zero-field axis is nearly perpendicular to $\hat{c}_2$, there are none where two axes are perpendicular to $\hat{c}_2$. Thus, the triplet state of the P_{890} special pair deviates significantly from C_2 symmetry despite the highly symmetric chromophore structure indicated by the x-ray data.

Luminescence studies of model systems [5,6] have shown that the interactions which mix the monomeric excited states in a symmetric dimer are much stronger for the singlet states than they are for the triplet states. This is because the ground-to-excited state transitions in the monomers are spin-forbidden for triplets [6]. The actual symmetry of the excited dimer state will depend upon the magnitude of this mixing term relative to any difference in the energies of the monomer states. Because the triplet states have a small mixing term, the symmetry of the dimer triplet will be very sensitive to differences in the local environments of the two monomers. We conclude that the observed asymmetry of triplet distribution in P_{890} most likely reflects an asymmetry in the protein environment immediately surrounding the special pair.

If an asymmetry exists in the protein environment of the primary donor, the donor cation of R. viridis might also be unequally shared on the special pair. The observed EPR linewidth of the cation (~11.5 gauss) is too broad to be a symmetrical dimer (~9.6 gauss), but too narrow to be a monomer (13.5 gauss). Thus, the cation EPR linewidth is also consistent with an asymmetrical spin distribution between the two chlorophylls of the special pair.

4. Acknowledgement

This work was supported by the Division of Chemical Sciences, Office of Basic Energy Sciences, Department of Energy under contract W-31-109-Eng-38.

5. References

1. J.R. Norris, R.A. Uphaus, H.L. Crespi, and J.J. Katz, "Electron Spin Resonance of Chlorophyll and the Origin of Signal I in Photosynthesis", Proc. Natl. Acad. Sci. 68, 625-628 (1971)
2. J. Deisenhofer, O. Kepp, K. Miki, R. Huber, and H. Michel, "X-Ray Structure Analysis of a Membrane Protein Complex. Electron Density Map at 3 Å Resolution and a Model of the Chromophores of the Photosynthetic Reaction Center from Rhodopseudomonas viridis", J. Mol. Biol. 180, 385-398 (1984)
3. P. Gast, M.R. Wasielewski, M. Schiffer, and J.R. Norris, "Orientation of the Primary Donor in Single Crystals of Rhodopseudomonas viridis Reaction Centers", Nature 305, 451-452 (1983)
4. P. Gast and J.R. Norris, "EPR Detected Triplet Formation in a Single Crystal of Reaction Center Protein from the Photosynthetic Bacterium Rhodopseudomonas sphaeroides R-26", FEBS Lett. 177, 277-280 (1984)
5. J.B. Birk, Photophysics of Aromatic Molecules, Wiley-Interscience, New York, 1970
6. D.P. Craig and S.H. Walmsley, Excitons in Molecular Crystals, Theory and Applications, W.A. Benjamin, New York, 1968, Chapter 1.

Triplet-minus-Singlet Absorbance Difference Spectroscopy of Photosynthetic Reaction Centers by Absorbance-Detected Magnetic Resonance

A.J. Hoff

Department of Biophysics, Huygens Laboratory of the State University, P.O. Box 9504, NL-2300 RA Leiden, The Netherlands

The technique of absorbance-detected magnetic resonance (ADMR) of triplet states and that of ADMR-monitored triplet-minus-singlet (T - S) absorbance-difference spectroscopy is discussed. Applications of T - S and linear dichroic T - S spectroscopy on photosynthetic reaction centers of bacterial and plant origin are reviewed.

1. Introduction

Recently Absorbance-Detected Magnetic Resonance (ADMR) has been introduced as a sensitive technique to measure triplet-minus-singlet (T - S) absorbance-difference spectra of photosynthetic reaction centers (RC). ADMR is a form of optically detected magnetic resonance (ODMR) of triplet states in 'zero' laboratory magnetic field. For RC of low fluorescence and phosphorescence yield, ADMR is the technique of choice to accurately measure the zero-field splitting and decay parameters of the reaction center triplet state. In addition, the technique is ideally suited to record T - S spectra with a sensitivity that is two to three orders higher than that in conventional flash absorbance difference spectroscopy. Employing a polarized microwave field, it has been possible to measure linear dichroic T - S spectra with high accuracy. This has allowed the determination of the angles that the transition moments of the pigments giving rise to the T - S spectrum make with two of the triplet dipolar axes, and consequently the angles between these transition moments themselves.

In this article we will present the ADMR and linear-dichroic (LD) ADMR techniques and we will highlight several of their applications on photosynthetic reaction centers.

2. The Reaction Center Triplet State

The reaction center is defined as the smallest isolated pigment-protein complex that can perform the primary photoreaction[1] $D_oA_oA_1 \xrightarrow{h\nu} D_o^+A_1^-A_1 \longrightarrow D_o^+A_oA_1^-$ with similar high yield and speed as in the intact photosynthetic membrane. When the primary acceptor A_1 is chemically prereduced or deleted from the RC, then electron transport beyond A_o is virtually blocked, and the

[1] D_o is the primary electron donor, A_o a transient intermediate electron-acceptor, A_1 the primary acceptor.

charges decay by recombination: $D_o^+A_o^- \rightarrow D_oA_o$ (1), $D_o^+A_o^- \rightarrow D_o^*A_o$(2), or $D_o^+A_o^- \rightarrow {}^3D_oA_o$ (3), where the asterisk denotes the singlet excited state and the superscript 3 the triplet state. In bacterial RC reaction (3) has a probability of about 15 % at room temperature and close to 100 % at cryogenic temperature (<20 K). Recombination to the excited or triplet state of A_o seems to be energetically forbidden.

3. Properties of the Triplet State

The triplet state contains two unpaired electrons, which sense each other's magnetic field via the dipole-dipole interaction. This lifts the degeneracy of the three triplet substates ($S = 1$, multiplicity $2S + 1 = 3$), whose energies then correspond to the eigenvalues of the traceless dipole-dipole tensor, viz. $-X$, $-Y$, $-Z$, $X + Y + Z = 0$. Conventionally, the dipolar splittings are expressed in the two independent zero-field splitting parameters $D = -3/2\ Z$ and $E = -\frac{1}{2}(X - Y)$. It can be shown (see e.g. [1]) that there is a magnetic transition dipole moment between any two of the three sublevels with eigenfunction $|T_i\rangle$ (i = x, y, z) given by $\gamma \langle T_x | S_y | T_z \rangle = i\gamma$ (cyclic) where S_y is the y-component of the spin angular momentum operator $\underline{S}$ and γ is the electronic gyromagnetic ratio. In a resonant microwave field $\underline{H}_1$ the transition probability is then given by $\gamma \langle T_z | \underline{H}_1 \cdot \underline{S} | T_y \rangle^2 = \gamma \underline{H}_{1x}^2$ (cyclic).

The triplet sublevels are normally populated via intersystem crossing (ISC) from the lowest excited singlet state with population probabilities p_i (i = x, y, z) that depend on molecular symmetry.For the radical recombination reaction (3) the p_i's cannot be given a priori,but they seem to be fairly close to each other [2]. Decay from the sublevels to the singlet ground-state proceeds via ISC with rates k_i that depend on molecular symmetry. Under continuous illumination,the equilibrium populations n_i of the sublevels are given by $n_i = c\ p_i/k_i$, where $\Sigma p_i = 1$ and c is a constant proportional to the light flux.

4. Absorbance Detected Magnetic Resonance, ADMR

The principle of ADMR is illustrated in Fig. 1, which depicts the triplet sublevels with equilibrium populations given by filled circles, and p_i's and k_i's indicated by arrows. Application of resonant microwaves leads to a redistribution of the populations of two sublevels: spins are transferred from a level with high n_i to one with low n_i. In our example this means that spins are transferred to a fast decaying level,and consequently the total triplet population $[T] = \Sigma n_i$ will decrease. As a consequence, the singlet ground-state population will increase by the same amount, as the population of the singlet

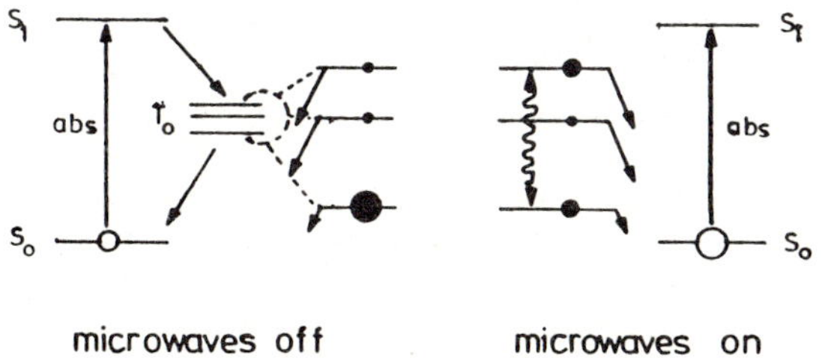

Fig. 1. Principle of absorbance-detected magnetic resonance (ADMR). Slanted arrows: population of probabilities and decay rates. Corrugated arrow: microwave transition

excited state is negligible under common light levels. This increase is of course reflected by the absorbance of the sample. Thus, scanning microwaves over the resonance frequency under conditions that spin-lattice relaxation is much slower than the microwave-induced transitions (in practice at temperatures around 2 K), one detects a resonance by monitoring the absorbance at a suitable wavelength (e.g. the $S_i \leftarrow S_o$ transition of D_o). To increase sensitivity.the microwaves are amplitude-modulated at a frequency compatible with the average decay rate of the sublevels,and the modulated absorbance is demodulated with a lock-in detector.

A schematic diagram of an ADMR set-up is shown in Fig. 2. Commonly, microwave irradiation is carried out in a helix by virtue of its broadbanded admittance. Helix and sample are inside a liquid helium bath cryostat equipped with at least two windows. Excitation is by a stable broadbanded light source, which in our set-up also serves as probe beam.

Decay rates are determined by monitoring absorbance response to a pulse of resonance microwaves. A nearly bi-exponential response results which yields the decay rates of the two sublevels connected by the microwaves [3].

For RC of a number of bacteria and of the two plant photosystems, the $|D|$ and $|E|$ values and the sublevel decay rates of ^3D_o have been accurately measured. This work has been reviewed recently [4] and will not be discussed here.

5. Triplet-minus-Singlet Spectroscopy

Introducing a triplet state on D_o has several consequences for the optical spectrum of RC. Firstly, the absorbance of D_o itself in the near-infrared will

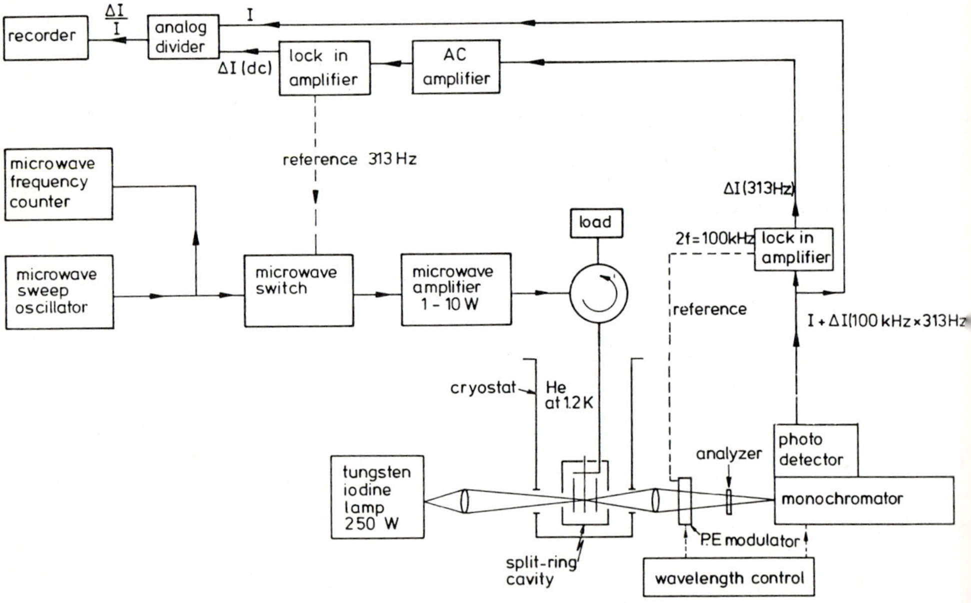

Fig. 2. Scheme of ADMR set-up, including attachments for linear dichroic measurements (see Fig. 3). From [6]

be bleached, as (bacterio)chlorophyll triplet states have litte absorbance in this region. Secondly, if D_o is a (B)Chl dimer, then the triplet state may be localized or delocalized (or very rapidly hopping on a timescale of say 10^{-13} s). In the former case, at any time one (B)Chl of D_o contains a triplet, the other does not. If the transition moments in the dimer are strongly coupled in the singlet state through electric dipole interaction, the optical spectrum will show exciton effects: bandsplits and/or shifts and intensity changes compared to the absorbance spectrum of monomeric (B)Chl. In the triplet state, the electrostatic coupling will generally be weaker or absent, because it then must operate through short-range exchange interaction. Thus, by generating ^3D_o, the original (exciton) band(s) disappear and (for a localized triplet) an absorbance band of monomeric (B)Chl will appear at a position that one expects to be much closer to the monomer Q_y absorption than the exciton component(s). For a delocalized triplet, of course, such a monomeric band does not appear. Thirdly, neighbouring pigments, as the two accessory BChl in bacterial RC or the purported Chl acceptor in RC of PS I, will sense different coupling with ^3D_o than with D_o, so that their absorbance bands (especially the Q_y bands) may shift, and possibly undergo a change in oscillator strength.

From the above it will be clear that the triplet state of D_o is a useful probe of the pigment interaction and thus of pigment geometry. It is a much less perturbing probe than the photoinduced D_o^+ cation, because there Coulombic interactions may lead to appreciable Stark shifts of the bands of neighbouring pigments, thereby obscuring details as appearing bands or bleachings of weak exciton components. Ambiguities in the interpretation of $D_o^+ - D_o$ absorbance difference spectra (redox ΔA spectra) may thus be resolved by a comparison with $^3D_o - D_o$ (triplet-minus-singlet, T - S) absorbance difference spectra.

With the ADMR technique it is a simple matter to record T - S spectra with high sensitivity and high resolution. By virtue of the fact that the total triplet state concentration is modulated by the resonant microwaves, lock-in demodulation yields a DC signal that is proportional to the absorbance change occasioned by the presence of the triplet state. Scanning the broad-banded transmitted light by a monochromator, while keeping the microwave frequency at resonance and dividing by the total light intensity at each wavelength directly yields the T - S spectrum [5]. (For the small intensity changes effectuated by the microwaves, the change in transmittance may be set equal to the change in absorbance).

6. Linear Dichroic Triplet-minus-Singlet Spectroscopy

The power of optical difference spectroscopy is much enhanced when use is made of polarized light. Linear dichroic (LD) spectra and LD-redox ΔA spectra of a great variety of RC have been published (see e.g. J. BRETON, this volume). Another technique is that of photoselection spectroscopy, where the sample is made anisotropic by, for example, selectively bleaching pigments with a beam of polarized light, which anisotropy is then probed with light polarized parallel or perpendicular to the excitation beam's electric vector. All these techniques have yielded a wealth of information on pigment orientation in the RC. Of course, the analysis of polarized spectra depends to some, and often a great, extent on the interpretation of the various features, e.g. is a feature part of a bandshift, a weak exciton component, an appearing band, an intensity change of an existing band, etc. Clearly, it would be of much use to be able to compare e.g. polarized redox ΔA spectra with polarized T - S spectra. Up to now such spectra have not been obtained with conventional flash spectroscopy owing to the inherent low sensitivity of that technique. It has been possible, however, to modify the ADMR technique such that LD-(T - S) spectra can be recorded with great accuracy.

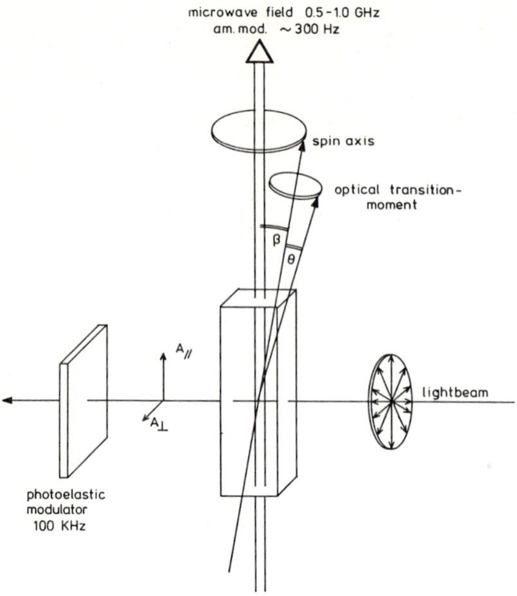

Fig. 3. Principle of linear dichroic ADMR. A horizontal beam of unpolarized white light passes the sample from the right to the left. The resonant microwaves that are vertically polarized introduce an ellipticity that is detected via a photoelastic modulator and an analyzer (not drawn). From [6]

LD-ADMR. The principle of LD-ADMR is based on the property that as mentioned in section 3 the magnetic transition dipole moment between the triplet sublevels is polarized. For example, the transition between the x- and z-level (Fig. 1) is polarized along the y-triplet spin axis. Hence, for a microwave field with the magnetic vector polarized along a certain direction, the transition probability for a particular molecule is proportional to $\cos^2\beta$ where β is the angle between the y-spin axis of this molecule and the direction of the microwave field. For $\beta = 90°$, the molecule will not respond to the microwaves. Thus, when polarized resonant microwaves are applied, out of the random distribution of triplet states an oriented selection is made that responds to the microwaves. When the sample is now probed with a beam of unpolarized light, the absorbance will depend on the direction of the electric vector with respect to the axis of the microwave field. In other words, the transmitted light becomes elliptically polarized. This ellipticity is detected via a photoelastic modulator, which sequentially transmits light in two perpendicular directions with a frequency of 100 kHz, and an analyzer (Fig. 3).

After double demodulation (the microwaves are again amplitude-modulated at a low frequency), a DC signal results that is proportional to $(T - S)_{\perp} - (T - S)_{//}$ where ⊥ and // refer to the axis of the microwave field. Scanning the monochromator yields a LD-(T - S) spectrum [6]. By virtue of the double modulation technique, the signal-to-noise ratio of this spectrum is comparable to that of the T - S spectrum. To facilitate comparison, a T - S spectrum is recorded simultaneously by demodulating only the low frequency microwave-induced modulations.

The LD-ADMR technique is very similar to regular ADMR, with the exception that the helix (in which the microwaves are more or less randomly polarized) is now replaced by a microwave cavity. For the relatively low frequencies of the ADMR transitions of ^3D_o it is of advantage to use a so-called split-ring cavity [7], Fig. 4, in which slots have been machined to admit excitation and probe light (in our set-up we use the same beam).

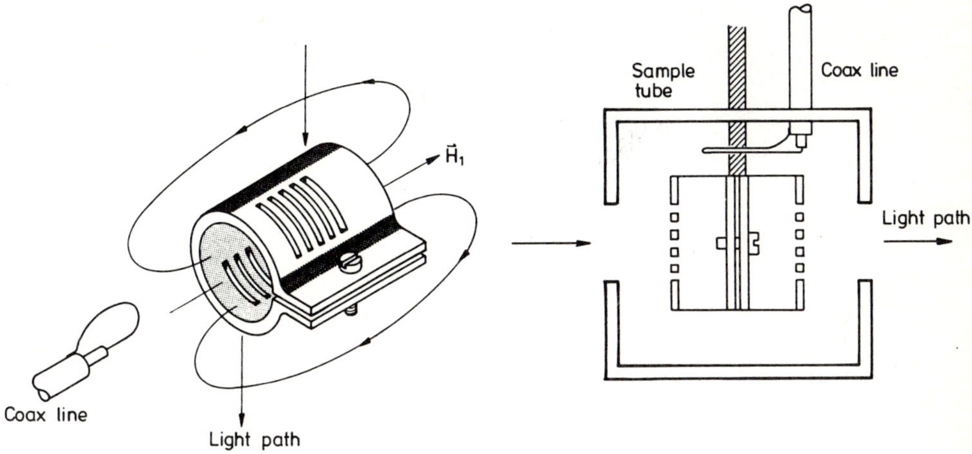

Fig. 4. Geometry of probe head with split-ring cavity for LD-ADMR measurements. From [6]

7. Analysis of LD-(T - S) Spectra

Since the LD-ADMR technique is analogous to that of photoselection we may use the same formalism to describe the results. Making the appropriate modifications it follows from [8], (1) - (4) that

$$A_{//} = A_z = 1/5(1 + 2\cos^2\alpha), \quad A_\perp = A_x = A_y = 1/5(2 - \cos^2\alpha) \tag{1}$$

where // and ⊥ denote parallel and perpendicular to the microwave magnetic field vectors, and α is the angle between the optical transition moment and the triplet spin axis.

The T - S spectrum is the sum of ΔA_z and ΔA_x, the LD-(T - S) spectrum is given by the difference of ΔA_z and ΔA_x. Hence the ratio R of an absorbance at wavelength λ in the LD-(T - S) to that in the T - S spectrum is given by

$$R_i\left[\frac{LD-(T - S)}{T - S}\right] = \frac{\Delta A_z - \Delta A_x}{\Delta A_z + \Delta A_x} = \frac{3\cos^2\alpha_i - 1}{3 + \cos^2\alpha_i} \tag{2}$$

where the subscript i refers to the x- or the y-triplet spin axis (which is selected by the microwave frequency). The angle α defines a cone that is circumscribed by the optical transition moment around the ith triplet-spin axis. Knowledge of two α_i's limits possible solutions of the direction of the transition moment to the intersections of two cones whose axes are at right angles. This yields two directions, and because transition moments are bidirectional vectors, two more that are symmetrically placed with respect to the former ones.

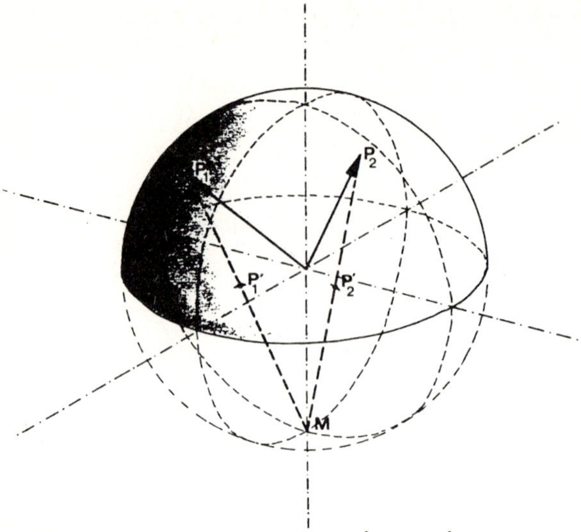

Fig. 5. The Wulf projection. P_1' and P_2' are the projection in the horizontal plane of the points P_1, P_2 on the unit sphere

An elegant way to visualize this is to use a Wulf projection (Fig. 5) [9]. All vectors have their origin in the center of a unit sphere. The intersections of the vector and the surface of the sphere are projected onto the midplane of the sphere by connecting them to one of the poles (for the upper halfsphere, the South pole). If the North-South axis is taken to be the x-triplet axis, vectors on cones with top angle 2α around this axis will give rise to concentric circles in the Wulf projection, whereas vectors on cones around the y-axis will project as curved lines. One may now calculate a net of circles and curves corresponding to R_x and R_y values and use this plot to determine the intersections of the cones. The standard Wulf net then directly gives the angles of the (four) possible solutions. This will be illustrated in section 9.

In the above, all angles are relative to the triplet spin axes. Although the precise location of those axes in the (B)Chl molecule, and even more pertinent, in the D_o aggregate is not known, relative angles between transition moments are of course easily determined from the Wulf plot, albeit with a fourfold ambiguity. Comparison with other polarized spectroscopies and with the known X-ray structure in the case of *Rps. viridis* is necessary to resolve this ambiguity.

8. T - S Spectra of Bacterial Reaction Centers

The very first T - S spectrum obtained by the ADMR technique [5] is shown in Fig. 6. Its accuracy and resolution is unparalleled. Several new features compared to those obtained earlier with flash spectroscopy [10] are apparent in the 830 nm region. It is clear that this region is composed of two bandshifts and one appearing band, whereas at 850 nm possibly a small bleaching is present. The bandshifts have been attributed to the two accessory BChl pigments, the appearing band to one of the BChl's in ^3D_o (i.e. a localized triplet state). The bleaching at 990 nm is of course due to D_o. It is significantly narrower than the corresponding bleaching in the redox ΔA spectrum. This is due to so-called *site selection*. The T - S spectra are

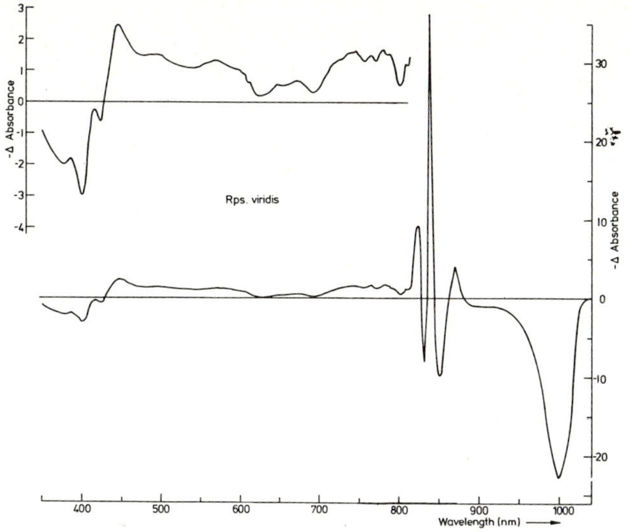

Fig. 6. The triplet-minus-singlet spectrum of Rps. viridis at 1.5 K. The 900 - 1000 nm region is enlarged in Fig. 7. From [5]

recorded at a fixed microwave frequency at the top of an ADMR resonance. It is known that these resonances are inhomogeneously broadened [11], i.e. they are composed of resonances of an ensemble of reaction centers with slightly different D and E values. Apparently, also, the optical transition is inhomogeneously broadened. This provides a direct explanation for the hitherto mysterious width of the long-wavelength absorbance band of bacterial RC's. If this bands results from exciton interaction within a dimer, then one would expect for symmetric dimers that the width of the exciton band is reduced by $\sqrt{2}$ compared to the monomeric linewidth [12]. In contrast, the long-wavelength band in RC's is always much broader than the monomeric Q_y transitions. From the ADMR and T - S spectra it follows that in frozen samples there exists a heterogeneity in the precise geometry of the dimer, which gives rise to band-broadening in both the optical and the ADMR spectra. It would be of considerable interest to see whether the remaining width in the T - S spectrum is purely due to homogeneous broadening or to further inhomogeneities unresolved by the ADMR technique. This could be decided by hole-burning or coherent photon echo spectroscopy at low temperatures.

The 830 nm region is practically insensitive to site-selection effects. One may surmise that these pigments are much less strongly coupled than the BChl's of D_o, and their spectra therefore less susceptible to site heterogeneity. Also, assuming exciton coupling, differences in precise relative orientation of the two BChl's of D_o will have an effect on the absorbance bands of D_o and on the zero-field parameters (and therefore on the ADMR spectrum) but leave the bands of the accessory pigments unaffected.

The bands in the T - S spectrum around 650 nm (the Q_x region) are much less distinct than in the redox ΔA spectrum. This is due to the fact that in contrast to $(B)Chl^+$ the triplet state of (B)Chl has considerable absorbance in this region.

An interesting feature of the long-wavelength bleaching is the presence of a shoulder at the long-wavelength side of the band. This shoulder was earlier observed in the low-temperature absorbance spectrum and attributed to charge-transfer (CT) interaction with an adjacent BChl [13,14]. The shoulder can be made more apparent when the microwave frequency is shifted somewhat within an ADMR transition (Fig. 7). From Fig. 7 it is also apparent that the exact location of the long-wavelength bleaching is rather sensitive to the precise microwave frequency. This is again a result of site selection, i.e. it reflects the heterogeneity of the RC's. There is a one-to-one correspondence with the shift in microwave frequency and the shift in the peak of the bleaching [15,16].

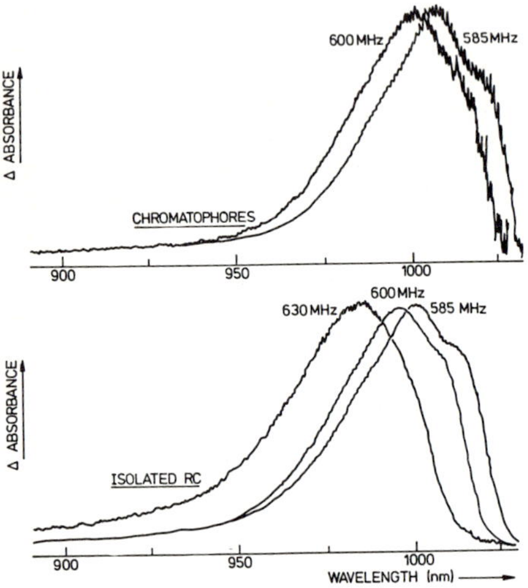

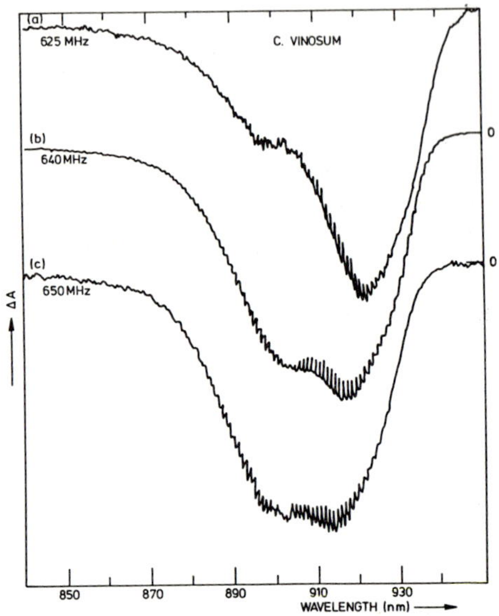

Fig. 7. The near-infrared bleaching in the T – S spectrum of Rps. viridis for different microwave frequencies within the |D| + |E| ADMR transition. The sign of the band is inverted with respect to Fig. 6 for ease of comparison with the Rps. viridis absorption spectrum. From [16]

Fig. 8. The long-wavelength bleaching of the T – S spectrum of reaction centers of Chromatium vinosum at frequencies within the |D| + |E| ADMR transition. The spikes are switching artefacts caused by sudden changes in the light intensity that falls on the photodetector when stepping the monochromator over the absorption band. From [19]

The above-mentioned shoulder is even more pronounced, and even predominates, in the T – S spectra of some other bacterial species. A striking example is displayed in Fig. 8 for Chromatium vinosum. The large oscillator strength of the shoulder in this case, and the fact that it is so sensitive to site-selection, makes it probable that it is due to RC heterogeneity rather than to CT interaction with a neighbouring pigment.

T – S spectra of a large number of BChl a and b containing purple bacteria and of green bacteria have now been recorded. It seems that those of the purple bacteria and of the green gliding bacterium Chloroflexus aurantiacus

[17] can be roughly divided into two classes: one which resembles the spectrum of *Rps. viridis* and one which is typified by *R. rubrum* (see below). There does not seem to be a correlation with the presence of either BChl *b* or BChl *a* [J.A. DIJKMAN and A.J. HOFF, unpublished results]. The spectrum of the green sulfur bacterium *Prosthecochloris aestuarii* is quite different from the above spectra [18].

9. LD-(T - S) Spectra of Bacterial Reaction Centers

Linear dichroic T - S spectra of several bacterial species have now been obtained [6,19]. The quantitative analysis described in section 7 is best illustrated with the spectra of *R. rubrum* (Fig. 9). One T - S spectrum and two LD-(T - S) spectra are displayed. The other T - S spectrum (at 465 MHz) is practically identical to that recorded at 668 MHz. The spectra have been recorded simultaneously. One notes the good signal-to-noise ratio of the LD spectra, which is comparable to that of the T - S spectrum.

The sign of the bands in either of the LD-(T - S) spectra relative to that of the T - S spectrum solely depends on the orientation of the transition moment with respect to the triplet spin axis (the x-axis for the 465 MHz transition, the y-axis for the 668 MHz transition) and not on whether the band is a bleaching or an appearing band as in the T - S spectrum.

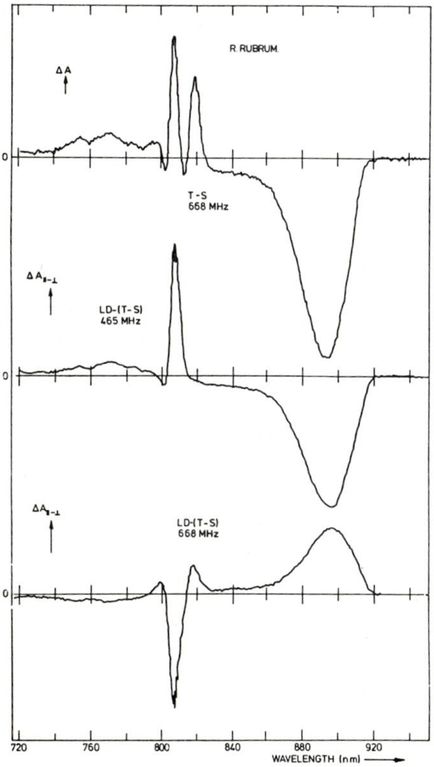

Fig. 9. T - S and LD-(T - S) spectra of reaction centers of *Rhodospirillum rubrum* at 1.5 K. The T - S spectrum at 465 MHz was virtually the same as the spectrum shown at 668 MHz. From [19]

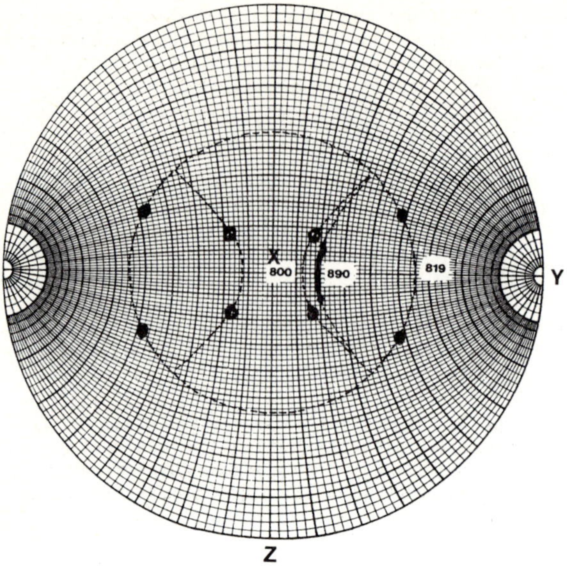

Fig. 10. Wulf plot of transition-moment orientation for R. rubrum as determined by LD-ADMR. The dots on the dashed circle correspond to the shift with positive lobe at 819 nm. Only one position of the 890 nm transition-moment is indicated. The triplet x-axis is projected at the center of the plot

Starting with the 890 nm band, we first note that it changes sign for the two LD spectra. Since the LD spectrum with the highest intensity at 890 nm (at 465 MHz) has the same sign as the T - S spectrum, the transition moment of the 890 nm band must make an angle smaller than 55° with the triplet x-axis and larger than 55° with the y-axis. In fact, R_x is close to 0.5, the theoretical maximum, whereas $R_y = 0.34$. (The precise values of $R_{x,y}$ cannot be directly obtained from the figure, as several calibration factors come in, such as lock-in gain, etc.). The intersections of the curves for those two values of $R_{x,y}$ in the Wulf net (Fig. 10) immediately give four possible equivalent positions, with $\alpha_x(890) = 18 - 20°$ and $\alpha_y(890) = 72°$.

The analysis of the 800 - 820 nm region is more complicated due to band overlap. There are good reasons, based on normal LD spectroscopy [9], to assume that the 819 nm band is due to a band-shift, analogous to that encountered in Rps. viridis. In Fig. 9b the 819 nm band is absent, and the 807 nm peak undistorted. This means that $R_x = 0$ and $\alpha_x(819) = 55°$. Furthermore, we see that there is little overlap between the 807 and 819 nm features, so the small peak at 819 nm in Fig. 9c can be fully attributed to the 819 nm shift. This yields $R_y = 0.37$ or $\alpha_y(819) = 26°$. There are four possible intersections in the Wulf plot, for which the conditions $\alpha_x = 55°$ and $\alpha_y = 26°$ are both satisfied (Fig. 10). Note that these angles make it unlikely that a significant amount of the bleaching at 810 nm is due to an exciton partner of the 890 nm, as the high and low-energy exciton components should be perpendicular.

The 807 nm band in the 465 MHz LD-(T - S) spectrum of Fig. 9b is undistorted, but the band in the T - S spectrum has to be corrected for band overlap with the negative lobe of the shift, whose positive lobe is at 819 nm. If we assume that the 819 nm peak intensity is undistorted, then we can construct the missing parts of the shift and subtract those from the 807 nm feature to give the undistorted 807 nm band. Carrying out similar procedures for the 668 MHz LD-(T - S) spectrum, we arrive at R_x and R_y values that yield angles of $\alpha_x(807) = 21°$ and $\alpha_y(807) = 74°$. The corresponding intersections in the Wulf net are indicated.

Except for the corrections for overlap of the 807 and 819 nm bands, the above analysis does not depend on assumptions regarding the nature of the bands. Hence it should be possible to check the above angles with data from other spectroscopic studies and X-ray crystallography. This would provide a check on our assignments, and would in the case of Rps. viridis resolve the remaining fourfold degeneracy in the positions in Fig. 10, because from such a comparison the relative angles between the transition moments can be deduced. Such a program is now underway for Rps. viridis.

10. T - S Spectra of Plant Photosystems I and II

The ADMR technique has also been applied to particles enriched in PS I and PS II [20,21]. The resulting T - S spectra are displayed in Fig. 11, together with the T - S spectra of Chl a in monomeric and dimeric form as obtained by flash spectroscopy [22,23]. The correspondence of the in vivo spectra to that

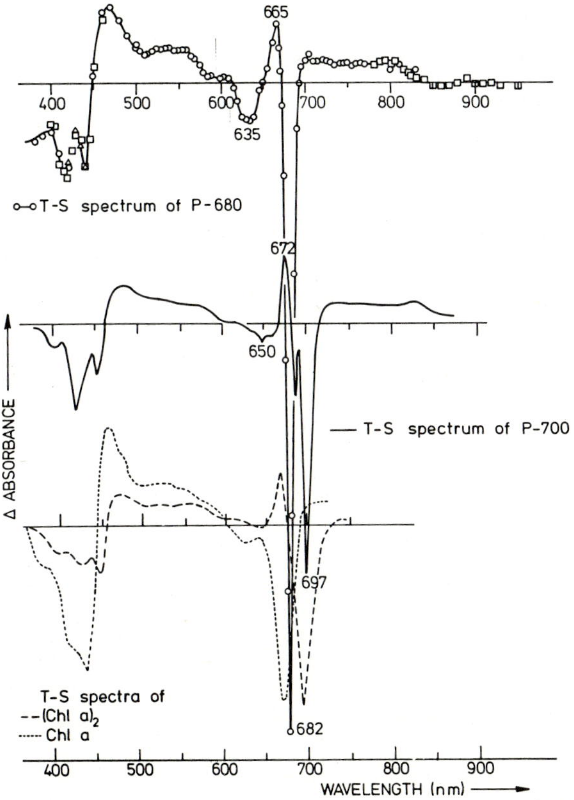

Fig. 11. T - S spectra of PS II (top) [21], PS I (bottom) [20], Chl a in pyridine [22] and $(Chl\ a)_2$ in methylcyclohexane [23]

of (Chl $\underline{a})_2$, and not to that of Chl $\underline{a}$, is remarkable. The red part of the in vitro dimer spectrum has been explained as a combination of the bleaching of the dimer Q_y band (that is shifted from 680 nm to 700 nm) and the appearance of an absorption band due to 'monomeric', i.e. non-coupled, Chl $\underline{a}$ [23]. This band arises because the covalently bonded (Chl $\underline{a})_2$ molecule is assumed to unfold when the triplet state is present, so that one ^{3}Chl $\underline{a}$ and one ^{1}Chl $\underline{a}$ molecule result that are not, or very weakly, coupled. The ^{1}Chl $\underline{a}$ molecule absorbs at about 680 nm, the ^{3}Chl $\underline{a}$ molecule has broad, unfeatured absorbance in this region.

On the basis of the analogy between the PS I and PS II spectra to that of (Chl $\underline{a})_2$ we concluded that in both these photosystems the primary donor is dimeric. In the PS I spectrum a small side band is present to the blue of the main bleaching at 700 nm. This has been interpreted as a shift of a neighbouring Chl $\underline{a}$ molecule that presumably acts as an intermediary acceptor [24]. In PS II particles, such a side band seems to be absent, in agreement with the generally held view that in PS II pheophytin $\underline{a}$ acts as intermediary electron acceptor. Note that in PS II the accepting Chl $\underline{a}$ absorbs around 670 nm [24], whereas the pheophytin $\underline{a}$ in PS II absorbs around 685 nm [25] so that a shift of this pigment would be hidden under the bleaching of the primary donor at 682 nm. (The value of the zero-field parameter D rules out a location of the triplet state on the pheophytin).

As a last remark we note that the shape of the T - S spectrum in the far-red of the green bacterium <u>Prosthecochloris aestuarii</u> [18] much resembles the shape of the T - S spectrum of PS I particles in the red region. This again evinces the strong similarity between PS I and the photosystem of green sulfur bacteria (see e.g. [26]).

11. Prospects

T - S and especially LD-(T - S) spectroscopy carried out with the ADMR technique is a powerful tool to gain insight in reaction center make-up. When properly calibrated on the pigment geometry of known RC, it should find wide application in the study of systems that as yet have not yielded to crystallization. In addition to structural data, the spectra offer information on pigment interaction, and as such provide an interesting testcase for theoretical calculations based on known RC structure.

Finally, pulsed ADMR combined with laser flash excitation will permit the observation of LD-(T - S) spectra at higher temperature. In this connection, it is of interest to note that for RC T - S spectra at higher temperatures may be obtained with another, new and highly sensitive technique, viz. magneto-optical difference spectroscopy (MODS) [27].

12. Acknowledgements

The author is indebted to Drs. H.J. den Blanken and R.F. Meiburg for their essential contributions to the work described in this paper, and to the other members of the Biophysics Department for their help and advice. The support from Dr. J. Schmidt of the Leiden Centre for the Excited States of Molecules is gratefully acknowledged. This work was carried out with the aid of the Nederlands Foundation for Chemical Research (SON), financed by the Netherlands Organization for the Advancement of Pure Research (ZWO).

References

1 J.H. van der Waals and M.S. de Groot: In: <u>The Triplet State</u> (A.B. Zahlan, ed., Cambridge University Press 1967), pp. 101-132
2 A.J. Hoff and H.G. de Vries: Biochim. Biophys. Acta <u>503</u>, 94 (1978)

3 W.G. van Dorp, W.H. Schoemaker, M. Soma and J.H. van der Waals: Mol. Phys. 30, 1701 (1975)
4 A.J. Hoff: In Light Emission by Green Plants and Bacteria (Govindjee, J. Amesz and D.C. Fork, eds., Academic Press New York 1986), in the press
5 H.J. den Blanken and A.J. Hoff: Biochim. Biophys. Acta 681, 365 (1982)
6 H.J. den Blanken, R.F. Meiburg and A.J. Hoff: Chem. Phys. Lett. 105, 336 (1984)
7 W.H. Hardy and L.A. Whitehead: Rev. Sci. Instrum. 57, 213 (1981)
8 A. Vermeglio, J. Breton, G. Paillotin and R. Cogdell: Biochim. Biophys. Acta 501, 514 (1978)
9 R.F. Meiburg: Doctoral thesis, University of Leiden (Leiden 1985)
10 V.A. Shuvalov and W.W. Parson: Biochim. Biophys. Acta 638, 50 (1981)
11 A.J. Hoff: Biochim. Biophys. Acta 440, 765 (1976)
12 R.P. Hemenger: J. Chem. Phys. 67, 262 (1977)
13 A. Vermeglio and G. Paillotin: Biochim. Biophys. Acta 681, 32 (1982)
14 V.A. Shuvalov and A.V. Klevanik: FEBS Lett. 160, 51 (1983)
15 H.J. den Blanken and A.J. Hoff: Chem. Phys. Lett. 98, 255 (1983)
16 H.J. den Blanken, A.P.J.M. Jongenelis and A.J. Hoff: Biochim. Biophys. Acta 725, 472 (1983)
17 H.J. den Blanken, H. Vasmel, A.P.J.M. Jongenelis, A.J. Hoff and J. Amesz: FEBS Lett. 161, 185 (1983)
18 H. Vasmel, H.J. den Blanken, J.T. Dijkman, A.J. Hoff and J. Amesz: Biochim. Biophys. Acta 767, 200 (1984)
19 A.J. Hoff, H.J. den Blanken, H. Vasmel and R.F. Meiburg: Biochim. Biophys. Acta 806, 389 (1985)
20 H.J. den Blanken and A.J. Hoff: Biochim. Biophys. Acta 724, 52 (1983)
21 H.J. den Blanken, A.J. Hoff, A.P.J.M. Jongenelis and B.A. Diner: FEBS Lett. 157, 21 (1983)
22 H. Linschitz and K. Sarkanen: J. Am. Chem. Soc. 80, 4826 (1958)
23 N. Perisiamy and H. Linschitz: J. Am. Chem. Soc. 101, 1056 (1979)
24 T. Swarthoff, P. Gast, J. Amesz and H.P. Buisman: FEBS Lett. 146, 129 (1982)
25 V.V. Klimov, A.V. Klevanik, V.A. Shuvalov and A.A. Krasnovsky: FEBS Lett. 82, 183 (1977)
26 A.J. Hoff: In Advances in Photosynthesis Research (C. Sybesma, ed.), Vol. I (Martinus Nijhoff/Dr. W. Junk, The Hague 1984), pp. 89-91
27 A.J. Hoff, E.J. Lous, K.W. Möhl and J.A. Dijkman: Chem. Phys. Lett. 114, 39 (1985)

ENDOR Studies of the Primary Donor in Bacterial Reaction Centers

W. Lubitz[1], F.Lendzian[2], M.Plato[2], K.Möbius[2], and E.Tränkle[3]

[1] Institut für Organische Chemie, Freie Universität Berlin, Arnimallee 14, D-1000 Berlin 33
[2] Institut für Molekülphysik, Freie Universität Berlin, Arnimallee 14, D-1000 Berlin 33
[3] Institut für Theorie der Elementarteilchen, Freie Universität Berlin, Arnimallee 14, D-1000 Berlin 33

1. Introduction

The structure of the primary donor P in bacterial photosynthesis is of considerable interest due to the unique role this species plays in the very first events of the charge-separation process in the reaction center (RC). It could be shown that P is a specialized bacteriochlorophyll (BChl) [1] which was proposed by NORRIS et al. [2] to be a BChl-dimer. This "special pair" model was a matter of considerable controversy in recent years. Different dimer models [3,4,5] and perturbed monomer models [6,7] have been proposed for the primary donor. The existence of a dimer has recently been demonstrated, at least for Rhodopseudomonas (Rp.) viridis RC's by an X-ray structure analysis of RC single crystals [8]. An alternative approach to the characterization of the primary donor in its cationic state is provided by EPR and electron nuclear double-resonance (ENDOR) spectroscopy [4,5, 9-12]. In this paper we summarize the most recent contributions of EPR and ENDOR in elucidating the electronic structure details of the primary donor cation radicals. In particular $P_{865}^{+\cdot}$ in RC's of Rp. sphaeroides and Rhodospirillum (R.) rubrum, and $P_{960}^{+\cdot}$ in RC's of Rp. viridis are studied. The isotropic hyperfine coupling constants (hfc's) extracted by ENDOR and TRIPLE resonance [13,14] are compared with values from advanced MO calculations, thereby relating the experimental values to structural details of the dimers.

2. Experimental and Theoretical Methods

Bacteriochlorophylls and reaction centers were isolated following procedures published elsewhere [4,5,15]. The biosynthetic isotopic labelling by D [4], ^{15}N [11], and ^{25}Mg [16] has been described. The BChl a and b cation radicals were generated in purified oxygen-free CH_2Cl_2/CH_3OH (6/1) mixtures using iodine and zinc tetraphenylporphyrin cation as oxidants, respectively. The cation radicals of the primary donors $P_{865}^{+\cdot}$ (Rp. sphaeroides R-26, R. rubrum G-9) and $P_{960}^{+\cdot}$ (Rp. viridis) were obtained by in-situ light-irradiation (UV and 10 cm water filter) of the respective bacterial RC's. The EPR/ENDOR/TRIPLE experimental set-ups are described elsewhere [13].

The theoretical method used for calculating s-spin densities is a Restricted Hartree-Fock procedure for doublet states ("half-electron-method") using the INDO approximation and parametrization [17] and a subsequent perturbation treatment

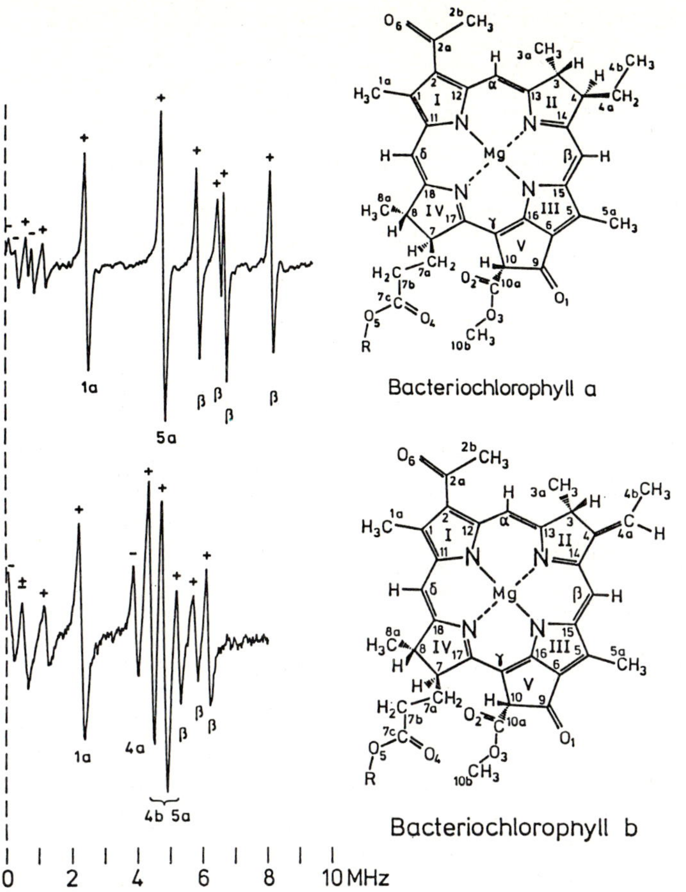

Figure 1 - Right: Molecular structures of BChl a and b, R = phytyl (R. rubrum has R = geranylgeranyl), numbering scheme; left: H Special TRIPLE spectra of the cation radicals in CH_2Cl_2/CH_3OH (6 : 1) at 220 K. Each line corresponds to one H hfc (Table 1), signs from General TRIPLE, assignments for β protons and methyl groups based on MO calculations (Fig. 2), for experimental details see [19]

of spin-polarization effects. The details of this RHF-INDO/SP-method will be described in a forthcoming paper [18]. The calculations were performed on a CRAY-1 M of the Konrad-Zuse Computer Center of Berlin. For a single BChl-dimer configuration (158 atoms, 416 valence orbitals) about 500 000 computer words and 20 minutes cpu time were required.

3. Result and Discussion

In Fig. 1 the liquid solution Special TRIPLE spectra of the BChl a and b cation radicals and their corresponding molecular structures are shown. The assignments shown in Fig. 1 were obtained from a comparison with MO calculations and, in the case of BChl a$^{+\cdot}$, corroborated by the use of selectively deuterated compounds and by relaxation studies [19]. BChl b differs from BChl a only at ring II where it carries an exocyclic double bond in conjugation to the bacteriochlorin π-system. One additional methyl hfc (4b) and a large negative α-proton hfc (4a) are measured instead of the β H hfc (pos. 4) found in BChl a$^{+\cdot}$ (Fig. 1). The ^{14}N-hfc's could also be determined for both species. In the case of BChl a$^{+\cdot}$ these values were confirmed by ^{15}N ENDOR (see Fig. 4, top) [11]. The isotropic proton and nitrogen hfc's are collected in Table 1. Furthermore, the spin density at the central magnesium could be determined for BChl a$^{+\cdot}$ from ^{25}Mg enriched material (a(^{25}Mg) = (-)0.3 MHz [16]).

The results of our MO calculations are given in Fig. 2 together with the experimental data for BChl a$^{+\cdot}$ and BChl b$^{+\cdot}$. The agreement between theory and experiment is good, not only for the various proton positions but also for the nitrogens and magnesium. The combined use of experimental and calculated spin densities therefore appears to work well for structure characterization of such large systems, and should consequently provide a sound basis for the comparison with the cation radicals of the primary donors in RC's.

In Fig. 3 the liquid-state Special TRIPLE spectra of BChl a$^{+\cdot}$ and $P_{865}^{+\cdot}$ in RC's of two different bacteria are shown

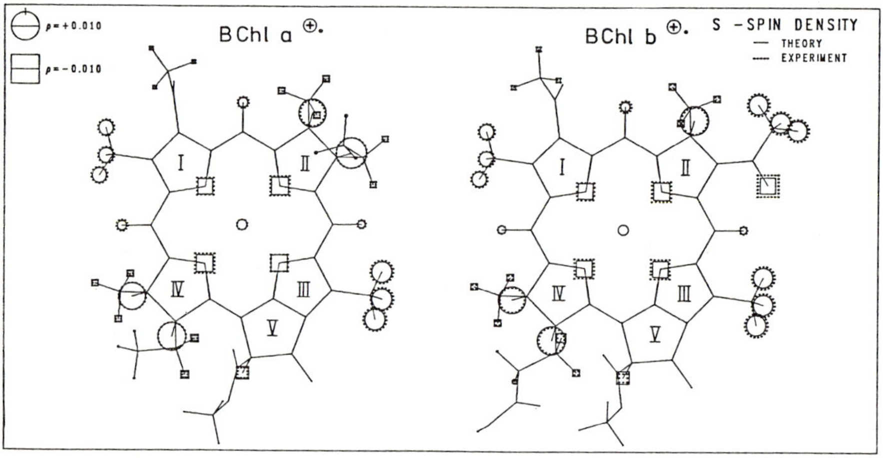

Figure 2 - Comparison of experimental (dotted) and calculated (solid lines) s-spin densities for BChl a (left) and BChl b (right) cation radicals. Their values are proportional to the area of the squares ($\rho < 0$) and circles ($\rho > 0$), respectively. For the calculation the RHF-INDO/SP method was used (see text). The full molecular structures were taken except for a truncation of the isoprenoid chain. Mostly, standard bond lengths and angles were employed [17], atomic positions for ring V, the methyl (1a, 5a) and acetyl groups (2) were determined by energy-minimization. Experimental values from isotropic hfc's (Table 1) by use of: Q(H) = 1420, Q(^{14}N) = 663, Q(^{25}Mg) = -224 MHz. Theoretical carbon and oxygen-spin densities were omitted for reasons of clarity.

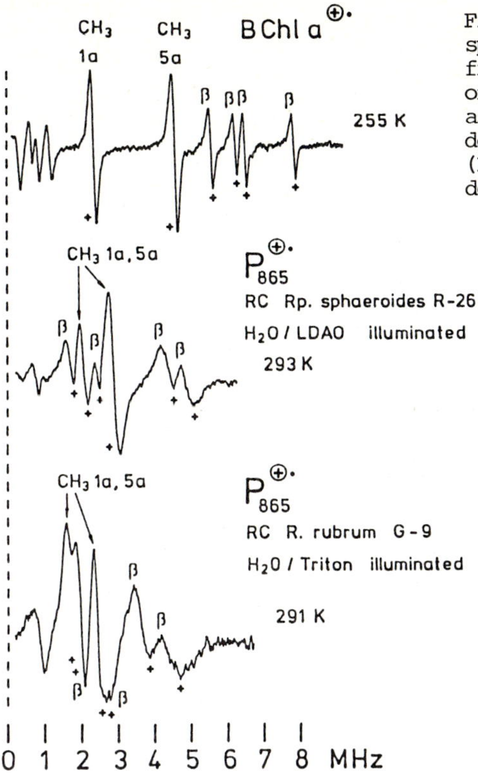

Figure 3 - H Special TRIPLE spectra of BChl $\underline{a}^{+\cdot}$ and $P_{865}^{+\cdot}$ (RC's from different bacteria). Signs of H hfc's from General TRIPLE, assignments based on specific deuteration and MO calculations (Figs. 2, 6). For experimental details see [4,5,19,20]

[4,5,20]. Although the spectra of the BChl $\underline{a}^{+\cdot}$ extracted from these two bacteria are undistinguishable, the corresponding spectra for $P^{+\cdot}$ show distinct differences. Different strains of the bacterium, e.g. <u>Rp. sphaeroides</u> R-26 and wild type, result in very similar $P_{865}^{+\cdot}$ spectra. An explanation could lie in a difference of the protein environment and/or a different anchoring of the dimer due to the different isoprenoid chains (Fig. 1) resulting in a slightly different dimer geometry.

All H hfc's (see Table 1) in $P_{865}^{+\cdot}$ are reduced in magnitude compared with those of BChl $\underline{a}^{+\cdot}$. Although individual reduction factors deviate strongly from 2, the average reduction factor for both bacteria is very close to this value. This indicates a delocalization of the unpaired electron (or the "electron hole") over <u>two</u> BChl $\underline{a}$ forming a supermolecule. Since there is no splitting of the ENDOR lines, e.g. of those assigned to the 1a and 5a CH_3 groups, the two halves are magnetically equivalent, indicating a C_2 symmetry of the dimer. The same conclusions were drawn from nitrogen hfc's of ^{15}N-BChl $\underline{a}^{+\cdot}$ and ^{15}N-$P_{865}^{+\cdot}$ in RC's of Rp. sphaeroides R-26 (Fig. 4), see Table 1 [11]. These conclusions from ENDOR/TRIPLE are in accordance with EPR simulations [4,5] and with a second moment analysis

Table 1 Hfc's [MHz] of the cation radicals of BChl _a_, BChl _b_, P_{865}, and P_{960}

nucleus (Pos.)	BChl $\underline{a}^{+\cdot}$ a)	$P_{865}^{+\cdot}$(R-26) b)	$P_{865}^{+\cdot}$(G-9) c)
β H (3,4,7,8)	+16.35 +13.47 +13.11 +11.76	+9.50 +8.60 +4.45 +3.30	+8.50 +7.50 +5.28 +3.95
CH_3 (1a,5a)	+9.62 +4.93	+5.60 +4.00	+4.85 +3.40
other H	+2.35, -1.64, +1.30, -0.55, -0.22	+1.40	+1.60
^{15}N (rings I-IV)	+4.40 +4.05 +3.25 +3.15	+2.60 +2.22 (+)1.61 (+)1.05	

nucleus (Pos.)	BChl $\underline{b}^{+\cdot}$ a)	$P_{960}^{+\cdot}$ d)
β H (3,7,8)	+12.45 +11.60 +10.55	+11.5 +8.5 +7.2
CH_3 (1a,5a,4b)	+9.70 +8.95 +4.70	+6.0 +3.4
α H (4a)	-8.02	±2.0
other H	+2.5, ±1.2 -0.4	±1.4
^{14}N (rings I-IV)	(-)3.3 (-)3.0 (-)2.3	

All data from ENDOR/TRIPLE in solution, a) 220 K (protons) and 230 K (nitrogens); note that for BChl $\underline{b}^{+\cdot}$ only three ^{14}N hfc's were obtained, for BChl $\underline{a}^{+\cdot}$ four ^{15}N hfc's were resolved but also only three ^{14}N hfc's (-3.16, -2.91, -2.28 MHz) [11]. b) RC's of Rp. sphaeroides R-26, 298 K [5, 11] . c) RC's of R. rubrum G-9, 293 K [4]. d) RC's of Rp. viridis, 293 K; note that the assignment is not safe (see text).

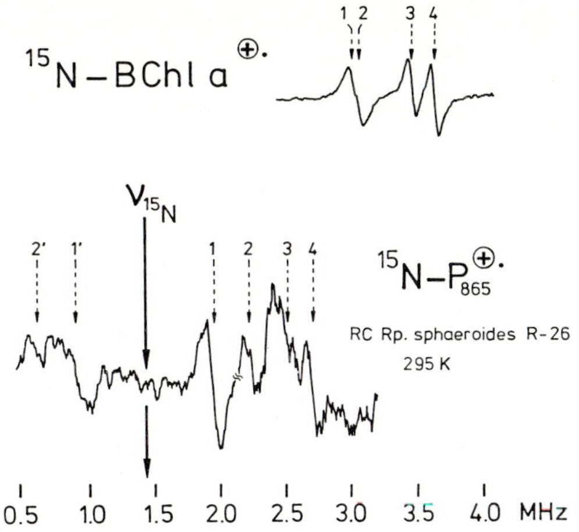

Figure 4 - High-frequency ^{15}N ENDOR lines (1 to 4) in ^{15}N labelled BChl $\underline{a}^{+\cdot}$ at 230 K (top) and ^{15}N ENDOR spectrum of $P_{865}^{+\cdot}$ in ^{15}N labelled RC's of Rp. sphaeroides R-26 (bottom). The frequency-scale holds for both spectra. For hfc's see Table 1. Figure adapted from [11].

of fully D and ^{13}C enriched $P_{865}^{+\cdot}$ and BChl $\underline{a}^{+\cdot}$ [22] and exclude the possibility of a perturbed monomeric model for $P_{865}^{+\cdot}$ [6,7].[1]

The structural analysis of the primary donor $P_{960}^{+\cdot}$ in RC's of Rp. viridis turned out to be more complicated. The EPR linewidth of this species, compared with that of BChl $\underline{b}^{+\cdot}$ *in vitro*, is not reduced by a factor of $\sqrt{2}$. This reduction factor was, however, found for $P_{865}^{+\cdot}$ compared with BChl $\underline{a}^{+\cdot}$ [1,2] and is also expected for a BChl $\underline{b}$ dimer. Nevertheless, solid-state ENDOR data have been interpreted on the basis of a dimeric nature of $P_{960}^{+\cdot}$, assuming a structural change of the component BChl $\underline{b}$ monomers upon dimer formation in the RC protein [12]. In Fig. 5 our liquid-state Special TRIPLE spectra of BChl $\underline{b}^{+\cdot}$ and $P_{960}^{+\cdot}$ are shown, the hfc's are presented in Table 1. Since specific deuteration studies have not been performed so far, final assignment of the lines in $P_{960}^{+\cdot}$ to the β protons and the three methyl groups (1a, 4b, 5a) cannot be given. The tentative assignment given in Table 1 is based on MO calculations, assuming a dimer with C_2 symmetry as in $P_{865}^{+\cdot}$. From the inspection of the spectra and hfs data, it is immediately clear that the spin densities in $P_{960}^{+\cdot}$ are - on the average - not reduced by a factor of 2. With the given assignment one obtains $\sim$ 1.6. However, with this assignment the EPR spectrum can be simulated quite satisfactorily. From the x-ray analysis of Rp. viridis RC single crystals, it is now known that P_{960} indeed involves two BChl $\underline{b}$ molecules in close contact [8]. So far, the analysis of the spin-density distribution of $P_{960}^{+\cdot}$ is not possible on the basis of the same model as used for $P_{865}^{+\cdot}$. This can be due to uncertainties in the assignments of the hfc's, or to an asymmetric nature of this dimer. Specific isotopic labelling (D, ^{15}N, etc.) will help to solve these questions in future studies.

----

[1] We have also obtained an H ENDOR spectrum of the triplet state of P_{865} at $\sim$ 10 K in "blocked" RC's 21 . A preliminary analysis of the 6 H-hfc's including signs shows that the two unpaired electrons in P^T_{865} are also delocalized over two BChl $\underline{a}$ molecules.

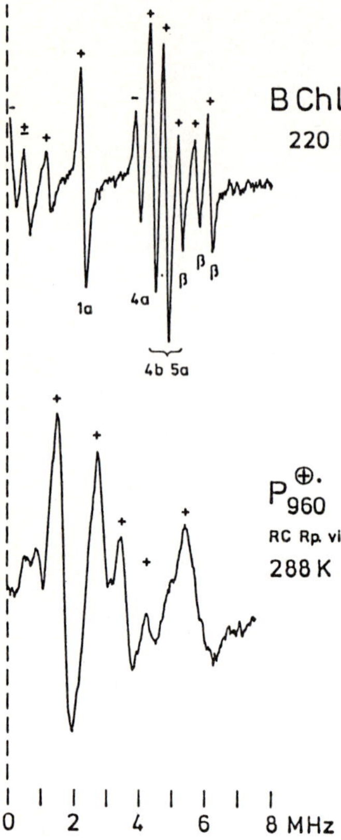

Figure 5 – H Special TRIPLE spectra of BChl $\underline{b}^{+\cdot}$ and $P^{+\cdot}_{960}$ (RC's of Rp. viridis), signs from General TRIPLE. Final assignments for $P^{+\cdot}_{960}$ are not yet possible (see Table 1).

The geometric relationship of the monomer components with respect to each other in the dimer can be obtained from the isotropic hfc's in conjunction with a reliable MO calculation. $P^{+\cdot}_{865}$, for example, can be described by a supermolecule with C_2 symmetry in which the "electron hole" is delocalized in a super MO extending over both BChl $\underline{a}$ halves. For distances > 3.5 Å (twice the van der Waals radius of π systems) the super MO's are symmetric or antisymmetric linear combinations $\psi^{\circ}_{s/a} = 1/\sqrt{2}(\psi_{M1} \pm \psi_{M1})$ of the wavefunctions of the non-interacting monomers M1 and M2. This case would result in an exact halving of all spin densities compared with the monomer. For shorter distances the increase of the resonance integral β, being proportional to the overlap integral $S_{\pi\pi}$, causes a distortion of the zero-order wave functions $\psi^{\circ}_{s/a}$ of the supermolecule due to first and higher-order perturbation terms. For a non-eclipsed dimer conformation (e.g. ring I over I or III over III) this will generally lead to distortions of the former zero-order spin densities, i.e. a halving of spin densities will not be found for all positions. The experimental results indicate that this situation actually exists in $P^{+\cdot}_{865}$ since several reduction factors deviate strongly from 2.

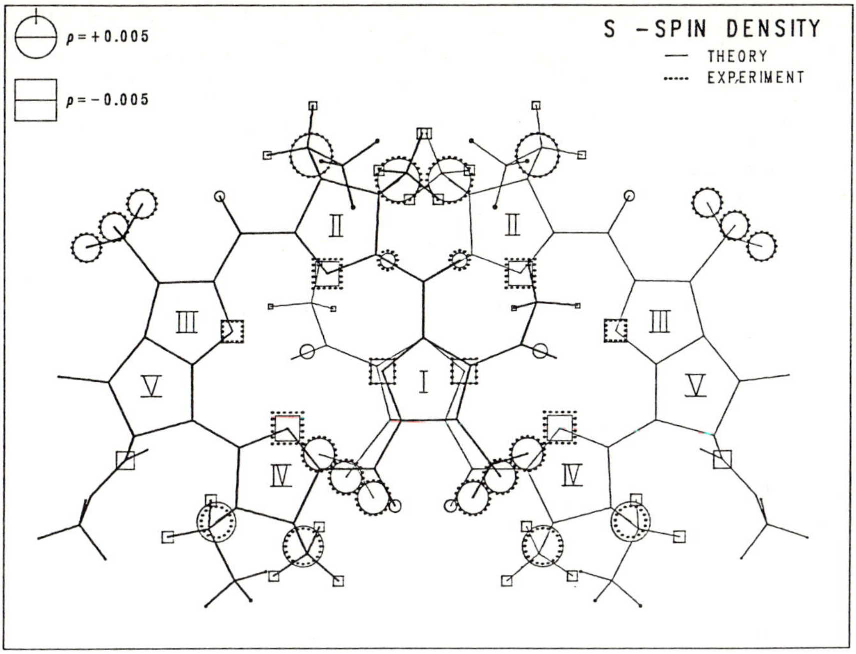

Figure 6 - Comparison of experimental (dotted) and calculated (solid lines) s-spin densities for the (BChl a)$_2^{+\cdot}$ without isoprenoid chain. Geometry of the dimer was optimized by energy-minimization using a truncated form (2 x 55 atoms) without neglecting critical steric interactions. For further geometrical details see text; scale: Mg-Mg distance in figure 5.9 Å. The two halves are distinguished by heavy and light lines, for numbering of rings compare Fig. 1. Representation of spin densities as in Fig. 2. Experimental proton data from $P_{865}^{+\cdot}$ (R. rubrum G-9), nitrogen spin densities from Rp. sphaeroides R-26 (Table 1).

We have used the above-mentioned RHF-INDO/SP approach to calculate energies and spin densities in various non-eclipsed BChl a dimer configurations with C_2 symmetry. Additionally, energy minimization was performed with respect to selected structural parameters involving the relative positions of the two halves of the dimer and distortions of the monomers themselves. A local energy minimum is obtained for a "face-to-face" overlap at rings I (horizontal Mg-Mg displacement ~ 6.o Å) with a distance of 3.1 Å between the centers of rings I. The acetyl group oxygens (pos. 2) are in close contact with the Mg atoms of the other BChl a halves (2.2 Å). The almost parallel planes (2° tilt, rings II approaching each other) of the two macrocycles are rotated against each other around an axis perpendicular to the planes by 30° (see Fig. 6). In its main features, the geometrically optimized structure of this isolated (BChl a)$_2^{+\cdot}$ is quite similar to that found for (BChl b)$_2$ in the RC protein environment by X-ray analysis [8]. The binding energy of (BChl a)$_2^{+\cdot}$ was calculated to be 2.2 eV. This value is, however, not very reliable, due to the well-known deficiencies of the INDO method in calculating absolute binding energies [17]. It could be shown that the major contribution to this value stems from the acetyl-Mg contact and not from the $\pi - \pi$ interaction. This structure is energetically favored over a dimer with overlap at rings III (without bridging ligands) which was proposed earlier on the basis of less sophisticated MO methods [4,5]. An interesting result of the calculations is that the dimerization process indeed leads to a strong change of the spin distribution in the dimer compared with the monomer, i.e. a new molecular species (supermolecule) is formed. Spin density in the dimer is concentrated in the overlap region (rings I) and decreases towards rings III/V. This leads, for example, to the fact that now the CH_3 hfc in ring I is larger than that in ring III, in contrast to the monomer where the opposite sense occurs.

In Fig. 6 the spin densities for (BChl a)$_2^{+\cdot}$ are shown for the configuration discussed above. The agreement between experimental and calculated values is quite good. A perfect agreement cannot be expected due to the unknown influence of the protein environment on $P_{865}^{+\cdot}$ which has not been accounted for in our dimer calculations. More detailed calculations, also on (BChl b)$_2^{+\cdot}$, are under way in this laboratory.

4. Conclusion

All results from different techniques taken together leave no doubt that the primary donor in bacterial RC's is a bacteriochlorophyll dimer. It is proposed that a dimer similar to that found in the X-ray analysis of *Rp. viridis* RC's [8] is also present in *Rp. sphaeroides* and *R. rubrum*. Furthermore, this supermolecule not only exists in the singlet ground-state but also in the cationic doublet and the excited triplet states, although the fine structural details could be changed in the process of charge-separation in the RC [23,24]. It is tempting to speculate why Nature uses a dimeric BChl as the primary phototrap in RC's. One reason could be that the "electron hole" in P^+ can be better stabilized by delocalization in a dimeric supermolecule than in a monomer. After charge-separation, a dimer also has more degrees of freedom to perform a slight structural change, e.g. caused by charge effects [23], by which the back reaction is slowed down and the efficiency of the forward electron transport is increased. Since the hfc's directly reflect the delocalized "electron hole" and are very sensitive to any kind of structural changes, ENDOR will be the method of choice for determining these constants in liquids, solids or even single crystals. In future, this technique will play an important role in the elucidation of similar structures in photosynthesis and other biological systems.

Acknowledgments

The authors want to thank the following colleagues who contributed to this work: H. Scheer and coworkers (Munich, FRG), A.J. Hoff (Leiden, NL), M.C. Thurnauer (Argonne, USA) and G. Feher and his collaborators (San Diego, USA). This work was supported by the Deutsche Forschungsgemeinschaft (Sfb 161).

References

1. J.D. McElroy, G. Feher, and D.C. Mauzerall: Biochim.Biophys. Acta 267, 363 (1972), and refs. cited therein.
2. J.R. Norris, R.A. Uphaus, H.L. Crespi, and J.J. Katz: Proc.Natl. Acad.Sci. USA 68, 625 (1971).
3. M.R. Wasielewski: in Light-Reaction Path of Photosynthesis, ed. F.K. Fong (Springer, Berlin 1982), pp. 234-276.
4. W. Lubitz, F. Lendzian, H. Scheer, J. Gottstein, M. Plato, and K. Möbius: Proc.Natl.Acad.Sci. USA 81, 1401 (1984).
5. W. Lubitz, F. Lendzian, H. Scheer, M. Plato, and K. Möbius: in Photochemistry and Photobiology, ed. A.H. Zewail (Harwood Academic, New York 1983) pp. 1057-1069.
6. P.J. O'Malley and G.T. Babcock: in Advances in Photosynthetic Research, ed. C. Sybesma (Nijhoff/Junk, The Hague 1984), Vol. 1, pp. 697-700.
7. P.J. O'Malley and G.T. Babcock: Proc.Natl.Acad.Sci. USA 81, 1098 (1984).
8. J. Deisenhofer, O. Epp, K. Miki, R. Huber, and H. Michel: J.Mol. Biol. 180, 385 (1984).
9. G. Feher, A.J. Hoff, R.A. Isaacson, and L.C. Ackerson: Ann.N.Y.Acad. Sci. 244, 239 (1975).
10. J.R. Norris, H. Scheer, and J.J. Katz: Ann.N.Y.Acad.Sci. 244, 260 (1975).
11. W. Lubitz, R.A. Isaacson, E.C. Abresch, and G. Feher: Proc.Natl. Acad.Sci. USA 81, 7792 (1984).
12. M.S. Davis, A. Forman, L.K. Hansen, J.P. Thornber, and J. Fajer: J.Phys.Chem. 83, 3325 (1979).
13. K. Möbius, M. Plato, and W. Lubitz: Phys.Rep. 87, 171 (1982)
14. H. Kurreck, B. Kirste, and W. Lubitz: Angew.Chem.Int.Ed. Engl. 23, 173 (1984).
15. R.K. Clayton and B.J. Clayton: Biochim.Biophys. Acta 501, 478 (1978).
16. F. Lendzian, K. Möbius, M. Plato, U.H. Smith, M.C. Thurnauer, and W. Lubitz: Chem.Phys.Lett. 111, 583 (1984).
17. J.A. Pople and D.L. Beveridge: Approximate Molecular Orbital Theory (McGraw-Hill, New York 1970).
18. M. Plato, E. Tränkle, W. Lubitz, F. Lendzian, and K. Möbius (1985), to be published.
19. W. Lubitz, F. Lendzian, H. Scheer, M. Plato, and K. Möbius: J.Am. Chem.Soc., submitted.
20. F. Lendzian, W. Lubitz, H. Scheer, C. Bubenzer, and K. Möbius: J.Am. Chem.Soc. 103, 4635 (1981).
21. F. Lendzian, H. van Willigen, S. Sastry, K. Möbius, H. Scheer, and R. Feick: Chem.Phys.Lett., in press (1985)
22. M.R. Wasielewski, J.R. Norris, H.L. Crespi, and J. Harper: J.Am. Chem.Soc. 103, 7664 (1981).
23. A.W. Warshel: Proc.Natl.Acad.Sci. USA 77, 3105 (1980).
24. D. Kleinfeld, M.Y. Okamura, G. Feher: Biochemistry 23, 5780 (1984).

ENDOR of Semiquinones in RCs from *Rhodopseudomonas sphaeroides*

G. Feher, R.A. Isaacson, M.Y. Okamura, and W. Lubitz‡

Department of Physics, University of California, San Diego, La Jolla, CA 92093, USA

I. INTRODUCTION

At the heart of photosynthesis are the electron-transfer reactions between the primary reactants. To understand these from first principles, a knowledge of both the spatial as well as the electronic structure of the reactants is essential. The exciting and beautiful work on the three-dimensional structure of reaction centers (RCs) in *Rps. viridis* by the Munich group will be discussed in a subsequent session. Here, we would like to discuss experiments designed to elucidate the *electronic structure* of the quinone acceptors, Q_A and Q_B in *Rps. sphaeroides* R-26.1. This work is still continuing; the presentation should, therefore, be taken in the spirit of a progress report.

The technique that we used in this work to explore the electronic structure is the Electron Nuclear Double Resonance (ENDOR). This method has been extensively used to investigate the structure of the primary donor, as will be discussed in the next paper by W. LUBITZ *et al.* ENDOR provided the proof that the primary donor is a specialized bacteriochlorophyll dimer [1-3]. The acceptors, the ubiquinones Q_A and Q_B, cannot be explored by ENDOR when they are in their native environment. The reason is the very broad EPR line of the radical which is caused by the magnetic interaction of the ubisemiquinone with the nearby paramagnetic high spin Fe^{2+} [4]. Recently, R. DEBUS in our laboratory was able, by a simple procedure, to replace Fe^{2+} with the diamagnetic metal Zn^{2+} thereby reducing the EPR linewidth by approximately 2 orders of magnitude [5]; this made the ENDOR studies possible.

In an ENDOR experiment, one induces NMR transitions and monitors them via the Electron Paramagnetic Resonance (EPR) signal. Since the magnetic moment of the electron is three orders of magnitude larger than that of the nuclei, one gains many orders of magnitude in sensitivity over regular NMR experiments. Furthermore, one often gains many orders of magnitude in resolution over regular EPR experiments.

In the ubisemiquinone anion radical, the electron is delocalized in the π-system of the ring and interacts with the various nuclei. The strength of this interaction (called hyperfine, hf, interaction) determines the NMR frequency and is related to the spin density and the electronic wave function of the unpaired electron. Thus, by properly identifying the NMR transitions with the corresponding nuclei, one can obtain a map of the spin density of the unpaired electron. Note that in this technique one focuses on the unpaired valence electron that is responsible for the primary photochemical reactions. In contrast, in X-ray crystallography the core electrons give rise to the observed diffraction pattern.

‡ Present address: Institut für Organische Chemie, Freie Universität Berlin, 1000 Berlin 33, West Germany

II. EXPERIMENTAL RESULTS

1. Model Systems

Before discussing the ENDOR spectra of quinones in RCs we will briefly review the ENDOR results obtained on model compounds, in particular the anion radicals of ubiquinone-10 (UQ-10) [6,7], 2,3-dimethoxy-5,6-dimethylbenzoquinone (*DDQ*) and 2,3,5,6 tetramethyl benzoquinone (*DQ*) [6,8] (see structures in Table I).

Basically, one observes two types of ENDOR spectra. In liquid solutions, when the molecular tumbling rate is fast compared to the hf interaction frequency, the anisotropic hf interactions are averaged out and a narrow, symmetric, line corresponding to the isotropic part, A_{iso}, is observed. For an axially symmetric hf coupling (hfc)-tensor A_{iso} is given by:

$$A_{iso} = \frac{1}{3}(2A_{\perp} + A_{\|}) \tag{1}$$

where $A_{\|}$ is the hfc when the magnetic field points along the axis of cylindrical symmetry and $A_{\perp}$ when the field is perpendicular to it. At lower temperatures (e.g., frozen solutions), the anisotropic hfc's are not averaged out and one obtains a more complicated spectrum from which $A_{\|}$ and $A_{\perp}$ can be deduced [7-10] (see Section II-2).

Liquid solution ENDOR spectra of $(UQ\text{-}10)^-$ in an aprotic medium (DME) are shown in Figure 1 for two different temperatures. At T = –3°C only three pairs of lines are clearly distinguished; they are attributed to the methyl, the methylene, and probably the methoxy protons. The isotropic hfc's were determined to be:

$$(\mathrm{UQ\text{-}10})^-(T = -3^\circ C): \quad +5.89 MHz\,(CH_3); \quad +2.92 MHz\,(CH_2); \quad -0.27 MHz. \tag{2}$$

The signs of the hfc's were obtained from General TRIPLE experiments [11]. When the temperature was lowered to –68°C, the following hfc's were obtained:

$$(\mathrm{UQ\text{-}10})^-(T = -68^\circ C):$$

$$+5.91 MHz\,(CH_3), \quad +3.72 MHz\,(CH_2), \quad +2.06 MHz\,(CH_2), \quad -0.29 MHz, \quad (+)0.06 MHz. \tag{3}$$

Note that *two* pairs of lines are now present for the CH_2 group, as previously observed by DAS *et al.* [6]. This effect is due to a freezing out of the rotational motion of the isoprenoid chain. The additional small splittings (0.06 MHz) is possibly due to the γ-proton on the isoprenoid chain. More work needs to be done to identify the two small splitting with certainty.

In a protic solvent (basic ethanol) the ENDOR spectrum of the anion radical $(UQ\text{-}10)^-$ at –3°C was similar to that in DME; the ^{1}H-hfc's were smaller by $\sim$ 2%. In deuterated ethanol the ENDOR spectrum remained unchanged, showing that all the observed hfc's belong to nonexchangeable protons on the quinone. At temperatures $<$ –3°C the radical in ethanol started to precipitate with a concomitant loss of the ENDOR signal.

For DQ^- in ethanol (–3°C) only one hfc for the four equivalent freely rotating methyl groups was detected with A_{iso} = +5.32 MHz. In DDQ^- under the same conditions the CH_3 hfc (+5.62 MHz) was larger than in DQ^- (+5.32 MHz) but smaller than in $(UQ\text{-}10)^-$ (+5.78 MHz). This shows that the spin density is influenced by both the methoxy groups and the isoprenoid chain (see Section III). A second hfc measured in DDQ^- (0.1 MHz) that was also present in deuterated ethanol, was tentatively assigned to the protons at the methoxy groups. This hfc is considerably smaller than that measured in $(UQ\text{-}10)^-$; this could be the result of the strong conformational dependence of this hfc.

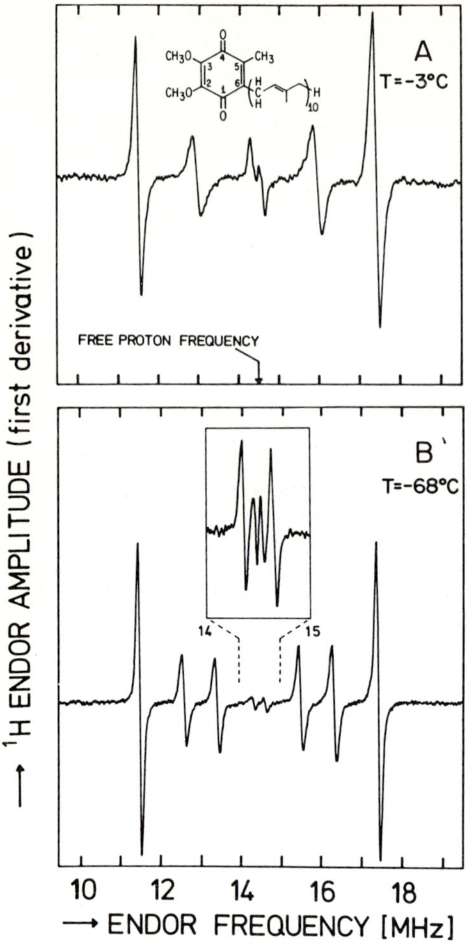

Figure 1. *Liquid solution ^{1}H-ENDOR spectra of $(UQ-10)^-$ in 1,2-dimethoxyethane (DME) at -3° C (A) and -68° C (B). Inset in A shows structure of UQ-10. Inset in B shows expanded region from 14 MHz to 15 MHz (T = -42° C). Experimental conditions: Anion produced by electrolysis, [UQ-10] = 2.5 × 10^{-4}M, microwave power (MW) 20 mW, radio frequency (RF) power 50 W, frequency modulation (FM) 100kHz, accumulation time 20 min.. Inset: MW power = 10 mW, RF power 30 W, FM 40 kHz.*

The anisotropic components of the hfc tensor for DQ^- were determined from frozen (ethanol) solution spectra to be $A_{\|}$ (CH_3) = 7.7 MHz and $A_{\perp}(CH_3)$ = 4.5 MHz. These values are in fair agreement with the reported values of 7.5. MHz and 4.4 MHz [8]. The $(UQ-10)^-$ spectrum in frozen solution gave hfc's of $A_{\|}$ = 8.4 MHz and $A_{\perp}$ = 5.3 MHz [7]. The unusually large anisotropy of the methyl group $\{(A_{\|} - A_{\perp})/A_{iso}\}$ of 57% for DQ^- and 49% for $(UQ)^-$ is due to the proximity of the oxygen (see Section III). An interesting feature of the spectra of immobilized semiquinones in frozen alcohol is the detection of exchangeable protons that are believed to bind to the carbonyl oxygens [7]. In $(UQ-10)^-$, the hfc's were determined to be $A_{\|}$ = 6.1 MHz and $A_{\perp}$ = 2.8 MHz [7]. These values show that the interaction is essentially dipolar in nature (i.e., $A_{iso} \simeq 0$). We have not detected this interaction in solution, at -3°C.

2. The ENDOR Spectrum of Q_A^- in Reaction Centers

Reaction centers from *Rps. sphaeroides* R-26.1 were extracted with lauryldimethylamineoxide (LDAO) and purified as described [12]. In all the experiments the Fe^{2+} was replaced by

Zn^{2+} as mentioned in the INTRODUCTION [5]. The metal content of the final preparation was determined by atomic absorption to be ~ 0.1 Fe/RC, and ~ 1 Zn/RC. The one electron reduction of the primary quinone, Q_A, was accomplished either photochemically by illuminating the sample at room temperature in the presence of a ten-fold excess of *cyt* c^{2+}, or by chemical reduction with dithionite. Following reduction, the samples were frozen quickly in liquid nitrogen.

The ENDOR spectrum from Q_A^- in reaction centers containing ~ 2Q/RC is shown in Figure 2. The central region of the spectrum has been expanded and is shown in Figure 2B. To obtain the exact hfc's, a knowledge of the line shapes is required. In solution spectra (Figure 1) one observes only the isotropic hfcs and the transition frequencies occur at the zero crossings of the derivative lines. In frozen solutions, the different orientations of the quinones cause a spread in resonance frequencies with some build-up of intensities at the principal values of the hf tensor This is illustrated for an axially symmetric tensor in Figure 3 (top). Experimentally one observes the derivative, $d\chi''/d\nu_E$, which exhibits prominent peaks along the principal axes of the tensor (bottom of Figure 3). The resonance position of the $A_\|$ component was determined by the procedure of WASSERMAN *et al.* [13]. The hfc's for $Q_A^- Q_B^=$ obtained from the spectra of Figure 2 are as follows:

$\mathbf{Q_A^-}$:	a-a′ : 0.09 MHz	d-d′ : 0.58 MHz	g-g′ : 2.6 MHz	j-j′ : 6.4 MHz
	b-b′ : 0.26 MHz	e-e′ : 1.14 MHz	h-h′ : 4.1 MHz	k-k′ : 7.2 MHz
	c-c′ : 0.47 MHz	f-f′ : 1.80 MHz	i-i′ : 4.7 MHz	l-l′ : 9.0 MHz.

Similar hfc's were obtained in RCs with ~ 1 Q/RC [14]. Thus the ENDOR spectra of Q_A^- and $Q_A^- Q_B^=$ are the same within experimental error.[1]

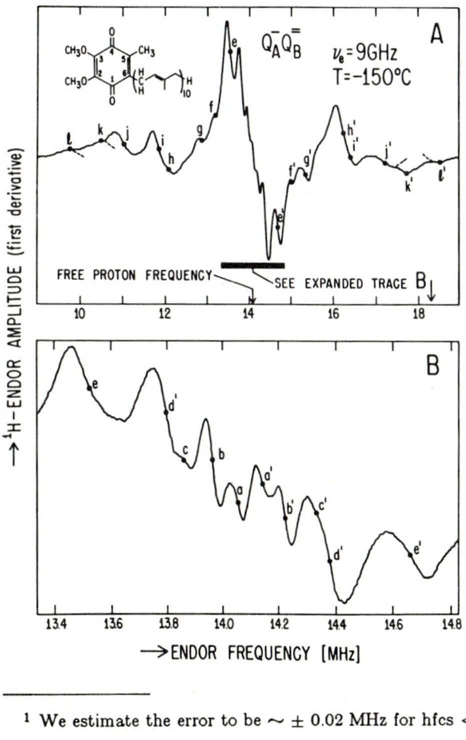

Figure 2. *¹H-ENDOR spectrum of Q_A^- (UQ−10)in Zn-substituted RCs from Rhodopseudomonas sphaeroides R-26.1. RCs (A^{800}_{1cm}=300) containing ~ 2Q/RC were reduced with 50 mM dithionite, 100 mM Tris in an EPR sample tube having an internal diameter of 0.2 cm. The central region has been expanded in the bottom trace. Experimental conditions: T = −150° C, ν_e = 9.3 GHz, MW 2 mW, RF power 100 W. FM modulation depth 100 kHz for full sweep and 40 kHz for expanded sweep. Magnetic field was set at center of EPR line. Sweep time 4s/sweep. One hundred sweeps were averaged.*

[1] We estimate the error to be ~ ± 0.02 MHz for hfcs < 1 MHz and ~ ± 0.1 MHz for hfcs > 2 MHz.

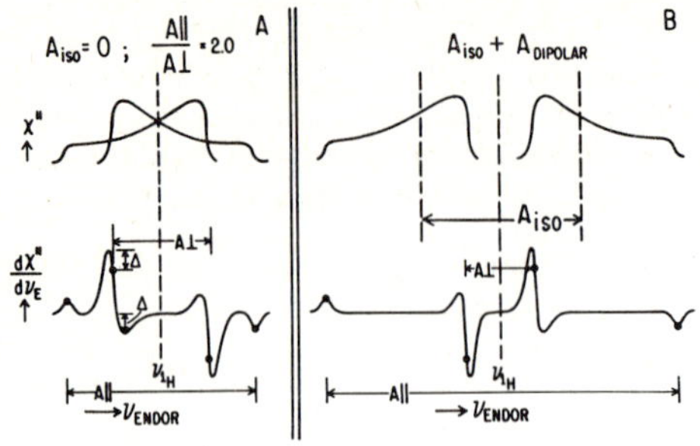

Figure 3. *ENDOR line shapes resulting from an axially symmetric hf tensor. A purely dipolar interaction ($A_{iso}=0$) is shown on the left. Experimentally we observe the derivative, $d\chi''/d\nu_E$ The procedure for determining the positions on the line [13] to obtain $A_\perp$ is indicated at the left bottom, i.e. the amplitude of the smaller peak Δ was substracted from the more intense peak.*

The most striking aspect of the spectrum is the large number of ENDOR lines compared with that obtained from model compound studies on semiquinone. The reason is that the quinone in the RC is embedded in a protein whose protons in the vicinity of the binding site can interact with the unpaired electron. These interactions give rise to so-called matrix ENDOR lines [15]. In addition, there is a class of protons that can exchange with the surrounding solution. In this category are, for instance, protons that form hydrogen bonds with the carbonyl oxygens on the quinones. Furthermore, there are the nonexchangeable protons on the quinone, e.g., methyl, methoxy, and methylene protons (see Figure 4).

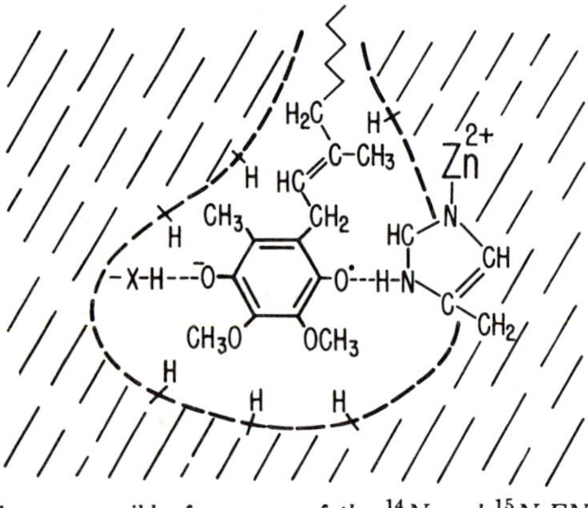

Figure 4. *Schematic representation of the quinone binding site, showing the different classes of protons that can interact with the unpaired electron on the quinone. An imidazole ring of histidine is shown to bind via an exchangeable hydrogen to one of the quinone oxygens as well as to Zn^{2+}. Such a structure was proposed by Bunker et al. (1982) (Biophys. J.* **37**, *539) and is in accord with recent X-ray data (Deisenhofer & Michel, private communication). The second quinone oxygen binds via another exchangeable hydrogen to an unknown moiety X (perhaps also a nitrogen). The histidine nitrogen is believed to be responsible for some of the ^{14}N and ^{15}N-ENDOR lines. Cross-hatched area represents protein.*

In order to obtain useful information from the ENDOR spectra it is necessary to identify each hfc with a particular proton. This has so far been partially accomplished by using the technique of isotopic substitution. However, before entering into the details of this procedure, let us first see whether there is a discernable difference between the ENDOR spectra of Q_A^- and Q_B^-.

3. Comparison of ENDOR Spectra from Q_A^- and Q_B^-

Although Q_A and Q_B are both UQ-10 in RCs from *Rps. sphaeroides* there must be a difference in their electronic structure since electron-transfer proceeds vectorially from Q_A^- to Q_B. This difference must be due to the surrounding protein ,which causes an inequality

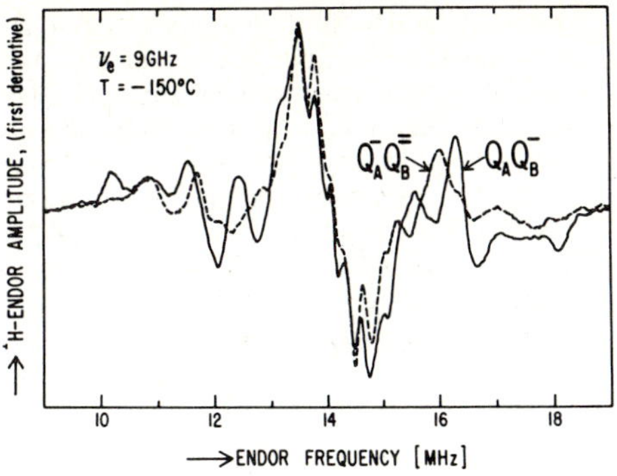

Figure 5. *Comparison of ¹H-ENDOR spectra of* $Q_A\, Q_B^-$ *(full line) and* $Q_A^- Q_B^=$ *(dashed line) in RCs from Rhodopseudomonas sphaeroides. The* $Q_A\, Q_B^-$ *spectrum has an admixture of* Q_A^- *(see text). Experimental conditions similar to those described in legend to Figure 2 (from Ref. [14]).*

in the redox potentials of Q_A^- and Q_B^- [16]. The difference in the electronic structure is reflected in the ENDOR spectra of Q_A^- and Q_B^- as shown in Figure 5. RCs in the $Q_A\, Q_B^-$ state were prepared by using a single short intense laser flash [17]. However, not all the reconstituted Zn/RCs have an active Q_B. Consequently, after the laser flash, some of the RCs are left in the Q_A^- state. Thus the spectrum labeled $Q_A\, Q_B^-$ has an admixture of Q_A^- in it. Furthermore, since Q_B^- was created by illumination (chemical reduction with dithionite produces $Q_B^=$) the concentration of RCs had to be low enough for light to penetrate. Thus, the maximum signal obtainable from Q_B^- was inherently lower than from Q_A^-. For these reasons the signal from Q_B^- has so far not been explored as extensively as that from Q_A^-. We proceed now to the identification of the individual ENDOR lines.

4. Exchangeable Protons

When bacteria are grown in a completely deuterated medium [18] but the RCs are isolated in an H_2O-based buffer, one expects all ¹H-ENDOR lines to be absent, except those arising from exchangeable protons. The results with such deuterated RCs are shown in Figure 6A. Two sets of prominent lines are observed in samples containing either Q_A^- or Q_B^-. When the RCs were equilibrated with D_2O these lines were eliminated. We have, therefore, assigned these lines to exchangeable protons. By performing a time-course study, we determined the half-time for the proton exchange with Q_A^- to be $\sim$ 3 hr (T = 20°C). The exchangeable protons most likely form hydrogen bonds with the two carbonyl oxygens on the quinones. The values of the hfcs obtained from Figure 6A are as follows:

$$Q_A^-:\ (1)\ A_\perp = 6.4\ \text{MHz} \qquad Q_B^-:\ (1)\ A_\perp = 5.0\ \text{MHz}$$
$$(2)\ A_\perp = 4.7\ \text{MHz} \qquad (2)\ A_\perp = 3.1\ \text{MHz}$$

$$A_\| = 9.0\ \text{MHz}. \tag{5}$$

The prominent lines were assigned to the perpendicular component, $A_\perp$, of the hyperfine tensor; the numbers (1) and (2) refer to the two oxygens. The smaller lines of the Q_A^- spectrum were assigned to $A_\|$ of the hyperfine tensor. It is, however, not clear at present whether they belong to oxygen (1) or (2). No corresponding lines have so far been observed for Q_B^-.[2] Model compound studies by O'MALLEY *et al* [7] indicated that the hyperfine tensor of exchangeable protons has axial symmetry (i.e., it is traceless with $A_\| = -2A_\perp$) (see Section II-1). If this conclusion holds for the RC, the observed component $A_\|$ must be associated with oxygen (2). It is interesting that the hfc's in Q_B^- are significantly smaller

[2] It is possible that the two hfc's obtained for Q_B^- belong to the same hfc tensor. In accordance with Figure 3 we then get $A_\| = 5.3$ MHz and $A_\perp = 3.1$ MHz. This would indicate that the two H-bonds in Q_B^- have the same strength as found for the *in vitro* model [7].

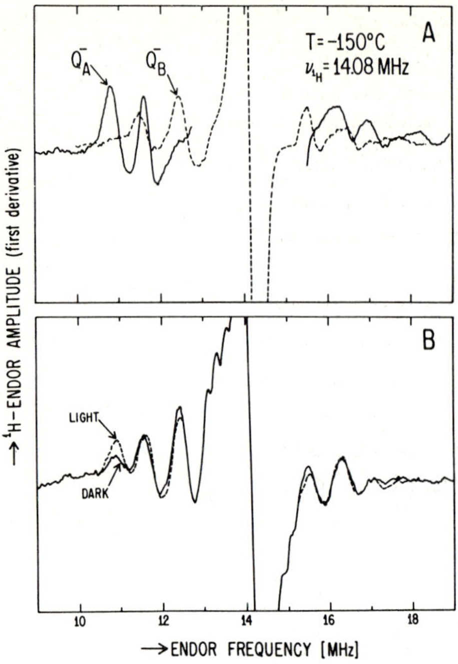

Figure 6. *^{1}H-ENDOR spectra of exchangeable hydrogens of Q_A^- and Q_B^- in RCs of Rhodopseudomonas sphaeroides.* **A:** *RCs were isolated in H_2O-based buffer from deuterated bacteria; Q_A^- was formed by illuminating the sample at 20° C for 5 s in the presence of a 10-fold excess of cyt c^{2+}. Q_B^- was formed by applying one short (~ 0.4 μs) laser flash in the presence of cyt c^{2+}.* **B:** *Nondeuterated RCs were reconstituted with deuterated UQ–10 and treated as described in the text. The broader central line (as compared with A) is due to the protons of the protein. The spectrum shows the presence of both Q_A^- and Q_B^- (see text). Upon illumination only the peak due to Q_A^- increased.*

than those found in Q_A^-. This implies that the hydrogen bonds in Q_B are weaker and could possibly account for the weaker binding of Q_B^- to the RC-protein [14].

The positions of the ENDOR lines were found to be temperature-dependent; e.g., for Q_A^- the hfcs were found to change from 6.4 to 5.8 MHz and from 4.7 to 4.3 MHz when the temperature was raised from –150°C to –20°C.

Since the ENDOR spectra from Q_A^- and Q_B^- are different, one can use them as an assay for the state of reduction of the quinones in the RC. An example of this is shown in Figure 6B. Undeuterated RCs were reconstituted with deuterated quinones in both the Q_A and Q_B sites. After reduction with dithionite,the ENDOR spectrum was characteristic of Q_A^- (they were presumably in the $Q_A^-Q_B^=$ state). After temporarily raising the temperature to –20°C, the spectrum shown in Figure 6B (solid line) was obtained at –150°C. The ENDOR signal due to Q_A^- was decreased and that due to Q_B^- was increased. $Q_B^=$ presumably had lost its two electrons at –20°C and had received the electron from Q_A^-. The mechanism of the loss of the two electrons is not understood at present. When the sample was illuminated at –150°C[3] the ENDOR signal corresponding to Q_A^- (see dashed line) increased. Thus ENDOR spectroscopy can be used to study the slow kinetics of charge transfers at low temperatures.

5. Hfc's of *CH_3*

The methyl protons are expected to give a strong ENDOR signal [8-10]. At –150°C the protons are freely rotating, giving rise to an approximately axial hfc tensor. To obtain the ENDOR transition from the methyl protons ,the interfering effects of the exchangeable protons (see lines i-i′ and j-j′ in Figure 2A) were eliminated by dialyzing RCs against D_2O

[3] At this temperature the electron-transfer from Q_A^- to Q_B does not occur within the time of the experiment.

TABLE I. HYPERFINE COUPLINGS [in MHz] OF METHYL PROTONS OF DIFFERENT QUINONES INCORPORATED IN RCs FROM *Rps. sphaeroides*

Quinone		Environment	$A_\perp$	$A_\parallel$	A_{iso}
UQ-10	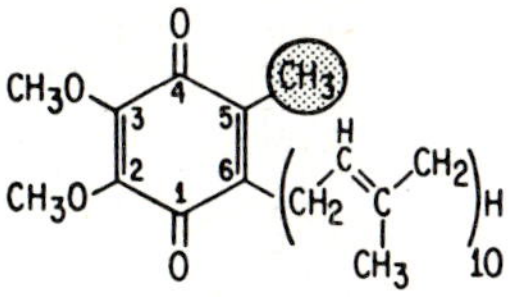	Q_A in RC	4.1	7.2	5.1
		EtOH[a]			5.8
		DME[a]			5.9
		Q_B in RC	4.7	7.9	5.8
DDQ	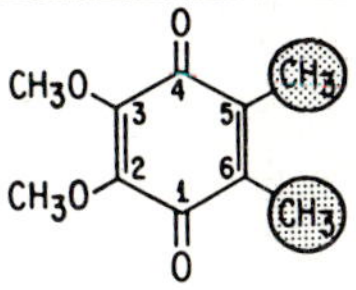	Q_A in RC	4.0	7.1	5.0
		"	5.6	9.1	6.8
		EtOH			5.6
DQ	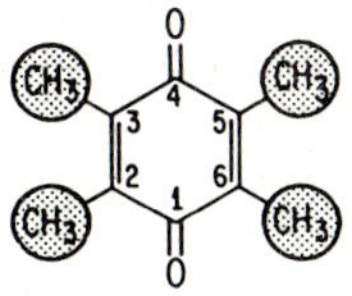	Q_A in RC	3.3	6.0	4.2
		"	5.2	8.5	6.3
		"	6.0	9.3	7.1
		Frozen EtOH[b]	4.5	7.7	5.6
		EtOH			5.3

a.) Similar values were found by DAS *et al.* [6].

b.) The values found by O'MALLEY and BABCOCK [8] were 4.4, 7.5 and 5.4 MHz, respectively.

[14]. The spectrum of the dialyzed sample exhibited two sets of prominent lines (data not shown) corresponding to h-h′ and k-k′ of Figure 2A. These were assigned to the two components $A_\perp$ and $A_\parallel$ of the methyl hfc tensor. The values of the hfc's for Q_A^- and Q_B^- are summarized in Table 1. The large anisotropy of the methyl group is in approximate agreement with the values found in isolated quinones [8] (see Section II-1). The isotropic hfc obtained from Eq. 1. is 5.1 MHz which is to be compared with values of 5.8 and 5.9 MHz found in solution (see Table 1). The discrepancy between these values may be explained by a redistribution of the electron density in $(UQ\text{-}10)^-$ due to hydrogen bonds to the carbonyl oxygens of different strengths (see Section II-4).

The isotropic hfc of the methyl group in Q_B^- is 14% larger than in Q_A^- and has the same value as that of $(UQ\text{-}10)^-$. This indicates an increase in spin density in Q_B^- at position 5 (see structure in Table I) and a predicted concomitant decrease at position 6. Such an equalization of the hfc's at positions 5 and 6 may be due to a weakening of the hydrogen bonds to the oxygens. This explanation is consistent with the smaller hfc's of the exchangeable protons in Q_B^-, as discussed in the previous section.

The redistribution of the electron density in the quinone ring, as well as the symmetry of the Q_A binding site, was further investigated in RCs in which the native UQ-10 was replaced by either DDQ or DQ. In DDQ^- we observed two sets of methyl hyperfine splittings (Table 1). One has approximately the same value as in $(UQ\text{-}10)^-$ (5.0 and 5.1 MHz, respectively), the second splitting is larger (6.8 MHz),presumably due to preferential hydrogen bonding to one of the oxygens. The average of the two hfc's in DDQ^- is 5.9 MHz, which is slightly larger than the value found in solution (5.6 MHz) (see Table I).

The ENDOR spectrum of DQ^- shows three sets of methyl splittings (see Table 1) Assuming that the smallest splitting is doubly degenerate, the average of the four isotropic hfc's is 5.4 MHz, in good agreement with the values found in model compounds (see Table I).

The above results can be used to draw conclusions about the symmetry of the quinone binding site. In DDQ^- one expects in the most general case, four sets of ENDOR transitions due to the methyl protons. Two of them are due to the inequivalence at positions 5 and 6. A different set of two transitions may be obtained if the site is asymmetric around the 1,4 molecular axis, and the quinone can be incorporated into the RC in two different ways. The fact that we see only two lines can be interpreted either by assuming that the site is symmetric around the 1,4 axis, or that the quinone is incorporated in only one way. To distinguish between these two possibilities, we use the results obtained from DQ^-. The fact that we obtained in this case three (instead of two) transitions shows that the site is asymmetric. Thus we conclude that the quinone binds to the RC in only one specific way.

So far we have considered the hyperfine tensor of the methyl protons to be axially symmetric. This is not exactly true, however, because of the preferential interaction with the oxygen along one axis. If one were to select molecules whose normal to their plane points along the external magnetic field, i.e., along the z direction (see Figure 7), one would observe a slightly different hyperfine interaction than when the field points in the plane, i.e., in the y direction. Since the RCs and hence the quinones are randomly oriented, one sees in general only an average of the perpendicular hf components. However, there is a procedure for selecting hf components in radicals having an anisotropic g value[4] [20]. It is based on the fact that if one sets the magnetic field off the center of the line, e.g. to lower magnetic fields, one observes molecules whose g value is higher. For $(UQ-10)^-$ this means that one preferentially selects molecules whose normal to their plane is parallel to the external magnetic field (z direction in Figure 7). By setting the magnetic field three to six Gauss from the center of the line to either higher or lower values we obtained

$$\frac{A_{zz}-A_{yy}}{\frac{1}{2}(A_{zz}+A_{yy})} \simeq 0.1. \tag{6}$$

Thus, the fractional difference between the two perpendicular hf component is approximately 10%. This is in fair agreement with theoretical calculations as will be discussed later.

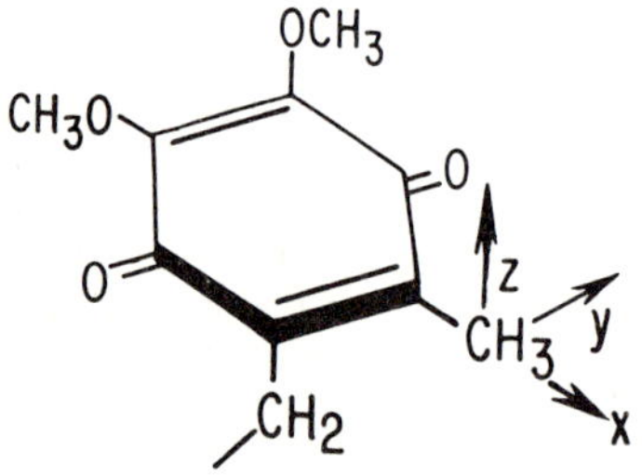

Figure 7. *System of axes used to interpret the components of the hyperfine tensor of the methyl protons.*

[4] The principal components of the g tensor of $(UQ-10)^-$ were determined to be $g_\perp = 2.0060\pm0.0002$ and $g_\parallel = 2.0023\pm0.0002$ [20].

6. Hfc's of CH_2

Unlike the methyl protons, the methylene protons of the isoprenoid chain are not expected to rotate freely. If, at low temperatures, the methylene protons are frozen in a distribution of configurations, one would expect broad ENDOR lines, since the hfc depends on the angle that the C-H bond makes with the p_z orbitals. At higher temperatures, the methylene protons may be partially mobile, giving rise to motionally narrowed ENDOR lines [e.g., see ref. 21]. Indeed at $-20°$ C a new set of ENDOR lines emerge. Figure 8 shows the expanded, high frequency, part of the spectrum. The two new lines were tentatively identified with the two hyperfine components of one of the methylene protons. Their values were found to be

$$\mathbf{Q_A^-}: \quad A\perp = 5.1 \text{ MHz}; \quad A\| = 8.4 \text{ MHz}$$
$$\therefore \; A_{iso} = 6.2 \text{ MHz}. \tag{7}$$

The deviation from axial symmetry was determined by magnetic field-selection as described in the previous section. A value similar to the one found for the CH_3 group (see Eq. 6) was found. The value of the isotropic hfc (obtained from eq. 1) is 6.2 MHz, which is close to the value found in DDQ^- for the methyl group at position 6. However, it is significantly larger than the value found in the model system studies (see Eqs. 2 and 3). To check whether the assignment of the methylene proton is correct, we plan to reconstitute RCs with UQ-0 (no isoprenoid chain). In these RCs the ENDOR lines assigned to CH_2 should be absent.

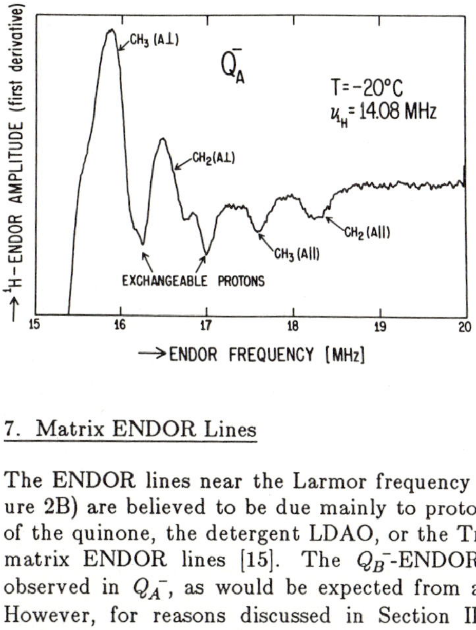

Figure 8. *High frequency part of the* ^1H-*ENDOR spectrum of* $Q_A^-(UQ-10)$ *in RCs from Rhodopseudomonas sphaeroides. Experimental conditions similar to those described in the legend to Figure 2, except* $T = -20°$ *C and RCs had 0.8 Q /RC*.

7. Matrix ENDOR Lines

The ENDOR lines near the Larmor frequency of free protons (small hf couplings) (see Figure 2B) are believed to be due mainly to protons associated with the protein in the vicinity of the quinone, the detergent LDAO, or the Tris buffer. They are commonly referred to as matrix ENDOR lines [15]. The Q_B^--ENDOR spectrum in this region differs from that observed in Q_A^-, as would be expected from a different environment of the two quinones. However, for reasons discussed in Section II-3 the spectrum of Q_B^- has not yet been analyzed in detail. In RCs isolated from completely deuterated bacteria that were dialyzed for one day against D_2O buffer, several of the matrix lines were still observed. These, presumably, are associated with the protons from LDAO or the Tris buffer.

So far we have not discussed the protons associated with the 2 methoxy groups. Their hfc's are expected to be small [22] giving rise to ENDOR lines in the region of the matrix ENDOR. By taking the difference between the spectra obtained from RCs reconstituted with deuterated $UQ-10$ and the native $UQ-10$, the ENDOR lines due to the methoxy protons may be obtained. Experiments along these lines are in progress.

8. Hfc's of ^{15}N and ^{14}N

In addition to ^1H-ENDOR, we also searched for ^{14}N and ^{15}N-ENDOR in Q_A^-, the latter in isotopically substituted RCs. Both spectra showed a number of lines in the vicinity of the respective nitrogen Larmor frequencies, indicating the presence of a nitrogen in the surroundings of $(UQ-10)^-$. An imidazole nitrogen of a histidine near the quinone binding site is expected from the model of HEARST and SAUER [23]. The nitrogen spectra are rather complicated and have not yet been fully analyzed.

9. Hfc's of ^{17}O

To obtain the hfcs of the unpaired electron with the oxygen nuclei, RCs were reconstituted with quinones in which the carbonyl ^{16}O was partially replaced by ^{17}O. Since the hf interactions with the oxygens are much larger than with the protons, they are resolvable in an EPR spectrum and one does not need to use ENDOR. When the isotopic substitution is small ($< 40\%$) the hyperfine structure will be dominated by quinones that have one ^{17}O and one ^{16}O (i.e., the likelihood that both oxygens are replaced by ^{17}O is small). The results for Q_A^- and Q_B^- are shown in Figure 9. The hyperfine lines in Q_A^- are split, indicating an inequivalence between the two oxygens.[5] The hyperfine lines in Q_B^- are broadened but not split, indicating that in this case the two oxygens are almost equivalent. This is consistent with the finding that the spin densities at carbon 5 and 6 are more similar to

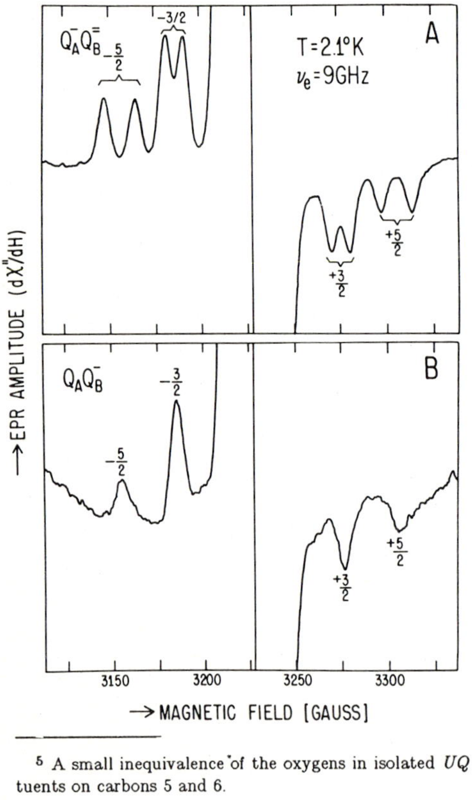

Figure 9. *EPR spectrum of Q_A^- (top) and Q_B^- (bottom) in RCs from Rhodopseudomonas sphaeroides reconstituted with isotopically (partially) substituted ^{17}O-ubiquinone-10. The substitution was accomplished by adding 5 mg of $UQ-10$ to 0.5 ml CH_2Cl_2, 0.4 ml methanol, 0.1 ml $H_2{}^{17}O$ (60% ^{17}O), 0.01 ml trifluoroacetic acid and stirring the mixture for 12 days at 23° C.* **A:** *Sample prepared with two ^{17}O-substituted quinones; Q_A^- was prepared as described in Figure 6A.* **B:** *RCs with one ^{16}O-quinone per RC were reconstituted with an ^{17}O-substituted second quinone; Q_B^- was formed as described in Figure 6A.*

[5] A small inequivalence of the oxygens in isolated $UQ-10$ can be expected due to the difference in the substituents on carbons 5 and 6.

each other in Q_B^- than in Q_A^- (see Table I) and with the weaker binding of the exchangeable protons (see Section II-4). The values of the ^{17}O hfc's determined from the spectra of Figure 9 are:

$$Q_A^-: \quad A_1 = 75 \text{ MHz}; \quad A_2 = 94 \text{ MHz}$$
$$Q_B^-: \quad A = 85 \text{ MHz}. \tag{8}$$

The hfc in Q_B^- is approximately the average of the two hfcs in Q_A^-. The values of the hfc's are much larger than those discussed in model compounds in solution at room temperature (e.g. $A \simeq 25$ MHz in p-benzoquinone [24]). The reason is that in the frozen state the large anisotropic components of the hf interaction are not averaged out and add to the isotropic part. Consistent with this notion, note that the shapes of the hf lines are characteristic of extrema of powder patterns (see Figure 3). The measured component ($A_{zz} = A_{||}$) is in approximate agreement with the theoretical estimate (see Section III).

III. THEORETICAL CONSIDERATIONS

For the calculation of the spin density distribution of the semiquinones we used an all valence electron self-consistent field (SCF) method based on INDO [25] and a restricted Hartree-Fock (RHF) approach (half-electron method) that included spin polarization effects. This method developed by PLATO *et al.* [26] proved to be successful in describing the spin density distribuiton in monomeric and dimeric bacterichlorophyll radicals [27]. For the calculation of the ubisemiquinone anion, a truncated isoprenoid chain ($-CH_2-CH=CH_2$) and standard bond lengths and angles were used [25]. Different geometries with various conformations of the methoxy groups and the 3-propenyl were tested. The calculated π- and s-spin densities for one of these geometries are given in Figure 10.

All π spin densities on the semiquinone are positive. Approximately 60% of the spin is located at the oxygens, $\sim$ 20% at C_5 and C_6 and $\sim$ 10% at C_2 and C_3 (the areas of the circles in Figure 10 are proportional to the spin densities). There is little spin density in the p_z orbitals of the substituents. The residual 10% are found in other s-type valence orbitals of the system. A small asymmetry of the spin distribution is caused by the different substitutions at C_5 and C_6.

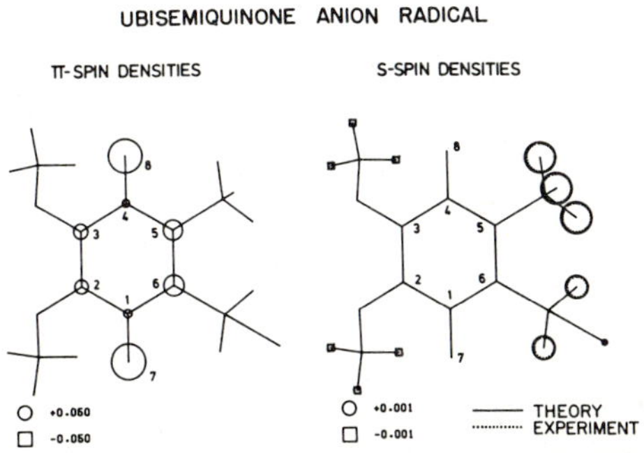

Figure 10. *π and s-spin densities calculated by the RHF-INDO method [26] for the ubisemiquinone anion with a truncated isoprenoid chain ($-CH_2-CH = CH_2$). For the molecular structure of UQ-10 see inset in Figure 1A. A perspective view of the skeleton is given; the following geometry was used in the calculation: the two methyls of the methoxy groups at C2 and C3 are pointing above and below the quinone plane; the twist angle is $\pm$ 45°. The atoms of the $-C_\alpha H_2-C_\beta H=C_\gamma H_2$ group at C6 are all in one plane, except for the methylene protons that point $\pm$ 60° away from the normal of the quinone plane. The H at C_β points away from the quinone (trans). The calculated s-spin densities of the CH_3 protons (full lines) were obtained by averaging the hfc's of all CH_3 protons (fast rotational motion). The experimentally determined spin densities were obtained from the measured isotropic hfc's (Section II-1) and the relation $A_{iso} = \rho^s Q$, with Q = 1420 MHz [27]. The values of the s-spin densities for carbons and oxygens were omitted for clarity.*

The calculated s-spin densities (solid lines) are compared in Figure 10 (right) to the experimental values (dotted) derived from the isotropic hfc's (see Section II-1). For the geometry chosen, the agreement between theoretical and experimental data is good. However, the hfc's of the methoxy and 3-propenyl substituents are strongly dependent on the geometry. This makes it difficult to correlate definitively the methoxy protons and the γ proton at the C_β of the isoprenoid chains with the measured hfc's.

In order to simulate the effect of hydrogen bonds on the spin density distribution, some preliminary calculations have been performed with one or two water molecules hydrogen-bonded to the oxygens (linear in-plane bond of 1.5 Å length). One water molecule symmetrically placed at each of the carbonyl oxygens *increases* the spin density at C_1 and C_4 but *reduces* it at the oxygens O_7 and O_8 and at C_5 and C_6. Consequently, the ^{1}H-hfc's of the CH_3 and the CH_2 group are reduced by 3.5 and 5.5% respectively. Such a trend was observed experimentally when changing from an aprotic (DME) to a protic (ethanol) solution (Section II.1). One water molecule attached to one of the oxygens (e.g. at O_8) increases one hfc (e.g. CH_2) and decreases the other one (e.g. CH_3). This finding is in accordance with the simple valence bond picture, and reflects the experimentally observed spin density shifts for Q_A^- with asymmetric hydrogen bonds (see Section II-5).

From the π-spin densities one can estimate the *anisotropic hfc tensors* for the various nuclei of the semiquinone using the method of McCONNELL & STRATHDEE [28]. For the rotating methyl group we obtained the principal values of +2.2, -0.9 and -1.3 MHz. Adding these to the calculated isotropic hfc of +7.2 MHz the complete hfc tensor becomes +9.4, +6.3, and +5.9 MHz. Averaging the 2 smaller components we obtain: $A_{||} = +9.4$ and $A_\perp = +6.1$ MHz. The anisotropy defined as $(A_{||} - A_\perp)/A_{iso}$ is 46% and the deviation from axial symmetry (see Eq. 6) is 7%. The magnitudes of both effects are in reasonable agreement with the experimental observations (see Section II-5); they are due to the proximity of the rotating CH_3 group to the carbonyl oxygen, which has a high π spin density ($\sim$ 0.3). The calculation without the inclusion of the nearby oxygen yields a more axially symmetric CH_3 tensor with an anisotropy of about 10% as is usually found for methyl groups in organic radicals [9].

The other carbonyl oxygen in the molecule influences the dipolar hfc tensor of the *methylene* protons in a similar way. The hfc's can become rather large and anisotropic; their values depending on the exact chain conformation. Analogous findings were obtained for the protons of the *methoxy* groups and for the γ-proton on the C_β of the isoprenoid chain. Although their values depend again critically on the geometry, the range of possible values was found to be restricted to the matrix region ($\leq$ 2.5 MHz).

An exact calculation of the dipolar hfc's of hydrogens in the proximity of the carbonyl oxygens is difficult due to the unknown geometry of the hydrogen bonds. Assuming that the =O---H bond is in the plane of the quinone ring and colinear with the O_7–O_8 direction [7], one can use the simple point-dipole formula[6]:

$$A_H = \frac{(g_e \mu_e)(g_H \mu_H)}{r^3} \rho(3\cos^2\Theta - 1) = \frac{79}{r^3} \rho(3\cos^2\Theta - 1) \text{ MHz} \tag{9}$$

to calculate the O-H distance. Since we have not determined $A_{||}$ for most of the transitions (see Eq. 5) we shall take only the values for $A_\perp$ and assume a purely dipolar tensor [7] (i.e. $A_{||} = -2A_\perp$). The O-H distances for Q_A^- and Q_B^- calculated from Eq. 9 are

$$Q_A^-: r_1 = 1.55 \text{ Å} \quad Q_B^-: r_1 = 1.69 \text{ Å} \quad (\text{or}^7: r_1 = r_2 = 2.0\text{ Å}) \tag{10}$$
$$r_2 = 1.78 \text{ Å} \quad r_2 = 1.97 \text{ Å}$$

[6] g_e, μ_e and g_H, μ_H are the electron and proton g-values and nuclear moments respectively, r is the distance in Å, ρ is the spin density at the oxygen (0.3) and Θ is the angle between the applied magnetic field and the line joining H and O.

[7] See footnote 2.

One should, of course, keep in mind the *ad hoc* nature of the assumptions made above and the approximation inherent in the point dipole treatment. More advanced calculations [e.g., see ref. 28] indicate that the hydrogen bond is probably not in the plane of the quinone.

The anisotropic tensor can also be estimated for the ^{17}O nuclei. Using ρ_0^π=0.3 we obtained principal values of -86, +43 and +43 MHz. Assuming an isotropic ^{17}O-hfc of -25 MHz [24] the complete tensor becomes: $A\|$ = -111 and $A\perp$ = +18 MHz. This estimate shows that in the solid state one ^{17}O hfc tensor component is large and should lead to a well resolved hf pattern of the EPR line, whereas the other one is small and will probably not be resolved. The estimated value of about 100 MHz is close to the measured values for Q_A^- (75 and 94 MHz) and Q_B^- (85 MHz) in RCs (see Section II-9).

IV. CONCLUSIONS

We have observed ENDOR spectra from Q_A^- and Q_B^- in RCs from *Rps. sphaeroides*. We were able by isotopic substitutions to identify the lines arising from the methyl, methylene and exchangeable hydrogens that are believed to bond to the carbonyl oxygens [29]. Several sets of lines corresponding to small hfc's (matrix ENDOR) were observed in the spectra of both Q_A^- and Q_B^-, indicating interactions with residues from the surrounding protein; ENDOR lines were also observed from nitrogens, most likely originating from the imidazole of histidine.

The quinone-binding site was found to be asymmetric with respect to both the 1-4 molecular axis (see inset in Figure 1A) and the axis perpendicular to it; the latter asymmetry makes the two carbonyl oxygens nonequivalent. In Q_B^- the nonequivalence of the oxygens was less pronounced than in Q_A^-. These conclusions were based on experiments performed with RCs in which the native quinone was substituted with *DQ* and *DDQ* (see Table I) and from the ^{17}O-hf splittings. The asymmetry of the two oxygens may be due to different strengths of the H bonds. Another possible cause is the presence of the positive charge on Zn^{2+}. Such a charge may stabilize Q^- in a hydrophobic environment, and may play a role in the electron-transfer process. The difference in the observed hfc's of Q_A^- and Q_B^- are presumably related to the change in redox potential that leads to the vectorial electron transport from Q_A^- to Q_B.

The ENDOR experiments with RCs were performed in the solid state (i.e., frozen solutions). Consequently, one observes both the isotropic and anisotropic components of the hfc tensor. We have also performed ENDOR experiments on model compounds in the liquid state in which only the isotropic hfc is observed. The isotropic hfc of the methyl group in the model compound differed from that found for Q_A^-. This is due to a reduction of the spin density at C_5 (with a concomitant increase at C_6). In contrast, the isotropic hfc obtained for Q_B^- was similar to that observed in the model compound, indicating a more symmetrical surrounding of the Q_B site. The isotropic hfc's of the methylene protons in Q_A^- and the model compound were different, suggesting a difference in the geometry of the isoprenoid chain.

Theoretical calculations were performed to obtain the π and s-spin densities of the semiquinones (see Figure 10). A SCF method based on INDO and restricted Hartree-Fock, including spin polarization was used [26]. Reasonable agreement between theory and experiment was obtained for the unusually large anisotropy of the methyl hfc and the deviation from axial symmetry. A simple point-dipole approach was used to estimate the distances of the exchangeable hydrogens to the oxygens in both Q_A^- and Q_B^-; they ranged from 1.55 Å to 1.97 Å (see Eq. 10).

A great deal of work remains to be done to complete the picture. The hfc's of the methoxy and the γ-proton of the isoprenoid chain have not been identified in the RC spectra, although a tentative assignment was made in the model compounds. The matrix ENDOR lines have not been explored in detail; in particular, the differences between the spectra of Q_A^- and Q_B^- have not been elucidated. The nitrogen ENDOR spectra need to be analyzed and correlated with specific residues of the RC protein. A study of the pH dependence of the spectra may shed light on the important problem of proton uptake by the semiquinones, which is understood only in part at present [30]. ENDOR spectra from metal-free RCs may provide a clue to the hitherto largely unknown function of the Fe^{2+} (or other divalent metals) [5]. ENDOR spectroscopy may also be used as a tool to explore conformational changes in RCs [31]. The ultimate goal is to correlate structure with function. To this end the electronic wavefunction obtained from this work should provide the necessary framework to calculate electron-transfer rates, which form the basis of photosynthesis.

V. ACKNOWLEDGMENT

We thank E. C. Abresch for the preparation of Reaction Centers and R. J. Debus for the replacement of Fe^{2+} with Zn^{2+} which made the present work possible. We thank M. Plato for the use of his computer programs and for his help with the calculations, G. T. Babcock for sharing with us his results on model compounds and to P. L. Loach for a gift of *DDQ*. This work was supported by grants from the National Science Foundation (PCM 82-02811) and the National Institute of Health (GM-13191). W. L. held a research scholarship from the Max Kade Foundation, New York.

References

1 J. R. Norris, R. A. Uphaus, H. L. Crespi, & J. J. Katz: *Proc. Natl. Acad. Sci. USA* **68,** 625-628 (1971).

2 G. Feher, A. J. Hoff, R. A. Isaacson, & L. C. Ackerson: *Ann. N. Y. Acad. Sci.* **244,** 239-259 (1975).

3 J. R. Norris. H. Scheer, & J. J. Katz: *Ann. N. Y. Acad. Sci.* **244,** 260-280 (1975).

4 M. Y. Okamura, R. A. Isaacson, & G. Feher: *Proc. Natl. Acad. Sci.* **72,** 3491-3495 (1975).

5 R. J. Debus, M. Y. Okamura, & G. Feher: Biophys J. (abstr.) **47,** 3a (1985).

6 M. R. Das, H. D. Connor, D. S. Leniart, & J. H. Freed: *J. Amer. Chem. Soc.* **92,** 2258-2268 (1970).

7 P. J. O'Malley, T. K. Chandrashekar, & G. T. Babcock: these Proceedings.

8 P. J. O'Malley and G. T. Babcock: *J. Chem. Phys.* **80,** 3912-3913 (1984).

9 H. Kurreck, M. Bock,N. Bretz, M. Elsner, H. Kraus, W. Lubitz, F. Mueller, J. Geissler, & P. M. H. Kroneck: *J. Am. Chem. Soc.* **106,** 737 (1984).

10 J. S. Hyde, G. H. Rist, & L. E. G. Eriksson: *J. Phys. Chem.* **72,** 4269 (1968).

11 K. Möbius, M. Plato, & W. Lubitz: *Phys. Rep.* **87,** 171-208 (1982).

12 G. Feher and M. Y. Okamura: in "*The Photosynthetic Bacteria,*" R. K. Clayton and W. R. Sistrom, eds., Chapter 19, 349-386 (Plenum, New York 1978).

13 E. Wasserman, L. C. Snyder, & W. A. Yager: *J. Chem. Phys.* **41,** 1763 (1964).

14 W. Lubitz, E. C. Abresch, R. J. Debus, R. A. Isaacson, M. Y. Okamura, and G. Feher: *Biochim. Biophys. Acta* (in press).

15 L. Kevan & L. D. Kispert: in "*Electron Spin Double Resonance Spectroscopy,*" p. 239 (J. Wiley, New York 1976).

16 C. A. Wraight: in "*Function of Quinones in Energy-Conserving Systems,*" B. L. Trumpower, ed., pp. 181-197 (Academic, New York 1982).

17 W. F. Butler, R. Calvo, D. R. Fredkin, R. A. Isaacson, M. Y. Okamura, & G. Feher: *Biophysical J.* **45,** 947-973 (1984).

18 R. C. Dougherty, H. L. Crespi, H. H. Strain, & J. J. Katz: *J. Amer. Chem. Soc.* **88,** 2854 (1966). J. J. Katz, R. C. Dougherty, H. L. Crespi, & H. H. Strain: *J. Amer. Chem. Soc.* **88,** 2856 (1966).

19 L. Kevan and L. D. Kispert: in "*Electron Spin Double Resonance Spectroscopy,*" p. 234 (J. Wiley, New York, 1976).

20 G. Feher, M. Y. Okamura, and J. D. McElroy: *Biochim. Biophys. Acta* **267,** 222-226 (1972).

21 G. E. Pake and T.L. Estle: in "*The Physical Principles of Electron Paramagnetic Resonance,*" p. 225 (W. A. Benjamin, Reading, Mass., 1973).

22 G. B. Rabold, R. T. Ogata, M. Y. Okamura, L. H. Piette, R. E. Moore, & P. J. Scheuer: *J. Chem. Phys.* **46,** 1161-1167 (1967).

23 J. E. Hearst and K. Sauer: *Z. Naturforsch.* **36c,** 421-424 (1984).

24 W. M. Gulick and D. H. Geske, *J. Am. Chem. Soc.* **88,** 4119 (1966).

25 J. A. Pople & D. L. Beveridge: *Approximate Molecular Orbital Theory* (McGraw-Hill, New York 1970).

26 M. Plato, E. Tränkle, W. Lubitz, F. Lendzian, & K. Möbius, to be published.

27 W. Lubitz, F. Lendzian, M. Plato, K. Möbius, & E. Tränkle: these Proceedings.

28 H. M. McConnell and J. Strathdee: *Mol. Phys* **2,** 129 (1959).

29 B. J. Hales and E. E. Case: *Biochim. Biophys. Acta* **637,** 291-302 (1981).

30 R. G. Prince and P. L. Dutton: in "*The Photosynthetic Bacteria,*" R. K. Clayton and W. R. Sistrom, Eds., Chapter 24, p. 439 (Plenum, New York 1978). A. W. Rutherford and C. W. Evans: *FEBS Lett.* **110,** 257 (1980). C. A. Wraight: *Israel J. Chem.* **21,** 384 (1981). D. Kleinfeld, M. Y. Okamura, & G. Feher: *Biochim. Biophys. Acta* **768,** 126 (1984) and (1985)(in press). P. Maroti and C. Wraight: *Biophys. J. (abstr.)* **47,** 5a (1985).

31 D. Kleinfeld, M. Y. Okamura, & G. Feher: *Biochemistry* **23,** 5780-5786 (1984).

Photoinduced Charge Separation in Bacterial Reaction Centers Investigated by Triplets and Radical Pairs

J.R. Norris, D.E. Budil, S.V. Kolaczkowski, J.H. Tang, and M.K. Bowman

Chemistry Division, Argonne National Laboratory, Argonne, IL 60439, USA

1. Introduction

A time domain of particular interest in the study of photosynthetic charge separation is from 5 to 100 ns after the initiation of primary photochemistry within the reaction center. This regime is especially significant to the magnetic resonance spectroscopist because, although charge separation has proceeded to a radical pair state, the electrons on the radicals remain close enough to interact with each other magnetically during this time. Such magnetic interactions largely govern the fate of the radical pair state when the charges recombine; however, the dynamic evolution of the state occurs too quickly to be directly observable with the EPR, ENDOR, or ESE (electron-spin echo) techniques. By using optically detected magnetic resonance (ODMR) and reaction yield-detected magnetic resonance (RYDMR), however, the time resolution necessary to observe short-lived paramagnetic states of the bacterial reaction center can be achieved, starting at about 5 ns.

In contrast to the EPR, ENDOR, and ESE methods, where a magnetic resonance spectrum is obtained by measuring absorption or emission of microwaves, the ODMR and RYDMR experiments rely upon the observation of an optical property of the system while it is being irradiated with resonant microwaves. For ODMR or RYDMR to be feasible, some spin-dependent branching pathway in the reaction scheme is required. In the case of the standard ODMR experiment (see, for example, references [1-6]) this branching occurs because each of the three zero-field spin sublevels of the triplet has a different decay-rate to the ground state. When resonant microwaves induce electrons occupying one spin sublevel to jump to another spin sublevel, the effective triplet lifetime changes. The amount of change in the triplet state is then detected by its optical absorption spectrum. Such experiments are performed at very low temperatures, ~2K, in order to avoid thermal equilibration of the spin sublevels which would "short-circuit" the microwave transitions. In our method, however, low temperatures are not required. Instead, the microwaves are sufficiently intense (1-8 kwatt) to cause spin state transitions at rates on the order of .1 ns^{-1}, which is competitive with many standard spin relaxation and triplet decay processes at room temperature.

The RYDMR version of ODMR [7-14] takes advantage of a different spin-dependent branching mechanism within the P^f state, namely radical pair intersystem crossing. The interpretation of P^f as a radical pair was initially founded on optical studies of the P^f state as well as standard EPR studies of the P^r triplet state [15-17]. According to the proposed mechanism, triplet molecules which are formed in blocked reaction centers originate from radical pairs which have their electronic spins correlated as a triplet. Thus, a microwave resonance which increases the population of the triplet sublevels of P^f should be detectable by measuring an increase in the optical absorbance of the P^r or carotenoid triplets which are formed after P^f. Since microwaves affect the observed yield of P^r and carotenoid triplets in the expected manner [12,13] the RYDMR experiment provides further, direct evidence of the radical pair nature of P^f.

A complementary type of magnetic experiment, which has been dubbed the MARY (magnetic field modulation of reaction yield) experiment [18], requires no oscillating magnetic fields. Instead, it depends upon the determination of the relative yield of triplet measured as a function of applied static magnetic field [19-24], an example of which appears in Figure 1. In other work, the applied field has been varied from near zero gauss to 50 kgauss [25,26].

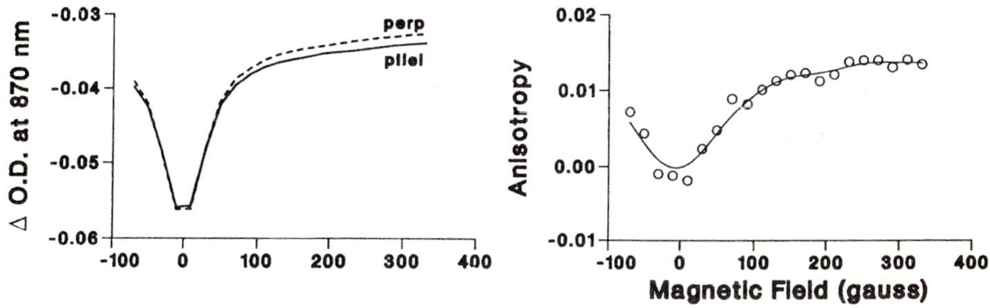

Figure 1. MARY spectra recorded with 870 nm light polarized parallel and perpendicular to the field direction with isotropic excitation. Right-hand curve: anisotropy parameter $(\Delta A_{\|} - \Delta A_{\perp})/(\Delta A_{\|} + 2\Delta A_{\perp})$ for each field.

We wish to illustrate the use of the RYDMR technique to study the magnetic resonance properties of P^f, P^r, and carotenoid triplets associated with the primary photochemical act of photosynthesis. We present evidence of a possibly new triplet state present during the primary stages of charge-separation which is detectable by room temperature ODMR.

2. Magnetic Anisotropy in the MARY Spectrum of P^f on a Nanosecond Timescale

Boxer et. al. [23,27] have reported orientation-dependence in the triplet yield in quinone-depleted reaction centers as measured at 870 nm on a microsecond timescale. A small anisotropy is observable at a 1 kgauss field, which they attribute to a electronic dipolar interaction or nuclear hyperfine anisotropy in the radical pair [27]. Theoretically, the observed dichroism at this low field could be accounted for entirely by hyperfine anisotropy, but this would require the maximal contribution of the hyperfine terms, which is an improbable case. The anisotropy measurements therefore suggest that some dipolar interaction (D) is present in P^f. However, Michel-Beyerle and co-workers are able to fit their nanosecond MARY results without inclusion of any D [24]; in addition, they have suggested that D must be very small in P^f, based upon calculations using the reported chromophoric structure in reaction center crystals of R. viridis [28].

One can rationalize these contradictory results because of the difference in timescale of their measurements. The orientation dependence of triplet yield measured on a microsecond timescale might be explained as an anisotropy in the rate of triplet decay as well as triplet formation. Anisotropic decay from a T_0-polarized triplet is to be expected if spin-orbit coupling contributes to the triplet decay mechanism. Chidsey et. al. have recently proposed a triplet intermediate with fast spin-orbit intersystem crossing [29]; the microsecond measurements cannot discount this as a source for triplet yield anisotropy at 1 kgauss. Thus, it is possible that the dichroism seen at low field does not reflect anisotropic interactions within P^f. In order to distinguish these two mechanisms and provide an independent determination of magnetic anisotropy in the radical pair, we have repeated the anisotropy measurements of Boxer at low (0-300 gauss) fields on a nanosecond timescale.

We demonstrate first that the photoselection we observe at 870 nm for P^f formation at zero field is comparable to previous measurements for $P^+ Q^-$ formation on a millisecond timescale [30]. The polarization ratio was measured using <15% saturating excitation pulses at 600 nm and 532 nm. The values measured immediately after a 5 ns laser flash are -.16 ± .02 and +.044 ± .002, for 600 nm and 532 nm, respectively. These numbers compare quite well to the ratios of -.15 and +.05 shown in Figure 4 of [30]. The polarization ratio decays to zero with a time-constant which depends upon buffer viscosity and detergent concentration; in aqueous buffer containing .03% LDAO at 298K, reaction centers undergo rotation with a correlation time of about 90 ns (data not shown).

The magnetic field contribution to dichroism at 870 nm can be determined by subtracting two photoselection measurements: one with the excitation light polarized parallel and one perpendicular to the direction of the field. Alternatively, the sample can be excited isotropically by rotating the excitation polarization to the magic angle relative to the field. Either method gives an equivalent result. Figure 1 shows the magnetically induced anisotropy measured 70 ns after the excitation pulse in a 60% ethylene glycol buffer at room temperature. The anisotropy increases from 0 at zero field to about .014 at saturating fields; a value somewhat less than the anisotropy reported by Boxer et. al at 1 kgauss on a microsecond timescale. Their observed anisotropy appears to be present at early times, confirming that it arises from anisotropic interactions in the radical pair rather than triplet decay processes.

3. RYDMR Detected by Carotenoid Triplets

In reaction centers which contain a carotenoid pigment, the first excited triplet of the carotenoid, car^T is seen in place of the state P^r which is observed in carotenoidless strains at room temperature [31]. A RYDMR spectrum of P^f can be detected in reaction centers of wild-type R. sphaeroides by observing either the 545 nm triplet-triplet absorption of the carotenoid or the bleaching of the ground state bands in the 420-490 nm region, as shown in Figure 2. Measurement at any characteristic car^T wavelength yields a spectrum equivalent to the P^r-detected RYDMR in reaction centers from the carotenoidless R-26 mutant [12]. This result lends support to the idea that car^T does not proceed directly from triplet P^f, since reaction rates large enough to make car^T formation predominant would result in significant changes in the width and microwave power-dependence of the RYDMR signal from those observed in R-26. Since the radical pair spectrum appears unaltered by the presence of the carotenoid, reaction centers which contain a carotenoid have proved particularly useful for RYDMR spectroscopy of P^f owing to the large, easily detected absorbance changes of that pigment.

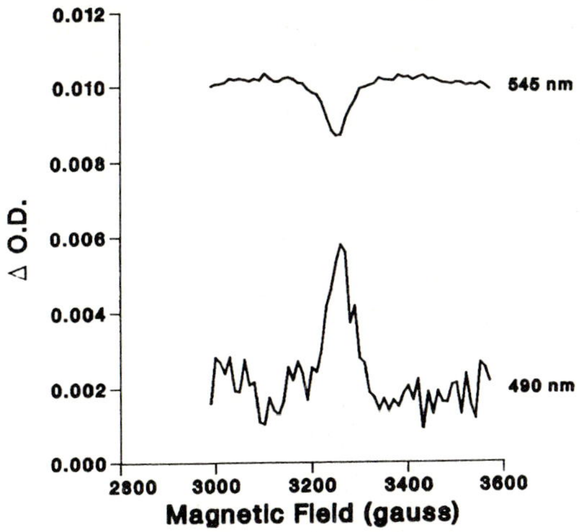

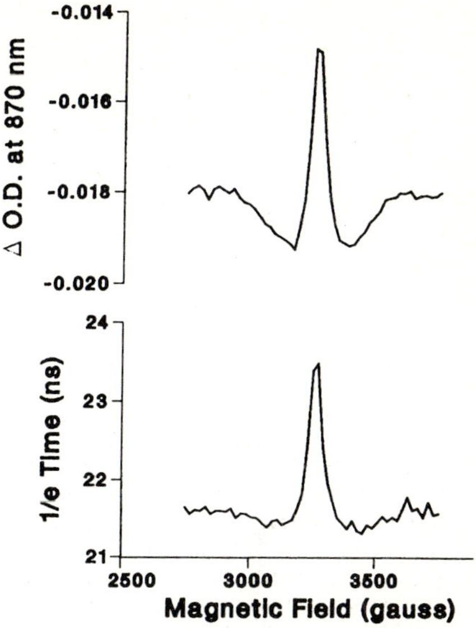

Figure 2. RYDMR detected at the 490 nm bleaching and the 545 absorption of carotenoid triplet in wild-type R. sphaeroides reaction centers.

Figure 3. Effect of high-power microwaves on 870 nm absorption and radical pair lifetime in quinone-depleted R-26 reaction centers.

4. Other Triplet States

In addition to the narrow (30 gauss) microwave absorption of the radical pair, RYDMR spectra frequently exhibit a broad background signal about 400 gauss wide [13]. This signal can be demonstrated not to arise from the radical pair. When P^f lifetime is plotted as a function of field [32], there is a feature which corresponds to the narrow microwave resonance, as shown in Figure 3. This is a result of the unequal decay rates for singlet and triplet P^f; when resonant microwaves populate the fast-decaying triplet states, the observed P^f lifetime decreases. There is no feature in the P^f lifetime spectrum which corresponds to the background signal, which suggests that it does not come from P^f. Furthermore, although the narrow signal disappears when the microwave pulse is delayed beyond the decay of P^f, the background signal remains [13]. Thus, the microwaves must be modulating a reaction or decay pathway, which either succeeds P^f in the kinetic scheme, or is a slower side reaction of photoexcitation.

The width of the background signal is suggestive of a bacteriochlorophyll triplet molecule; indeed, at high microwave powers, it is very near the value of $|2D|$ measured for P^r [17]. Boxer has recently proposed a mechanism for P^r decay at room temperature [33] which could in principle account for an ODMR spectrum from P^r. The spin-dependent branching in this mechanism occurs because only the T_0 spin state of the P^r triplet can decay to ground state by thermal repopulation of P^f, assuming slow spin-lattice relaxation. However, there is some evidence, presented in Figures 4 and 5, which suggests that P^r is not the source of the background signal.

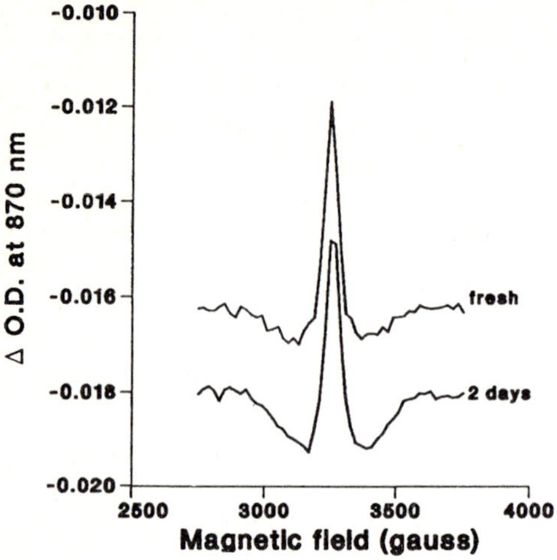

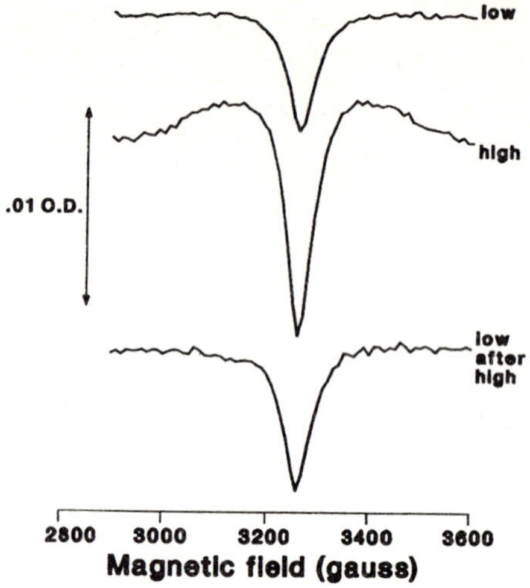

Figure 4. Appearance of background signal with sample age in R-26 quinone-depleted reaction centers under equivalent conditions of illumination and microwave irradiation.

Figure 5. Background signal observed in wild-type reaction centers as a function of actinic light flux, showing a reversible appearance at excitation intensities near saturating.

The broad signal does not always appear under experimental conditions where P^r is present. More specifically, it is absent in fresh quinone-depleted samples excited at low laser intensities, and appears irreversibly with sample age (Figure 4) and reversibly with high intensities of excitation light (Figure 5). This would indicate that P^r per se is not responsible for the background signal; however, the primary donor is implicated to some extent because the background is detectable at the 870 nm bleaching.

The background signal is observed in RYDMR of carotenoid-containing strains of R. sphaeroides as well as the R-26 mutant. The optical difference spectrum of the broad background observed in wild-type reaction centers closely parallels that of car^T; otherwise it resembles the background observed from R-26 in its width, microwave power-dependence, requirement for intense actinic light, and its persistence when the microwaves are delayed after P^f decay. The signal appears to arise from some additional triplet state which is present in R-26 reaction centers, but quenched by the carotenoid in wild-type. It seems unlikely that this triplet is P^r, in light of the rapid rate at which P^r is presumed to transfer its excitation energy to car^T at room temperature [34]. Such a fast P^r decay would compete very strongly with any microwave transitions within the P^r manifold, and the signal intensity and power dependence would be expected to be dramatically different for the carotenoid-containing reaction center.

We propose that the background signal is a room-temperature ODMR signal from a triplet state which has not previously been observed in photoexcited reaction centers. Although the identity of the triplet is unclear, the conditions required to observe it are suggestive. The dependence on high light flux could indicate some multiply excited state of the reaction center, but it is also possible that the triplet is formed in damaged or altered reaction centers. The signal might arise either from alteration of the decay rate of a triplet which is subsequently trapped by P^r or car^T, or by microwave inhibition of triplet-triplet annihilation processes [35,36] in multiply excited reaction centers. To our knowledge, this is the first triplet ODMR spectrum recorded at room temperature.

If the observed background signal reflects the presence of an additional sample-dependent triplet in reaction centers, it affords a reasonable explanation for the variability of absolute triplet quantum yield measurements among different laboratories [24,29,37] and within the same laboratory [18,24,29]. Chidsey et. al. invoked such a triplet state to explain the disparity of their quantum yield measurements with theoretical values [29]. The width of the background signal could be useful in ruling out some of the possibilities these authors suggest for the identity of that triplet state.

It should be emphasized that, since the broad ODMR signal does not appear in fresh samples, additional triplets may not be present in the de-excitation of "native" reaction centers. Alternatively, damage to the reaction center could modify the decay processes of a "native" triplet intermediate, rendering it visible by ODMR. In either case, the time-resolved ODMR technique provides a more rigorous assay for the integrity of a given reaction center preparation, and may prove useful in resolving discrepancies in quantum yield measurements.

5. Acknowledgement

This work was supported by the Division of Chemical Sciences, Office of Basic Energy Sciences, Department of Energy under contract W-31-109-Eng-38.

6. References

1. W. G. van Dorp, W. H. Schoemaker, M. Soma and J. H. van der Waals, "The Lowest Triplet State of Free Base Porphyrin. Determination of Its Kinetics of Populating and Depopulating from Microwave-Induced Transients in the Fluoresence Intensity", Mol. Phys. **30**, 1701-1721 (1975).
2. R. H. Clarke, R. E. Connors, J. R. Norris and M. C. Thurnauer, "Optically Detected Zero-Field Magnetic Resonance Studies of the Photoexcited Triplet State of the Photosynthetic Bacterium Rhodospirillum rubrum", J. Am. Chem. Soc. **97**, 7178-7179 (1975).
3. R. H. Clarke and R. E. Connors, "Optically Detected Zero-Field Triplet State Magnetic Resonance in Photosynthetic Bacteria", Chem. Phys. Lett. **42**, 69-72 (1976).
4. A. J. Hoff, "Kinetics of Populating and Depopulating of the Components of the Photoinduced Triplet State of the Photosynthetic Bacteria Rhodospirillum rubrum, Rhodopseudomonas sphaeroides (Wild Type) and its Mutant R-26 as measured by ESR in Zero Field", Biochim. Biophys. Acta **440**, 765-771 (1976).
5. A. J. Hoff and H. G. de Vries, "Electron Spin Resonance in Zero Magnetic Field of the Reaction Center Triplet of Photosynthetic Bacteria", Biochim. Biophys. Acta **503**, 94-106 (1978).
6. H. J. den Blanken and A. J. Hoff, "High-Resolution Optical Absorption-Difference Spectra of the Triplet State of the Primary Donor in Isolated Reaction Centers of the Photosynthetic Bacteria Rhodopseudomonas sphaeroides R-26 and Rhodopseudomonas viridis Measured with Optically Detected Magnetic Resonance at 1.2K", Biochim. Biophys. Acta **681**, 365-374 (1982).

7. Y. Ruedin, P. A. Schnegg, C. Jaccard and N. Aegerter, "EPR Optical Detection of F Centre Pairs in Alkali Halides I. Pumping Cycle Kinetics and Characteristics of the Resonances", Phys. State Sol. (b) **54**, 565-576 (1972).
8. E. L. Frankevich and A. I. Pristupa, "Magnetic Resonance of Excited Complexes with Charge-Transfer Revealed by Fluorescence at Room Temperature", Soviet Physics. JETP Lett. **24**, 362-365 (1976).
9. O. A. Anisimov, V. M. Grigoryants, V. K. Molchanov and Yu. N. Molin, "Optical Recording of the EPR Signal of Short-Lived Radical Pairs at Superlow Concentrations", Dokl. AN SSSR **248**, 380-382 (1979).
10. O. A. Anisimov, V. M. Grigoryants, V. K. Molchanov and Yu. N. Molin, "Optical Detection of ESR Absorption of Short-Lived Ion-Radical Pairs Produced in Solution by Ionizing Radiation", Chem. Phys. Lett. **66**, 265-268 (1979).
11. Yu. N. Molin, O. A. Anisimov, V. M. Grigoryants, V. K. Molchanov and K. M. Salikhov, "Optical Detection of ESR Spectra of Short-Lived Ion-Radical Pairs Produced in Solution by Ionizing Radiation", J. Phys. Chem. **84**, 1853-1856 (1980).
12. M. K. Bowman, D. E. Budil, G. L. Closs, A. G. Kostka, C. A. Wraight and J. R. Norris, "Magnetic Resonance Spectroscopy of the Primary State, P^f, of Bacterial Photosynthesis", Proc. Natl. Acad. Sci. USA **78**, 3305-3307 (1981).
13. J. R. Norris, M. K. Bowman, D. E. Budil, J. Tang, C. A. Wraight and G. L. Closs, "Magnetic Characterization of the Primary State of Bacterial Photosynthesis", Proc. Natl. Acad. Sci. USA **79**, 5532-5536 (1982).
14. M. R. Wasielewski, J. R. Norris and M. K. Bowman, "Time-Domain Magnetic Resonance Studies of Short-lived Radical Pairs in Liquid Solution", Faraday Discuss. Chem. Soc. **78**, 279-288 (1984).
15. W. W. Parson and R. J. Cogdell, "The Primary Photochemical Reaction of Bacterial Photosynthesis", Biochim. Biophys. Acta **416**, 105-149 (1975).
16. J. Fajer, D. C. Brune, M. S. Davis, A. Forman and L. D. Spaulding, "Primary Charge Separation in Bacterial Photosynthesis: Oxidized Chlorophylls and Reduced Pheophytin", Proc. Natl. Acad. Sci. USA **72**, 4956-4960 (1975).
17. M. C. Thurnauer, J. J. Katz and J. R. Norris, "The Triplet State in Bacterial Photosynthesis: Possible Mechanisms of the Primary Photo-Act", Proc. Natl. Acad. Sci. USA **79**, 3270-3274 (1975).
18. W. Lersch, A. Ogrodnik and M. E. Michel-Beyerle, "On the Influence of Microwaves and Static Magnetic Fields on the Recombination of Radical Ions in Reaction Centers of Photosynthetic Bacteria", Z. Naturforsch. **37a**, 1454-1456 (1982).
19. R. E. Blankenship, T. J. Schaafsma and W. W. Parson, "Magnetic Field Effects on Radical Pair Intermediates in Bacterial Photosynthesis", "Biochim. Biophys. Acta" **461**, 297-305 (1977).
20. H.-J. Werner, K. Schulten and A. Weller, "Electron-Transfer and Spin Exchange Contributing to the Magnetic Field Dependence of the Primary Photochemical Reaction of Bacterial Photosynthesis", Biochim. Biophys. Acta **502**, 255-268 (1978).
21. R. Haberkorn, M. E. Michel-Beyerle and R. A. Marcus, "On Spin-Exchange and Electron-Transfer Rates in Bacterial Photosynthesis", Proc. Natl. Acad. Sci. USA **76**, 4185-4188 (1979).
22. A. J. Hoff, "Magnetic Field Effects on Photosynthetic Reactions", Quart. Rev. Biophys. **14**, 599-665 (1981).
23. S. G. Boxer, C. E. D. Chidsey and M. G. Roelofs, "Magnetic Field Effects on Reaction Yields in the Solid State: An Example from Photosynthetic Reaction Centers", Ann. Rev. Phys. Chem. **34**, 389-417 (1983).

24. A. Ogrodnik, H. W. Kruger, H. Orthuber, R. Haberkorn and M. E. Michel-Beyerle, "Recombination Dynamics in Bacterial Photosynthetic Reaction Centers", Biophys. J. **39**, 91-99 (1982).
25. C. E. D. Chidsey, M. G. Roelofs and S. G. Boxer, "Effect of Large Magnetic Fields and g-Factor Differences on Triplet Population in Photosynthetic Reaction Centers", Chem. Phys. Lett. **74**, 113-117 (1980).
26. S. G. Boxer, C. E. D. Chidsey and M. G. Roelofs, "The Use of Large Magnetic Fields to Probe Photoinduced Electron-Transfer Reactions: An Example from Photosynthetic Reaction Centers", J. Am. Chem. Soc. **104**, 1452-1454 (1982).
27. S. G. Boxer, C. E. D. Chidsey and M. G. Roelofs, "Anisotropic Magnetic Interactions in the Primary Radical Ion-Pair of Photosynthetic Reaction Centers", Proc. Natl. Acad. Sci. USA **79**, 4632-4636 (1982).
28. J. Deisenhofer, O. Epp, K. Miki, R. Huber, and H. Michel, "X-ray Structure Analysis of a Membrane Protein Complex. Electron Density Map at 3Å Resolution and a Model of the Chromophores of the Photosynthetic Reaction Center from Rhodopseudomonas viridis", J. Mol. Biol. **180**, 385-398 (1984).
29. C. E. D. Chidsey, C. Kirmaier, D. Holten, and S. G. Boxer, "Magnetic Field Dependence of Radical-Pair Decay Kinetics and Molecular Triplet Quantum Yield in Quinone-Depleted Reaction Centers", Biochim. Biophys. Acta. **766**, 424-437, (1984).
30. A. Vermeglio, J. Breton, G. Paillotion, and R. Cogdell, "Orientation of Chromophores in Reaction Centers of Rhodopseudomonas sphaeroides: A Photoselection Study", Biochim. Biophys. Acta, **501**, 514-530 (1978).
31. R. F. Cogdell, T. G. Monger, and W. W. Parson, "Carotenoid Triplet States in Reaction Centers from Rhodopseudomonas sphaeroides and Rhodospirillum rubrum", Biochim. Biophys. Acta, **408**, 189-199 (1975).
32. M. R. Wasielewski, C. H. Bock, M. K. Bowman and J. R. Norris, "Controlling the Duration of Photosynthetic Charge Separation with Microwave Radiation", Nature **303**, 520-522 (1983).
33. S. G. Boxer, C. E. D. Chidsey, L. Takiff, and R. Goldstein, "Magnetic Field Effects on the Lifetime of the Triplet State in Photosynthetic Reaction Centers: Evidence for Thermal Re-population of the Primary Radical Pair", Biophys. J., **45**, 215a (1984).
34. C. C. Schenck, P. Mathis, and M. Lutz, "Triplet Formation and Triplet Decay in Reaction Centers from the Photosynthetic Bacterium Rhodopseudomonas sphaeroides", Photochem. Photobiol., **39**, 407-417 (1984).
35. W. Steudle, and J. U. von Schuetz, "Phosphorescence and Delayed Fluorescence Detected Magnetic Resonance at Zero Field on the 1:1 CT-Crystal Anthracene/TCNB", J. Lumin. **18/19**, 191-195 (1979).
36. J. U. von Schuetz, W. Steudle, H. C. Wolf, and V. Yakhot, "Time-Correlated Delayed Fluorescence ODMR. Dynamics of Triplet Exciton Annihilation in Anthracene-tetracyanobenzene Crystals", Chem. Phys., **46**, 53-65 (1980).
37. C. C. Schenck, R. E. Blankenship and W. W. Parson, "Radical-Pair Decay Kinetics, Triplet Yields and Delayed Fluorescence from Bacterial Reaction Centers", Biochim. Biophys. Acta, **680**, 44-59 (1982).

Spin Dipolar Interactions of Radical Pairs in Photosynthetic Reaction Centers

A. Ogrodnik, W. Lersch, and M.E. Michel-Beyerle

Institut für Physikalische und Theoretische Chemie, Technische Universität München, D-8046 Garching, F. R. G.

J. Deisenhofer and H. Michel

Max-Planck-Institut für Biochemie, D-8033 Martinsried, F. R. G.

INTRODUCTION

The optical detection of recombination dynamics and its dependence on external magnetic fields gives access to kinetic and structural features of short-lived radical pairs (RPs). In reaction centers (RCs) of purple bacteria this RP consists of the cation of the bacteriochlorophyll dimer, $(BC)_2^{\pm}$ and the anion of the bacteriopheophytin, BP^- and is formed within a few picoseconds by electron-transfer from the singlet excited dimer state. In quinone-depleted RCs at room temperature the RP is stabilized for approximately 10 ns. The initially formed RP is singlet-phased as its precursor state and recombines via the rate k_S to form the ground state species. This is illustrated in the following kinetic scheme:

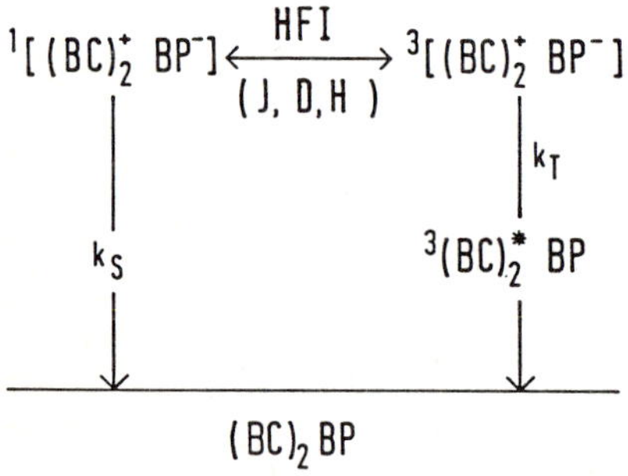

Hyperfine interaction HFI of the two electron spins with their different nuclear spin environments can change the multiplicity of the RP spin state. This leads to a RP triplet state, thus providing a new channel of recombination forming the dimer triplet state with the rate k_T. Such singlet-triplet transitions, however, can be hindered by exchange and spin dipolar interaction (J and D) between the radical electrons as well as by the recombination rates, which are responsible for the lifetime broadening of the RP states.

Two methods are suited to provide values for these structural and kinetic parameters. Both make use of the fact that the extent of singlet-triplet mixing, and hence the relative amounts of singlet and triplet recombination products, can be manipulated by external magnetic fields. The magnetic field shifts the triplet levels due to Zeeman interaction, and thus changes the hyperfine induced coupling of each of them with the singlet state. The first method therefore studies the influence of the magnetic field on the recombination yields, and will be referred to as MARY (MAgnetic field modulation of Reaction Yields). At strong static fields when the triplet states are split off far enough and two of them are completely isolated from the singlet state, one can reconnect these again by application of microwaves under specific resonance conditions. Besides tracing the resonance feature by changing the magnetic field, one can alter the extent of coupling by varying the microwave power. In the second method these effects again are monitored on the recombination yields, hence it is called Reaction Yield Detected Magnetic Resonance (RYDMR). Both methods are equivalent in principle /1/. In MARY- as well as in RYDMR-experiments time-resolved laser spectroscopy may be utilized to probe the time evolution of the triplet yield and of the decaying RP concentration.

Time-resolved absorption data on the decay of RPs and the rise of triplets $^3(BC)_2^*$ together with the magnetic field dependence thereof have been accumulated for quinone-depleted RCs of Rp. sphaeroides R26 and rigorously analysed /2/. The experimental quantities could be fitted with the two rates k_S and k_T and the exchange interaction J without any further parameters i.e. spin dipolar interaction. From MARY and RYDMR experiments other groups have attempted to extract D, yielding values of $|D|$=50G /3/ and $|D|$=75G /4/. Taking these large D values into account, close face-to-face positions of the RP constituents have been assumed. Comparison of data acquired in different laboratories is somewhat difficult due to the pronounced sensitivity of RP recombination dynamics to high light intensities, and also due to preparative conditions /5/.

It is the intention of this paper

i) to calculate the dipolar interactions of radical pairs in RCs of Rp. viridis on the basis of the x-ray structure data /6/ and

ii) to discuss the implications of this interaction on MARY spectra.

Calculation of spin dipolar interaction from crystal structure data

In a very simple model - certainly appropriate for distant radical pairs - we calculate the spin dipolar interaction tensors /7/ by averaging over the individual dipolar contributions of spin pairs localized at the various nuclear positions of the π-systems weighted by their spin density /8-12/. The tensor is then diagonalized yielding eigenvalues X,Y,Z. Its eigenvectors determine the orientation of the charge distribution relative to the RC. Because the tensor is traceless, the interaction can be expressed by the zero-field splitting parameters $D = 3/2 \cdot Z$ and the $E = (X-Y)/2$. The magnitude of D essentially refers to the average of the inverse cube of the radius of the charge distribution, and to its deviation from spherical symmetry. Positive values result from oblate, negative ones from prolate charge distributions. E quantifies the deviation from axial symmetry.

Fig.1 shows the D-values of various pairs of radicals. To test the sensitivity to the exact spin distribution we also calculated these values assuming constant spin density throughout the π-systems of the molecules. The deviations were less than 10 %. In the case of $(BC)_2^+BC^-$ direct interactions could perturb the assumed spin density distribution significantly. At the same time, the changes of $1/r^3$ when integrating over all nuclear coordinates are by far strongest in this case.

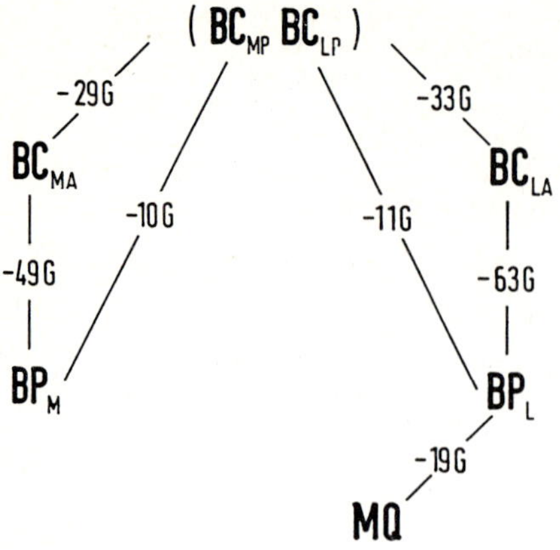

Fig.1: Calculated spin dipolar interaction parameter D for radical pairs formed of the various constituents of the RC: BC Bacteriochlorophyll, BP Bacteriopheophytin, MQ Menaquinone. Indices L and M refer to pigments near the L and M protein chains respectively. Index P assigns the special pair, A the accessory pigments.

Thus, somewhat larger deviations might have to be tolerated. The values of interest, because of their implication on experimental data, however, are D = -11 G for $(BC)_2^+BP^-$ and D = -19 G for BP^-MQ^-.

In case, structural similarity holds for RCs of Rp.viridis and Rp. sphaeroides R26, the small value for the RP $(BC)_2^+BP^-$ is in contradiction to the propositions, 50 G /3/ and 75 G /4/. From EPR-measurements on MQ reconstituted in RCs of Rp.sphaeroides R26 and from the temperature-dependence of the electron-transfer rate from BP to the MQ a value for the dipolar interaction in $BP^- MQ^-$ of |D|=10 G has been estimated /13/. The difference between this value and the calculated one is within accuracy of estimate. It could also reflect different binding properties for menaquinone in Rp. viridis and Rp. sphaeroides.

Finally, we want to mention that E values for all pairs are extremely low (<<0.5G), as one would expect for radical pairs. Even in the case of nearest neighbours $(BC)_2^+BC^-$ E was less than 10% of D. From analysis of MARY-data, it has been concluded, that E should be on the order of the maximum possible value E=1/3·D /14/. Such a deviation from axial symmetry certainly does not apply to any reasonable RP configuration in RCs, and above conclusion has been withdrawn /4/, since it was based on the wrong assumption that $k_S = k_T$.

Can spin dipolar interaction be extracted from MARY data?

In order to gain some insight on how spin dynamics are influenced by spin dipolar interaction, we performed simulations of MARY-spectra. Computations of MARY-spectra allowing for spin dipolar interaction have to include averaging over all orientations of RCs relative to the magnetic field. Running a free fit of all experimental data with a complete set of parameters, including dipolar interaction, is therefore very time consuming and has not been carried out yet. For the simulations in a two proton model /15/, we have adopted the following generally accepted set of parameters: k_S=0.06 ns^{-1}, k_T=0.6 ns^{-1}, J=+/-16G, and the hyperfine coupling constants: A_1=9 G and A_2=13 G.

In Fig. 2 experimentally accessible quantities are depicted for various values of the zero-field splitting parameter D:

a) The triplet yield at zero-magnetic field is constant up to D-values of -20 G, before it decreases with further splitting of the triplet sublevels. For an axially symmetric dipole tensor with E=0 there are two degenerate triplet levels T_x and T_y with spacing $\Delta E=J+D/3$ from the singlet level and a third triplet level T_z with the spacing $\Delta E=J-2/3\cdot D$ (for clarity, additional levelshifts due to hyperfine interaction are not discussed). The mixing efficiency of the states coupled by hyperfine interaction is reduced proportionally to $1/\Delta E^2$.

In case the signs of J and D are opposite and for small values of D, dipolar interaction shifts the pair of degenerate levels T_x and T_y closer to the singlet, while the loan level T_z is shifted away. Thus the resulting decrease in mixing efficiency of the singlet state with T_z is overcompensated by the increased coupling with the pair T_x, T_y. An expected rise in triplet yield with D, however, does not occur, because of the lifetime broadening of the levels. In case, J and D have the same signs, T_z approaches the singlet level instead. At larger D, relevant triplet levels cross the singlet level and eventually split off totally. T_z moves twice as fast with increasing D as T_x and T_y. Consequently, in case J and D have the same sign, the triplet yield starts to decrease already at lower values of D. Alternatively a decrease of triplet yields can result from splitting due to exchange interaction or by the value of k_T relative to both k_S and the hyperfine interaction. One should bear in mind however, that for sufficiently large values of k_T (in this case for $k_T > 1.0\ ns^{-1}$) the triplet yield starts to decrease again with k_T further increasing /16/.

b) In Fig.2b the ratio of the triplet yield at 600G and in zero field is shown in its dependence on D. The magnetic field causes Zeeman splitting of the triplet sublevels. For magnetic fields large compared to the hyperfine interaction, only the T_0 state can mix with the singlet state. The energy splitting between these two states now depends on the orientation of the dipole tensor axis relative to the magnetic field. For E=0 the splitting is $\Delta E = J-(\cos^2\theta - 1/3)\cdot D$. θ is the angle between the z-axis of the dipole tensor and the magnetic field. For random orientation of RCs, as has been assumed, this splitting spreads from $J-2\cdot D/3$ to $J+D/3$. While at high D values the triplet yield at zero-field is reduced, there always is a certain portion of RCs oriented in such a way as to be in resonance at high fields, if D is large enough compared to J. The modulation then becomes less pronounced. Similarly, weaker modulation will occur /16/, when kinetic conditions favour the triplet channel as at low temperature /17/, while this time the triplet yield rises. These counteracting trends of D and kinetics on the relative behaviour of the triplet yield and its modulation, however are obscured by the influence of J. Time resolved measurements of these complex, time integral quantities might give additional information.

c) The magnetic field dependence of the triplet yield shows an unresolved level crossing resonance structure at small fields. The half width at half maximum of this feature has a weak dependence on D up

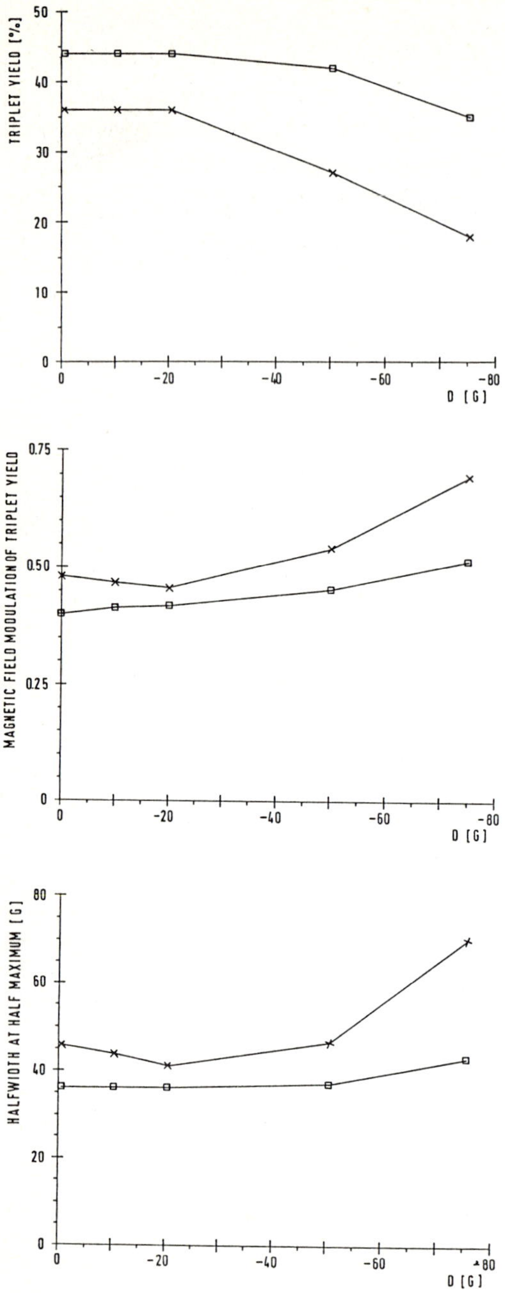

Fig.2:

a) Triplet yield,

b) Magnetic field modulation of the triplet yield,

c) Halfwidth at half maximum of the magnetic field dependence of the triplet yield

calculated for increasing values of D, 100 ns after excitation.

□ J=+16 G, x J=-16G.

to D=-50G as shown in Fig.2c. In this regime the halfwidth is essentially determined by the lifetime broadening of the triplet levels, as a consequence of the high rate k_T.

The orientation dependence of the S-T-splitting at high magnetic fields leads to an orientation dependence of the triplet yield. In an absorption measurement using polarized light to detect the triplet yield, a photoselective anisotropy therefore is expected. The triplet induced absorption change with the polarization of probing and excitation beam parallel to the magnetic field ($\Delta A_{\parallel}$) is different from the one with the polarizations perpendicular to the field ($\Delta A_{\perp}$). The anisotropy may be quantified by $a = (\Delta A_{\parallel} - \Delta A_{\perp}) / (\Delta A_{\parallel} + 2\cdot\Delta A_{\perp})$.

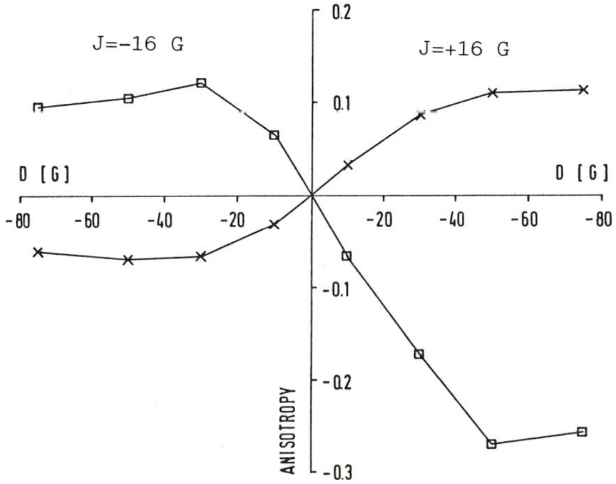

Fig.3: Anisotropy of the triplet yield for various values of D. On the right-hand side J=+16 G, on the left-hand side J=-16G.
□ : $\alpha = 0°$, ×: $\alpha = 90°$. Magnetic field = 600 G.

Calculated values of this anisotropy with increasing D are shown in Fig.3. Of course it depends on the orientation of the transition dipole moments responsible for excitation and probing with respect to the triplet axis. Here it was assumed that probing and excitation occurs at the same transition. The angle between this transition moment and the triplet z-axis is α. The sign of the anisotropy value is different for $\alpha = 0°$ and $\alpha = 90°$. For angles in between the anisotropy decreases in magnitude. With this set of parameters it approaches zero at $\alpha = 62°$. This deviates somewhat from the magic angle, because the triplet yield is not related to the projections of the transition moments on the triplet z-axis in a simple way.

An anisotropy value of 0.035 at 1000 G has been measured for RCs of <u>Rp.sphaeroides</u> R26 /4/. Because of the angular dependence of the anisotropy, one can only deduce a lower limit for the magnitude of D, as long as α is unknown. Fig. 3 suggests this limit to be less than 10G. For <u>Rp.viridis</u> one can obtain an angle of α=72° for the 960 nm transition from the structure data. Together with D=-10G a calculated anisotropy value of 0.023 would be compatible with that measured for Rp. sphaeroides R26.

Most interestingly, the sign of the anisotropy is related to the relative signs of D and J. For instance, if D and J have the same sign, and if transition moments and triplet z-axis are parallel, the anisotropy must be positive. Assuming angles α for Rp.sphaeroides to be larger than 62°, similarly to Rp.viridis, this would implicate a positve exchange interaction in the RP $(BC)_2^+BP^-$ of this species, as indeed has been proposed in /4 /. Efforts to estimate the sign of J from RYDMR and MARY data without considering anisotropy, however, led to contradictory results /3/.

In our calculations, anisotropic contributions of the hyperfine interaction have been neglected. They can be discerned by time resolved measurements /18/.

The kinetic properties, in particular the high recombination rate k_T in the triplet channel of the RP-state in photosynthetic RCs dominate the magnetic field-dependence of various observables: any possible splittings of states (D and J) are obscured by lifetime broadening of the RP states. Fig. 4 demonstrates that well-resolved level crossing resonances can be expected in principle by MARY, provided that suffitiently slow kinetics allow the necessary resolution. In the less favourable case we have to deal with in actual RCs the dipolar interaction tensor and its orientation contribute together with various other kinetic and interaction parameters in a rather complex way to the various observables of magnetic field-dependent measurements on the RP state and its products.

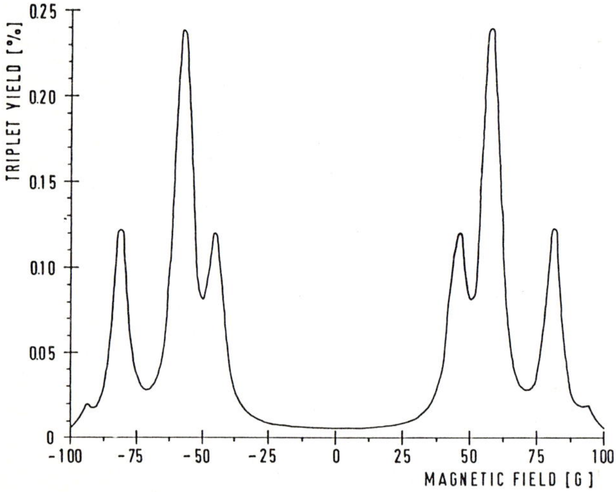

Fig.4: Simulated MARY-Spectrum at 1 μs after excitation. Set of parameters:
A_1=1G, A_2=2G, k_S=0.001 ns^{-1}; k_T=0.01 ns^{-1}; J=-70G, D=-50G, E=-10G

Conclusions

Since the dipolar interaction of RPs allows to infer on the distance of the radical constituents, it would be attractive to determine D in MARY or RYDMR experiments. Is such an undertaking feasible? Although orientational anisotropy appears as a new observable, when dipolar interaction is operative, it can only utter a lower limit of $|D|$. For values of D small compared to the rates (i.e. $|D| < 34$ G for the set of parameters used in Fig. 2) the observables show almost no dependence on D and there seems to be no chance of its determination. D grows in magnitude as the radicals come closer. However, the rates even increase faster with distance, after the radicals have come close enough. Thus, there might be an intermediate range of D-values, for which it is possible to extract D for a certain combination of rates and J, from a complete set of time-resolved data including anisotropic measurements. The answer to the question above can therefore not be given in a general way, but must be obtained by careful analysis of the behaviour of an all-parameter free fit for each individual case. However, it should be kept in mind that further complications might arise from

i) a mechanism of recombination involving hopping between various sites /19/. The structure data for RCs of Rp.viridis, at least, suggest the BC monomer to act as an intermediate electron acceptor.

ii) perturbations on BP by the presence of reduced quinone. In RCs of Rp.viridis, where the menaquinone on the active branch is too tightly bound to be removed, the RP has to be stabilized by blocking electron transfer via reduction of the menaquinone.

From the nuclear coordinates obtained in x-ray structure analysis, it is easy to calculate the dipolar interaction D. Such an independent determination of D is especially favourable in view of the complexity of MARY- and RYDMR-experiments, since it helps to increase the precision of the obtainable exchange interaction and spin-dependent recombination rates. Both of these quantities are suited to elucidate the mechanism responsible for slow recombination, the necessary condition for high efficiency of charge separation.

Acknowledgement

This work was supported by the Deutsche Forschungsgemeinschaft, SFB 143.

References:

/1/ Lersch W., M.E. Michel-Beyerle. 1983. Chem.Phys.Lett. 78: 115

/2/ Ogrodnik A., H.W. Krüger, H. Orthuber, R. Haberkorn, M.E. Michel-Beyerle, H. Scheer. 1982. Biophys.J. 39: 91

/3/ Norris J.R., M.K. Bowman, D.E. Budil, J. Tang, C.A. Wraight, G.L. Closs. 1982. Proc. Natl. Acad. Sci. USA 79: 5532

/4/ Chidsey C.E.D., C. Kirmaier, D. Holten, S. Boxer. 1984. Biochim. Biophys.Acta 766: 424

/5/ Lersch W., A. Ogrodnik, M.E. Michel-Beyerle. in: Photochemistry and Photobiology ed. A.H. Zewail. Vol 2 p.986. Harwood academic publishers. 1983.

/6/ Deisenhofer J., O. Epp, K. Miki, R. Huber, H. Michel. 1982. J.Mol.Biol. 180:385

/7/ Carrington A., A.D. McLachlan. Introduction to Magnetic Resonance. p.116. Chapman and Hall, 1980.
/8/ Lendzian F., Thesis 1982. Freie Universität Berlin
/9/ Petke J.D., G.M. Maggiora, L.L. Shipman, R.E. Christoffersen. 1981. Photochem.Photobiol. 31:243
/10/ Petke J.D., G.M. Maggiora, L.L. Shipman, R.E. Christoffersen. Photochem.Photobiol. 33: 663
/11/ Petke J.D., G.M. Maggiora, L.L. Shipman, R.E. Christoffersen. Photochem.Photobiol. 32: 661
/12/ Matsunaga Y., 1960. Bull.Chem.Soc.Jap. 33: 1436
/13/ Okamura M.Y., R.A. Isaacson, G. Feher. 1979. Biochim.Biophys.Acta 546: 394
/14/ Boxer S.G., C.E.D. Chidsey, M.G. Roelofs. 1982. Proc.Nat.Acad.Sci. USA 79: 4632
/15/ Lersch W., Diplomarbeit 1983. Technische Universität München
/16/ Haberkorn R., M.E. Michel-Beyerle. 1979. Biophys.J. 26: 489
/17/ Ogrodnik A. 1983. Thesis. Technische Universität München.
/18/ Lersch W., M.E. Michel-Beyerle. 1984. Chem.Phys.Lett. 107: 522
/19/ Haberkorn R., M.E. Michel-Beyerle, R.A. Marcus. 1979. Proc.Nat. Acad.Sci.USA 76: 4185

Protein/Lipid Interaction of Reaction Center and Antenna Proteins

J. Riegler, W.M. Heckl, J. Peschke, M. Lösche, and H. Möhwald

Physik-Department, E 22, Technische Universität München, D-8046 Garching, F. R. G.

1. Introduction

Being a membrane-bound process photosynthesis depends on the functional interaction of membrane-bound or associated proteins [1]. For molecules of the photosynthetic membrane carrying out the primary process of energy and electron-transfer it is therefore essential to study protein/lipid and lipid mediated protein/protein interaction.

A meaningful study of these interactions requires the establishment of well-defined model systems. We therefore built up model membranes of synthetic phospholipids into which the antenna proteins B800-850 and the reaction center of R. sphaeroides 2.4.1 were reconstituted. These proteins offer the advantages that their structure is rather well known and that the structure/function relationship can be studied kinetically via optical techniques.

Information on the interactions occurring predominantly in the polar head group region of the membrane are expected from measurements on monolayers at the air/water interface. Therefore,these are investigated by thermodynamical, fluorescence microscopic and spectroscopic measurements.

To conclude on interactions within the complete bilayer membrane, especially on elastic forces arising from a mismatch in the hydrophobic parts of protein and phospholipid membrane, vesicles of phosphatidyl cholines with varying chain lengths were used. These contained reaction centers in different amounts and were studied thermodynamically, optically and by electron microscopy.

2. Experimental

Reaction centers (RC) and the B800-850 light-harvesting complex (LHCP) were grown and isolated in the laboratory of Prof. H. Scheer. The lipids DL-α-dilauryl phosphatidic acid (DLPA), DL-α-dimyristoyl phosphatidic acid (DMPA), DL-α-dilaurylphosphatidylcholine (DLPC), DL-α-dimyristoylphosphatidylcholine (DMPC) and DL-α-dipalmitoylphosphatidylcholine (DPPC) were purchased from Fluka, Buchs, CH. DL-α-dipentadecanoylphosphatidylcholine (DPePC) and the fluorescent dye DL-α-dipalmitoyl-nitrobenzoxadiazol-phosphatidylethanolamine (DP-NBD-PE) was provided by Avanti Polar Lipids, Inc., Birmingham, AL.

Monolayers at the air/water interface were prepared by first spreading the phospholipid in form of a chloroform/methanol (3:1) solution. Before use,the 0.1% LDAO solution containing the protein was diluted with millipore water by a factor of 20. 50 µl of this detergent (LDAO) solution were then deposited in 30 small droplets. This procedure yielded a homogeneous protein distribution in the fluid lipid phase,as observed by fluorescence microscopy with LHCP. Working with inert gas atmosphere, the native fluorescence spectrum of LHCP can be preserved over hours.

The fluorescence microscope and spectrometer to study monolayers at the air/water interface has been described elsewhere [2]. Excitation of LHCP was achieved into the carotenoid bands (440 nm - 490 nm), emission was detected via a SIT TV camera or, after wavelength selection by a grating monochromator (Jobin-Yvon H 20 V) via a peltier-cooled red-sensitive photomultiplier (RCA C31034) and a photon-counting device, respectively.

Protein reconstitution into vesicles was achieved by detergent dialysis. By this the 0.1% detergent solution was diluted by 6 orders of magnitude, after mixing large vesicles with RC and twice centrifuging (2000g, 10 minutes) the suspension. The influence of possible residual detergent on the phase-transition was shown to be negligible by reference experiments. The protein/lipid ratio given below was determined via material inweight used for preparation. Flash spectroscopy showed that about 35% of the RC were functional within the membrane. Thus the concentration may be up to a factor of 3 smaller than given below.

Phase-transition curves were measured via static light scattering. Tests performed at selected samples using density measurements and fluorescence depolarisation of diphenylhexatriene showed the same results. The curves presented were obtained by cooling the samples (30°C/hour). Corresponding heating runs showed a slight hysteresis (0.5°C), but the relative shifts in transition-temperature on protein incorporation were identical.

Freeze-etch electron micrographs were obtained after annealing the sample for more than 30 minutes at the desired sample temperatures. After shock cooling in liquid propane (77 K) the replica was prepared by platinum shadowing (30 Å, 45°) and by carbon deposition (200 Å) and investigated in a Phillips EM400.

3. Results and Discussion

3.1 Monolayer experiments

Figure 1 shows the pressure/area diagram of a DLPA monolayer containing LHCP in various amounts. One clearly observes the break in the slope of the diagram corresponding to the fluid/gel phase-transition. On protein incorporation, the transition pressure is apparently reduced and the

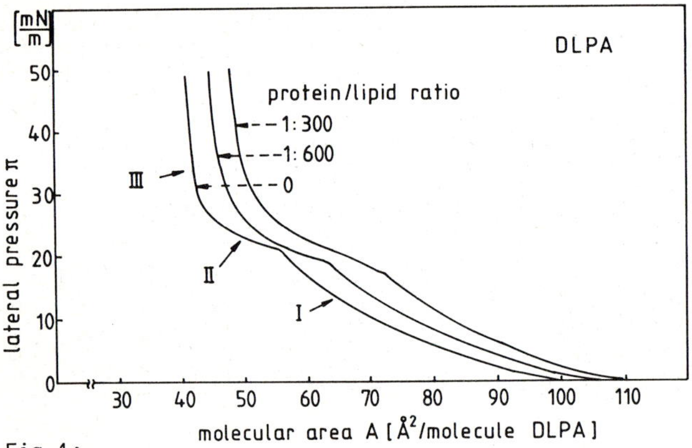

Fig.1:
Pressure-area diagrams of DLPA monolayers containing LHCP in different protein/lipid ratios. The ratios given in the figure are calculated assuming complete incorporation. T=21°C, pH=5.5

transition is broadened. This can be understood qualitatively as an increase in ordering of the fluid phase due to protein-incorporation: Lipids near the protein are condensed, thereby orienting the hydrocarbon chains. This may lead to nucleation centers for crystallization and/or to an entropy decrease favouring the establishment of the ordered gel phase.

The fact that the reduction in transition pressure is observed for LHCP reconstituted into the acidic lipid DLPA but not for LHCP in the uncharged lipid DMPE [3] suggests that there is a preferential interaction between the negatively-charged phospholipids and the positive surface charges of the protein. Indeed, an inspection of the amino acid sequence of the two subunits of B800-850 reveals that at one end of long hydrophobic segments (positively charged) lysines reside [4]. As the hydrophobic segments may be ascribed to the transmembrane α-helices, one can rationalize a Coulombic attraction between lysines and phosphate head groups due to their possible close proximity.

Information on lateral distribution of chromophores is obtained by fluorescence microscopy. Fig.2a shows a fluorescence micrograph of a DLPA monolayer containing the dye DP-NBD-PE. In this case, a pressure of 20 mN/m is established to maintain coexistence between fluid and gel phase. Due to low dye solubility the solid phase appears dark [5]. It was recently shown that, enforced by electrostatic interactions, solid domains have finite and unique size and repel one another, thus forming hexagonal patterns [6].

Using this technique, observing LHCP fluorescence, one can verify the establishment of a homogeneous protein distribution in the fluid lipid or a defined inhomogeneous distribution in the coexistence region. The latter is observed from Fig.2b, indicating that the protein partitions preferentially into the fluid phase.

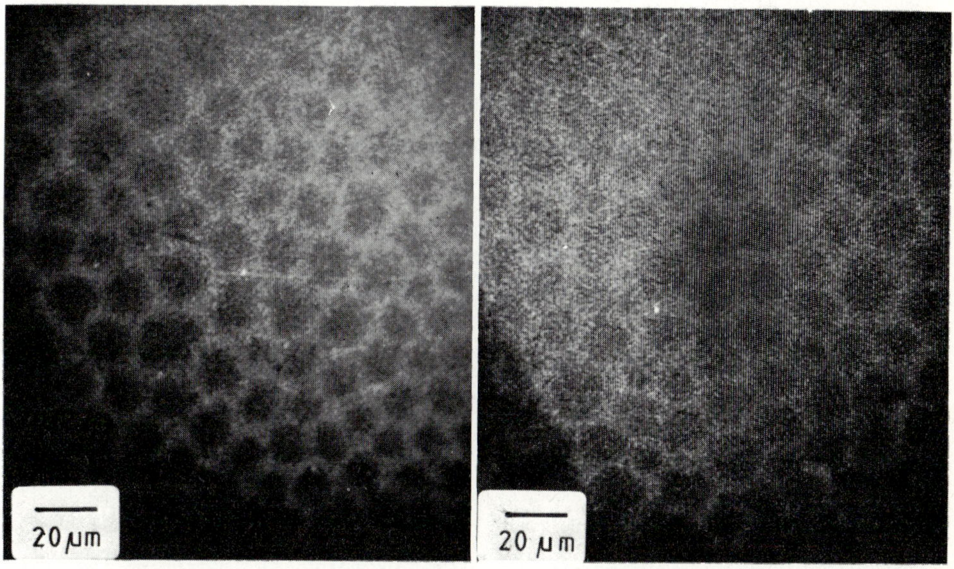

Fig.2:
Fluorescence micrographs of DLPA monolayers in the fluid/gel phase coexistence region
Fig.2a (left-hand side): Fluorescence emission of a dye probe
Fig.2b (right-hand side): Fluorescence of LHCP embedded in a concentration of 1.5 mol% in the monolayer

An inhomogeneous fluorescence distribution would also be observed if the LHCP fluorophore would partition in any phase with equal probability but if its fluorescence quantum yield depends on lipid enviroment. This is indeed not the case, because the fluorescence yield does not significantly depend on phase-state but drastically on surface pressure.

The latter conclusion can be deduced from Fig.3 comparing the surface pressure and the total fluorescence intensity of a DLPC/LHCP monolayer as a function of molecular area. In contrast to DLPA the DLPC monolayer does not undergo any phase-transition under these experimental conditions. Hence, this system is well-suited to study the influence of a pressure change without affecting the phase-state. The fluorescence reveals a pronounced maximum followed by an abrupt decrease on exceeding the pressure of about 20 mN/m. This change is reversible, and not due to a squeezing out of the protein into the subphase. The latter would cause a change in the slope of the pressure/area diagram, in contrast to the findings.

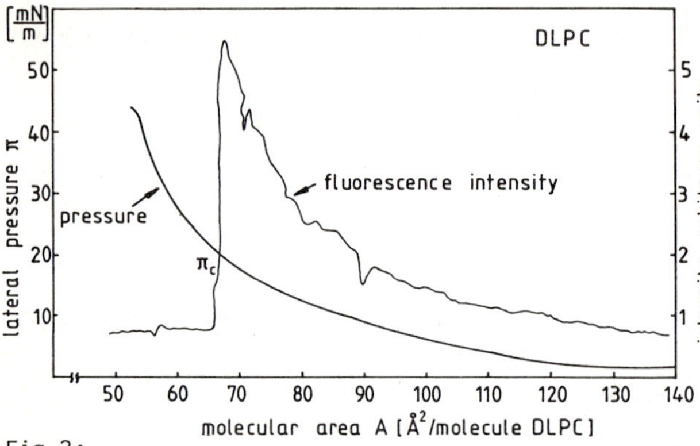

Fig.3:
Surface pressure and fluorescence intensity (at 860 nm) as a function of molecular area for DLPC monolayers containing 0.6 mol% LHCP. T=21°C, pH=8.

It is interesting to speculate on the abrupt fluorescence change which is obviously due to a structural rearrangement of the protein. Experiments to conclude on this are under way; possible explanations that are also discussed in this volume are:

- The structural change may lead to a disruption of the energy-transfer from carotenoids to the "dimeric" trap [7]. As in the experiments excitation was performed into the carotenoid bands; this could cause the fluorescence decrease.
- Increase of vibronic coupling of the fluoresent state to a lower level excited state could enhance thermal relaxation processes [8].

Concluding the section on monolayer experiments,we should add that it is possible to reconstitute natural LHCP/RC interaction in these systems. By incorporating RC into the LHCP/lipid monolayer, we were able to establish energy-transfer between the proteins by measuring the LHCP fluorescence quenching [3].

3.2 Vesicle experiments

Reconstitution of a membrane protein into the lipid bilayer causes a distortion, because generally the hydrophobic portions of protein and

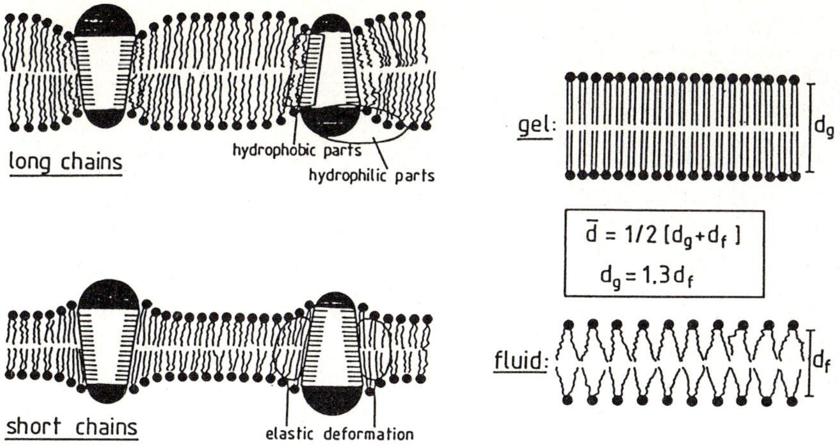

Fig.4:
Sketch of elastic deformations caused by protein incorporation into the bilayer.

bilayer disagree (Fig.4) [9]. The degree of mismatch depends on the lipid phase, as it is known that the membrane thickness is reduced by about 30% on going from the gel to the fluid phase [10]. Hence, one expects that a protein with length of hydrophobic parts (d_{pr}) longer than a characteristic mean value d will favour the gel phase, whereas it will favour the fluid phase in the opposite case (c.f. Fig.4). d depends on lipid chain length, and is the arithmetic mean of the thicknesses of the hydrophobic regions of the membrane in the gel and fluid phase [9]. Correspondingly, for a lipid with short acyl chains, the protein stabilizes the gel phase and the transition temperature is increased. If, on the other hand, the lipid exhibits long chains, the transition-temperature is reduced.
These expectations, based on elastic models, are indeed met: Fig.5 shows measurements of the gel/fluid phase-transitions of diacylphosphatidylcholines with varying chain lengths. All samples contain RC in about equal molar ratio of 1/3000. Protein reconstitution generally causes a broadening of the transition curves and a change in transition-temperature. This change is an increase of about 10°C for the shortest lipid DLPC where the transition-temperature is expected below 0°C for the pure lipid (Fig.5a). There is still a slight increase observable for DMPC (Fig.5b), but a decrease for DPPC (Fig.5d). No change in transition-temperature is observed for the lipid with the pentadecanoyl chain length (Fig.5c) indicating that this length d fits best with the hydrophobic part of the protein.

From neutron diffraction data d = 28 Å may be derived for this lipid [11]. Hence an average length of the α-helices of 19 amino acid residues is determined to be in accordance with elasticity theory. This length has indeed been assumed in searching for α-helical segments after determining the amino acid sequences of the L and M subunits [12,13]. These units comprise the major membrane integral parts of the protein, each possessing 5 α-helices, whereas the H unit is oriented towards the water phase, being anchored within the membrane by merely 1 or 2 α-chains. The length determined above resembles the value reported on another well-known transmembrane protein, the bacteriorhodopsin, where α-helices containing 20 hydrophobic amino acids were reported [14].

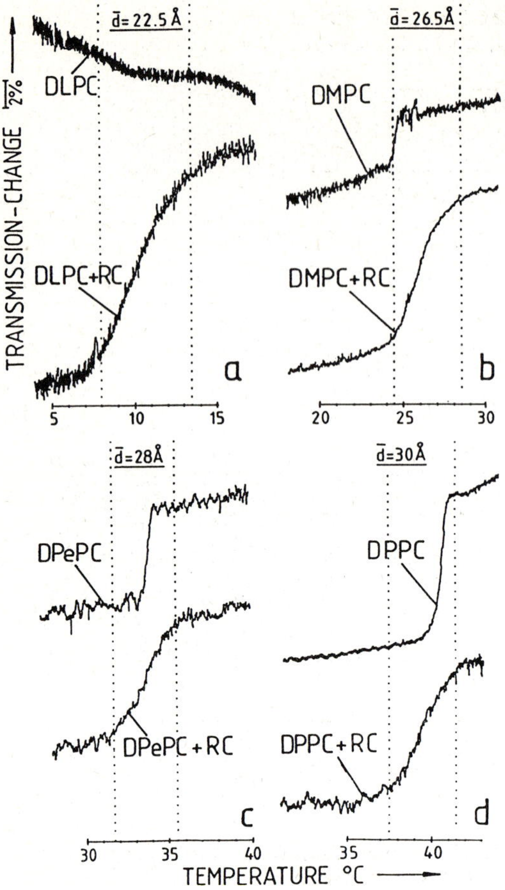

Fig.5:
Transmission change due to light scattering for diacylphosphatidylcholine vesicles containing no RC and RC in a protein/lipid ratio 1 : 3000; pH=7. Fig.5a: DLPC, 5b: DMPC, 5c: DPePC, 5d: DPPC.

From an inspection of the sketch in Fig.4 it is additionally obvious that the elastic energy connected with the membrane distortion may be minimized by protein aggregation [15]. (For dissimilar proteins also an elasticity-mediated repulsion may result.)
Conditions for patching would be [9]:

(1) Large mismatch between protein and membrane thickness and/or
(2) high stiffness of the membrane.

The first condition is fulfilled for RC reconstituted in the fluid phase of DLPC. Yet for this case we observe a homogeneous protein distribution on the membrane surface (Fig.6a). Apparently the membrane is not stiff enough to enforce protein aggregation. The second condition is fulfilled for proteins in the solid lipid phase. For this membrane the freeze-etch electron micrographs of Figs.6b, c show a heterogeneous protein distribution on the membrane surface. However, in the case of solid DMPC one normally observes protein/protein distances of more than 100 Å. There is apparently no strong interprotein attraction. The absence of intermolecular contact over short distances may then be ascribed either to kinetic hindrance of patch formation or to an energy barrier opposing the annealing of those defects surrounding each protein.

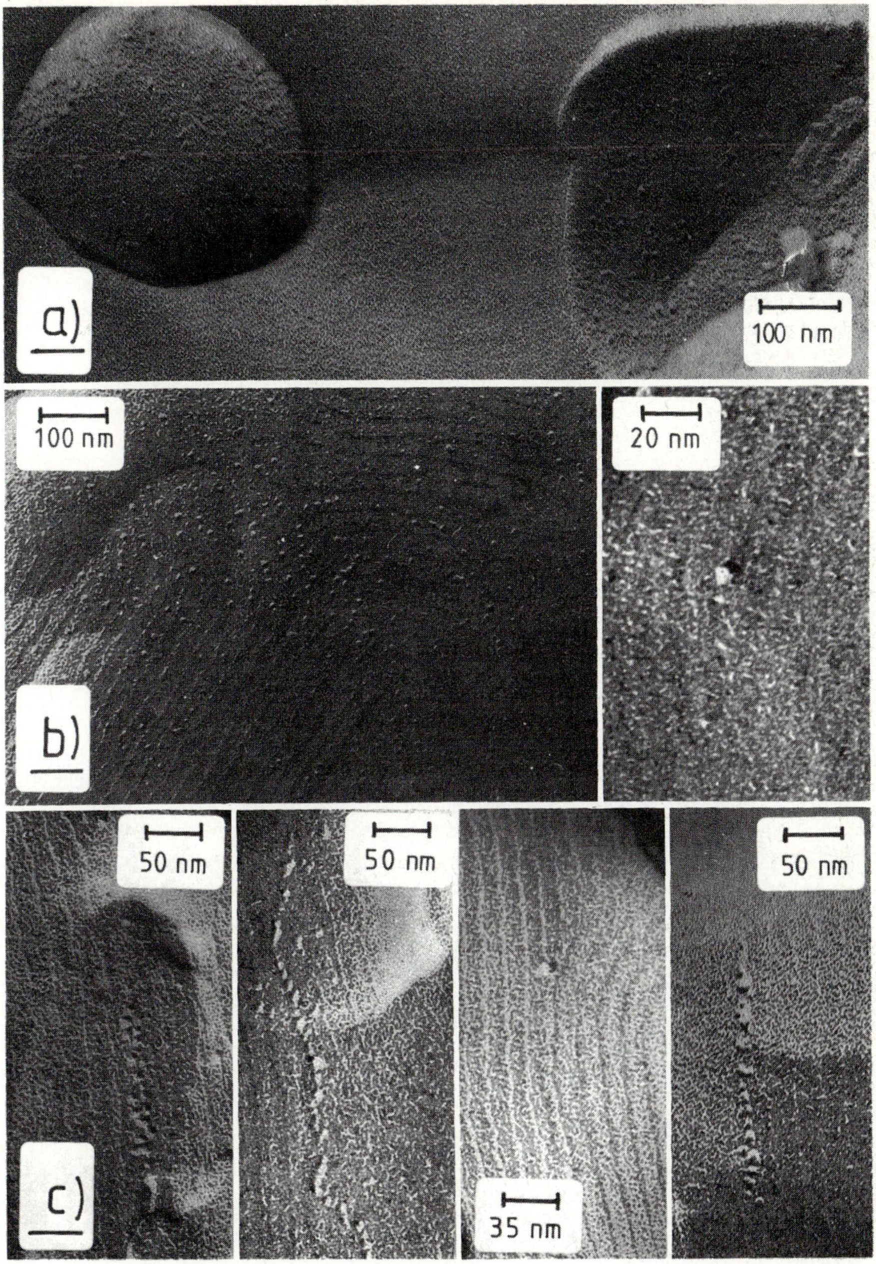

Fig.6:
Freeze-etch electron micrographs of RC containing vesicles
6a: DLPC in fluid phase; T=24°C
6b: DMPC in solid "ripple" phase.
6c: DPPC in solid "ripple" phase.

The only slight mismatch of membrane thickness (30 Å) and d_{pr} (28 Å) also explains why RC are arranged on the planar part of the surface that forms the ripple phase at this temperature. For RC in solid DPPC, however, the membrane is considerably thicker (34 Å) than the protein. This obviously causes stronger segregation of the protein along defect lines (Fig.6c) or at defect points.

The aspect of elasticity-induced local enrichment of proteins is also of possible importance for interactions between proteins. E.g. preliminary experiments indicate that the energy - transfer from LHCP to RC reconstituted in vesicles occurs very efficiently. This demonstrates that the microscopic arrangement of proteins may on one hand influence the efficiency of photosynthesis. On the other hand. the measurement of functional parameters may also be used to study the structure of the apparatus [16].

4. Conclusions

We have shown that protein functions like fluorescence and energy - transfer may depend on protein/lipid interactions. These interactions determine many membrane-bound processes, not only photosynthesis. On the other hand proteins derived from the photosynthetic apparatus and reconstituted into model membranes provide a structurally very well-defined system. Its study therefore allows conclusions on models describing electrostatic and elastic features of protein/lipid interactions.

5. Acknowledgements

We are grateful for many advice, discussions and for the protein material provided by H. Scheer. We profited from helpful discussions with E. Sackmann and from the skilful technical assistance of C. Fahn. This work was supported by the Deutsche Forschungsgemeinschaft through SFB 143 and Mo 283/9.

6. References

[1] Clayton, R.K.: "Photosynthesis: physical mechanisms and chemical patterns" JUPAB Biophys. Series, Cambridge Univ. Press, 1980

[2] Lösche, M. and Möhwald, H., Rev. Sci. Instr. 55, 1968 (1984)

[3] Heckl, W.M., Lösche, M., Scheer, H. and Möhwald, H., subm. to Biochim. Biophys. Acta

[4] Cogdell, R.J. and Thornber, J.P., FEBS Lett. 122, 1 (1980)

[5] Lösche, M. and Möhwald, H., Europ. Biophys. J. 11, 35 (1984)

[6] Fischer, A., Lösche, M., Möhwald, H. and Sackmann, E., J. Physique Lett. (Paris) 45, L-785 (1984)

[7] Cogdell, R.J., presented in this conference report

[8] Pearlstein, R., presented in this conference report

[9] Mouritsen, O.G. and Bloom, M., Biophys. J. 46, 141 (1984)

[10] Janiak, M.J., Small, D.M. and Shipley, G.G., Biochemistry 15, 4575 (1976)

[11] Lewis, B.A. and Engleman, D.M., J. Mol. Biol. 166, 211 (1983)

[12] Williams, J.C., Steiner, L.A., Feher, G. and Simon, M.I., Proc. Natl. Acad. Sci. USA 81, 7303 (1984)

[13] Williams, J.C., Steiner, L.A., Ogden, R.C., Simon, M.I. and Feher, G., Proc. Natl. Acad. Sci. USA 80, 6505 (1983)

[14] Engelmann, P.M., Goldmann, A., and Steitz, T.A., Methods Enzymol., 88, 81 (1981)
[15] Owicki, J.C., Springgate, M.W. and McConnell, H.M., Proc. Natl. Acad. Sci. USA, 75, 1616 (1978),
Owicki, J.C. and McConnell, H.M., Proc. Natl. Acad. Sci. USA 76, 4750 (1979)
[16] Riegler, J., Peschke, J. and Möhwald, H., Biochem. Biophys. Res. Commun. , 125, 592 (184)

The Architecture of Photosystem II in Plant Photosynthesis Which Peptide Subunits Carry the Reaction Center of PS II?

A. Trebst and B. Depka

Lehrstuhl für Biochemie der Pflanzen, Ruhr-Universität Bochum, Postfach 102148, D-4630 Bochum 1, F.R.G.

Photosystem II in the thylakoid membrane of plant chloroplasts catalyzes oxygen evolution and plastoquinone reduction. Enriched photosystem II particles contain a reaction center P680, pheophytin, Fe, plastoquinones Q_A and Q_B, Mn (for the oxygen-evolving system), core antenna chlorophyll a and a cytochrome b559 of unknown function. The protein complex consists of 47, 44, 34, 32 and 10 kD peptide subunits, probably in ratios of 1 : 1. Experiments will be presented with trypsinated photosystem II particles, still capable of oxygen evolution and Q_A reduction, that do not support the notion in the literature that the 47 kD peptide is the subunit carrying the reaction center. Instead it is suggested that the reaction center of PS II consists of a complex of the 34 and 32 kD peptide. The amino acid sequence and particularly their predicted membrane folding is highly homologous to the L and M subunit of the bacterial reaction center. It is therefore very suggestive that the principle architecture of photosystem II is quite comparable to the photosystem of purple bacteria.

1 Introduction

Photosynthetic electron flow from water to NADP in the thylakoid membrane of chloroplasts is catalyzed by three integral protein complexes of several peptide subunits each: they are the PS II complex, the PS I complex and the cytochrome b/f-complex. Several peripheral peptides are attached to them and mobile carriers connect the three integral complexes.

Membrane fractions highly enriched in the photosystem II complex and devoid of photosystem I and the cytochrome b/f-complex have been prepared in a number of ways [1,2]. They contain the reaction center P680, the primary donor, pheophytin, the acceptor quinones Q_A, Q_B, Mn, Fe, cytochrome b559 and about 30 to 40 core antenna chlorophyll a (for review see [3,4]). Depending whether peripheral peptides of 33, 24 and 17 kD size are still attached or not, such particles may or may not evolve oxygen [5]. A fourth integral complex, the light-harvesting chlorophyll complex LHCP (containing chlorophyll a and b and several peptides of about 22-25 kD, see gels below) is enriched in PS II particles. Recently also this LHCP complex has been removed from photosystem II by octylglucoside sucrose density centrifugation [6] or a similar detergent [7].

Some of the peptide subunits of photosystem II have been correlated with specific functions. The 32 kD peptide is considered to be a herbicide binding and from there the Q_B binding peptide [8] . The 44 kD peptide is carrying several core antenna chlorophyll a. It can be removed without loss of primary photochemistry [9] . The 34 kD peptide is involved on the donor side of PS II [10] , but recently it has been shown to be also involved in triazine binding [11] and therefore to function on the acceptor side. We assume this 34 kD peptide to be the product of the D-2 gene (see below). Particularly important is, of course, the peptide subunit of the reaction center chlorophyll itself. A number of groups have put forward evidence that suggested that the 47 kD peptide carries the reaction center of photosystem II, i.e. the special chlorophyll P680, pheophytin and Q_A. The evidence is based on fluorescence [12,13] , thermoluminescence [14] , the DCP-DCPIP reaction activity [15,16] , pigment composition and P680 spectral changes [7,13,16-18]).

We wish to cast doubts on this assignment. By trypsination of photosystem II we can degrade the 47 kD peptide without loss of electron transport from water to Q_A. We propose, but cannot prove, that a complex of the 34 and 32 kD peptide carries the reaction center.

2 Oxygen Evolution in Trypsinated Photosystem II Particles

Recently the clarification of the role of the three peripheral peptides that are attached to the integral peptides of photosystem II on the oxygen-evolving side has been much advanced (for review see [5]). Two smaller subunits (24 and 17 kD) can be replaced by calcium ions in oxygen evolution, whereas the 33 kD (the so-called Murata peptide) cannot, shown by several groups. The 33 kD peptide has been shown to be directly attached to the membrane on the donor side of photosystem II and is responsible for keeping the manganese properly bound to the membrane on the donor side of P680. The calcium stimulation of the water splitting system in stripped photosystem II particles is also experimentally important, as it makes it possible to purify for the reaction center of photosystem II by following oxygen evolution rather than using an artificial donor system. Of course, this demands that the Murata 33 kD peptide remains intact and attached to the membrane. This is possible even in trypsination experiments, if run at low pH.

A first conclusion in such enriched oxygen-evolving photosystem II particles is that the DPC-DCPIP reaction does not necessarly reflect electron flow through the reaction center. As Table 1 is to indicate, in oxygen-evolving PS II particles neither ferricyanide nor DCPIP are good acceptors for Q_B [19]. However, a lipophilic quinone (MDBQ) is [20] . In tris treated particles the peripheral peptides of oxygen evolution systems are displaced

Table 1 Comparison of acceptor systems for photosystem II
The DPC-DCPIP reaction is an artefact in enriched PS II systems, as the rate of oxygen evolution in a ferricyanide reduction is much higher.

	rates in 1e/h/mg chlorophyll		
electron acceptor	ferricyanide	DCPIP	
electron donor	water	water	diphenyl-carbazide = DPC
PS II particles			
control	120		
control + MDBQ	408	102	110
tris treated	10	4	136
trypsin treated	120		
trypsin treated + $CaCl_2$	310	21	34

MDBQ = methylenedioxy-dimethyl-p-benzoquinone

from the membrane. The donor DCP restores the low rate in the DCPIP system [21]. If, however, Q_A becomes exposed by trypsination of PS II particles and ferricyanide is now an acceptor at the Q_A site [22] , oxygen is evolved. Also photosystem II preparations can be treated with trypsin (at low pH according to INOUE, RENGER et al. [19]) and oxygen evolution with ferricyanide still proceeds at high rates (Table 1). In the presence of calcium ions the rates are very high in this system, whereas the rates remain low in the DCPIP system. Therefore DCPIP is not an acceptor of the Q_A site either, and therefore the small rate induced by DPC has to be catalyzed by a route parallel to the one from water to Q_A-ferricyanide. We suggest that an antenna chlorophyll carrying subunit (the 47 kD peptide?) may be converted to induce the reaction.

We have studied the subunit structure of trypsinated photosystem II particles in more detail following oxygen evolution rather than an artificial donor system. After trypsination of triton particles (those according to [1]) at low pH (following the procedure of VÖLKER et al. [19]) as well as by trypsination of octylglucoside particles - that are devoid of the LHCP complex [6] - one obtains enriched photosystem II particles that are still able to evolve oxygen (Table 2). Even after the trypsinated particles have been washed, oxygen evolution is still intact, calculated on a chlorophyll basis. Calcium ions are, however, essential for electron flow in the water to ferricyanide reaction.

Table 2 Oxygen evolution in detergent-treated photosystem II preparations after trypsination at low pH

	µ electrons transported from water to ferricyanide per mg chlorophyll x h in the presence of 10 mM Ca-ions
triton particles [1]	120
triton particles + MDBQ	444
trypsinated triton particles	336
trypsinated particles washed with 0.5% triton	192
trypsinated particles washed with octylglucoside	408
octylglucoside particles [6] + MDBQ	317
trypsinated octylglucoside particles	414

The reaction may be followed spectroscopically (FeCy reduction) or by an oxygen electrode. For the subunit composition see Figs. 1 and 2.

Ferricyanide in trypsinated thylakoids of photosystem II particles is a reduction through Q_A, now exposed to the hydrophilic environment. This has been established particularly by RENGER [22] . This reaction is also DCMU insensitive, which is a basis for assigning the Q_B site to the 32 kD peptide, degraded by trypsin [23] . The time of trypsin treatment in our experiments of Table 2 can be extended beyond 20 min. without loss of photosynthetic activity, if run at pH 6.5, as originally shown by VÖLKER et al. [19]. A 2 min. trypsination is already sufficient to show the degradation of larger peptides. The SDS gels in Fig. 1 show the subunit composition of trypsinated photosystem II particles in comparison with controls. The Murata 33 kD peptide is seen on all gels (our reference peptide) as well as a band-carrying chlorophyll at about 90 kD (this band is

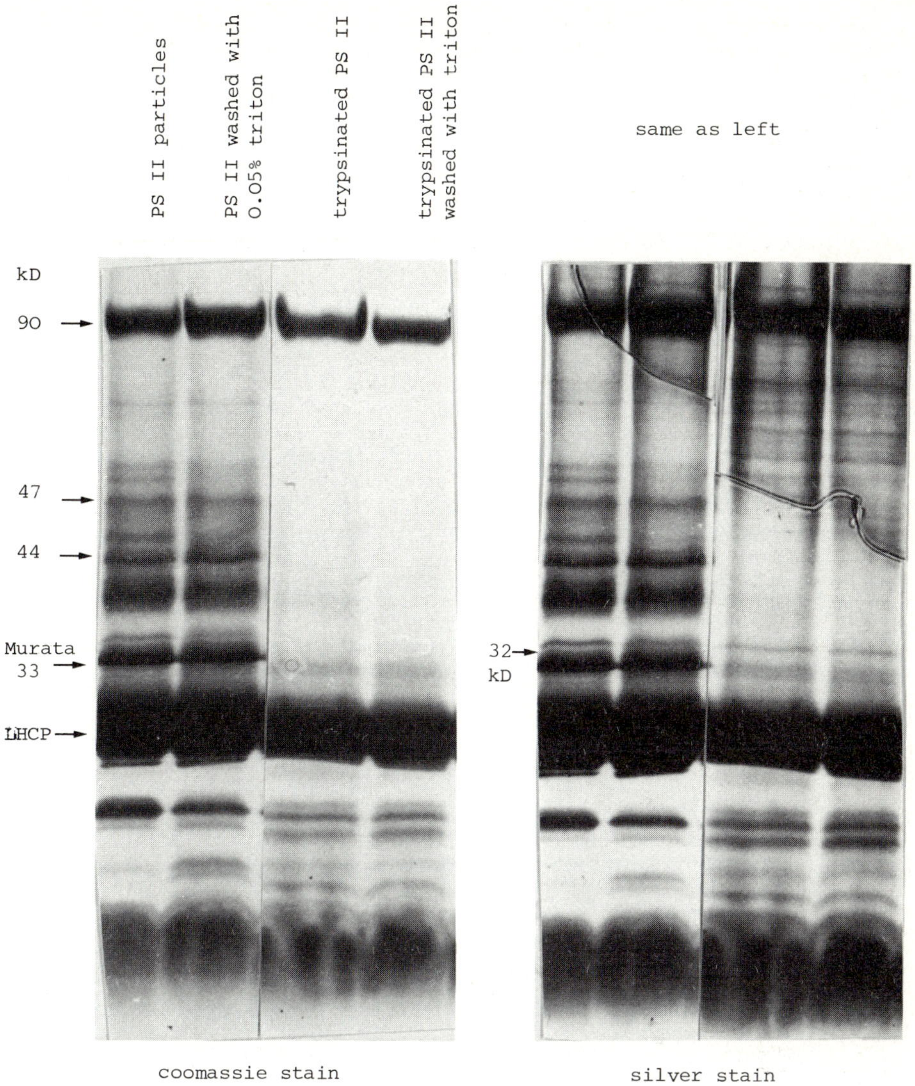

Fig. 1 Subunit peptides of photosystem II particles
showing the loss of the 47 and 44 kD peptides after trypsination
The particles according to [1] with 300 µg chlorophyll in 3 ml were trypsinated at pH 6.5 with 200 µg trypsin (Merck) for 20 min. at 20°, trypsin inhibitor added, spun down, washed with the detergent where indicated, again spun down, solubilized with LiDS, then separated at room temperature by acrylamide electrophoresis. The gel was first stained with coomassie blue (left) and then with silver nitrate (right). Photosynthetic activity is given in Table 2.

usually assigned to photosystem I. We do not have any PS I activity in this preparation,and therefore doubt this assignment. Probably it is a complex of subunits of PS II). The light-harvesting complex LHCP is slightly smaller after trypsin treatment in the triton particles, as is known. The most interesting phenomenon is that the 47 and 44 kD peptides of photosystem II are both clearly attacked by trypsin treatment of the membrane, as they are barely seen on the gels. Particularly, when the trypsin treated particles were washed further with triton, the gels show only a shade of these peptides above the Murata 33 kD peptide. On the other hand,a number of low molecular weight peptides appear now below the light-harvesting peptides after the trypsin treatment.

As is known, coomassie blue does not stain the 32 kD herbicide binding peptide. It does show up in silver stain. The same gel in Fig. 1 after silver stain on top of the coomassie blue stain does show an additional band just above the Murata peptide. By azidoatrazine labeling (not shown on the gels in Fig. 1) this band is indeed identified as the 32 kD herbicide-binding peptide. It should be mentioned that we rarely succeed in picking up the 32 kD peptide in PS II particles or even in thylakoids by silver stain (experienced also by other investigators). Then, of course, it is not seen in the trypsinated samples prepared from those. This makes positive identification of the 32 kD peptide very difficult,and this is true even more for the second 34 kD integral peptide.

In the trypsin-treated triton particles the LHCP complex is still present (although attacked by trypsin). Recently INOUE et al. [6] described a method to remove the LHCP by octylglucoside sucrose gradient centrifugation. We find this method very effective (Fig. 2). Trypsination of such PS II particles before or after octylglucoside treatment also leads two particles still very active in oxygen evolution in a ferricyanide system. Again,as the gels in Fig. 2 show, there is only a shade of the 47 and 44 kD peptides left in these particles.

These experiments do not support the notion that the 47 kD peptide is the reaction center subunit of PS II. It would rather assign to this peptide the role of a core antenna-carrying peptide, alike the accepted function for the 44 kD peptide, that can be removed by trypsin and detergent treatment. Other smaller subunits (better: a complex of these) would have the function of the reaction center peptides, the complex perhaps being stabilized by the 47 kD subunit. Our results, however, do not prove this, either,and we do not wish to ignore all of the experiments (quoted above) in favour of the 47 kD peptide. We could explain our results,for example,that we split the 47 kD peptide by trypsin on hydrophilic membrane-exposed segments, but that the integral, membrane buried, parts of this peptide remained stuck together without loss of function. Only in the SDS gel electrophoresis would the degraded segments split apart. However, we think this possible for a stabilizing function of the 47 kD peptide, but not for the primary function itself.

3 Homology Between the 32 and 34 kD Peptides of Photosystem II to the Bacterial Reaction Center Peptides

Although we like to question the assignment of the 47 kD subunit as the reaction center peptide, we have no experimental proof which peptides make up the reaction center. But strong considerations should be given to the similarities in structure and function of photosystem II and the bacterial photosystem,and particularly to the peptide subunit structure of both. The overall pigment and redox carrier composition is very much the same (keeping in mind, of course, the obvious differences in function, for review see [3,4]). Particularly intriguing is the amino acid sequence and

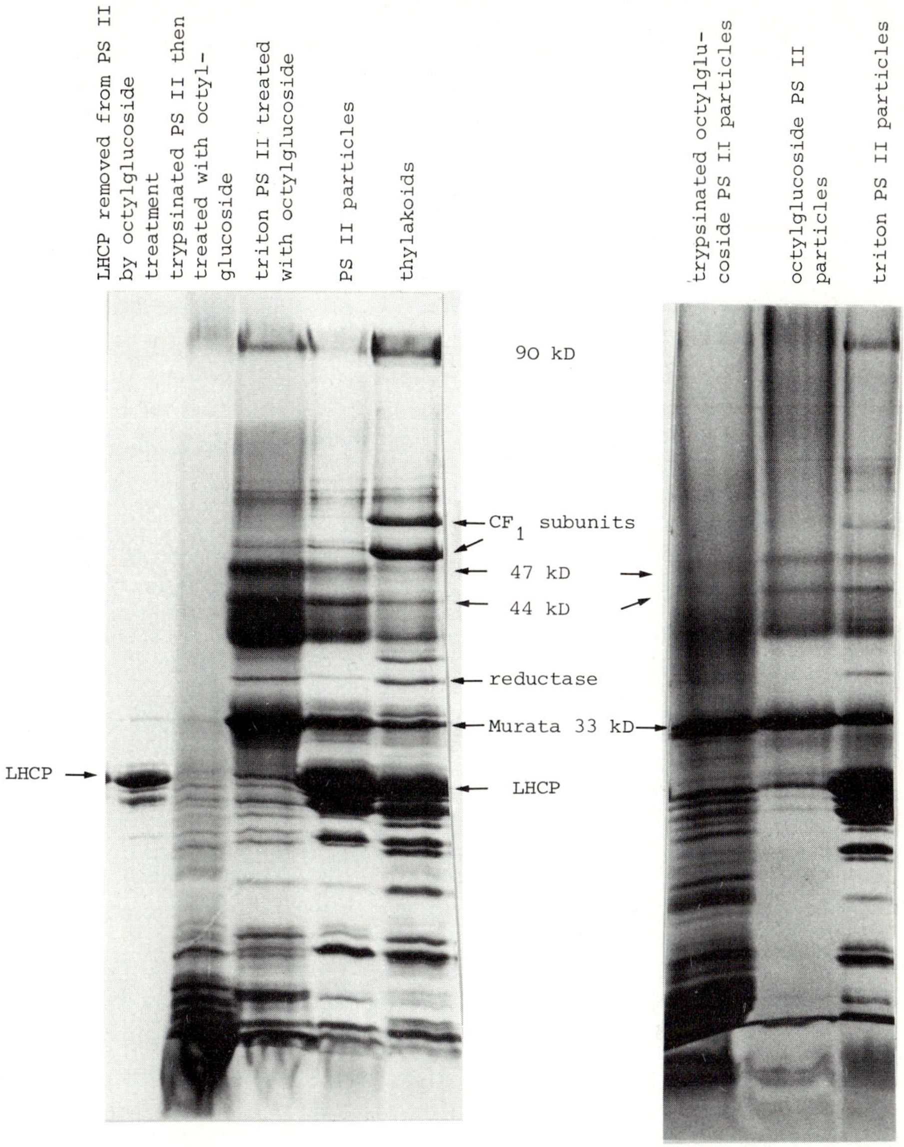

Fig. 2 Subunit peptides of photosystem II particles prepared with octylglucoside according to INOUE [6] (i.e. devoid of LHCP)
Trypsination before or after the octylglucoside centrifugation leads to a loss of the 47 and 44 kD peptides without loss of electron flow. Octylglucoside sucrose density centrifugation according to [6]. Trypsination and other conditions as in Fig. 1. Photosynthetic activity is given in Table 2.

```
                                                             transmembrane
                                                                helix I
D-1        MT--AILERRESESLWGRFCNWITSTENRLYIGWFGVLMI-PTLLTATSVFIIAFIAAPPVDIDGIREPV-SGSLLYGNNIISGAI
L                     MALLSFERKYRVPGGTLIGGSLFDFWVGPFYVGFFGVTTIFFATLGFL-LILWGAAMQGTWNP

D-2        MT-IAVGKFTKDEKDLFDSMDDWLRRDRFVFVGWSG-LLLFPCAYFALGGWFTGTTFVTSWYTHGLASSYLEGCNFLTAAVSTPAN
M    MAEYQNFFNQVQVAGAPEMGLKEDVDTFER--TPAGMFNILGWMGNAQIGPIYLGIAGTVSLAFGAAWFF-TIGVWYWYQAGFDP      FI

     parallel                    transmembrane                      transmembrane
       helix                        helix II                          helix III
D-1  IPTSAAIGLHFYPIWEAASVDEWLYNGGPYELIVLHFLLGVACYMGREWELSFRLGMRPWIAVAYSAPVAAATAVFLIYPIGQGSFSDGMPLGI
L        QLISIFPPPVENGLNVAALDKGGLWQVITVCATGAFCSWALREVEICRKLGIGFHIPVAFSMAIFAYLTLVVIRPMMMGSWGYAFPYGI

D-2  SLA-HSLLLLWGP-EAQGDFTRWCQLGGLWAFVALHGAFALIGFMLRQFELARSVQLRPYNAIAFSGPIAVFVSVFLIYPLGQSGWFFAPSFGV
M    FMRDLFFFSLEPPPAEYGLAIAPLKQGGVWQIASLFMAISVIAWWVRVYTRADQLGMGKHMAWAFLSAIWLWSVLGFWRPILMGSWSVAPPYGI

     parallel                transmembrane                                parallel
       helix                    helix IV                                    helix
D-1  SGTFNFMIVFQAEH-NILMHPFHMLGVAGVFGGSLFSAMHGSLVTSSLIRETTENESANEGYRFGQEEETYNIVAAHG      YFGRLIFQY
L    WTHLDWVSNTGYTYGNFHYNPFHMLGISLFFTTAWALAMHGALVLS     AANPVKGKTMRTPDHEDT                YFRDLMGYS

D-2  AAIFRFILFFQGFH-NWTLNPFHMMGVAGVLGAALLCAIHGATVENTLFEDGDGANTFRAFNPT-QAEETYSMVTANR       FWSQ-IFGV
M    FSHLDWTNQFSLDHGNLFYNPFHGLSIAALYGSALLFAMHGATILA                VTRFGGERELEQIVDRGTASERAALFWRWTMGFN

                    transmembrane
                       helix V
D-1  ASFNNSRSLHFFLAAWPVVGIWFTALGISTMAFNLNGFNFN-QSVVDSQGRVIN--TWADIINRANLGMEVMHERNAHNFPLDLAAIEAPSTNG
L    VGTLGIHRLGLLLALNAVFWSACCMLVSGTIYFDL              WSDWWYWWVNMPFWAD                          MAGGING

D-2  AFSNK-RWLHFFMLFVPVTGLWMSALGVVGLALNLRAYDFVSQEIRAAEDPEFE--TFYTKNILLNEGIRAGW-AAQDQFHENLIFPEEVLPRGNAL
M    ATMEGIHRWAIWMAVMVTLTGGIGILLSGT                    VVDNWYVWA                       QVHGYAPVTP
```

Fig. 3 Homology between subunits of photosystem II and of the bacterial photosystem
The amino acid sequences and hydropathy plots (indicating hydrophobic helices) between the 32 and 34 kD (products of the D-1 and D-2 gene) subunits of PS II from tobacco and spinach and the L (light) and M (medium) subunits of the photosystem of Rh. capsulata [24] , indicate a close similarity in their folding through the membrane. Five transmembrane and three parallel helices are indicated,as well as the conserved histidines in helices IV and V involved in binding functional groups [29] .

(predicted) protein-folding homology between the L and M sbunit of the bacterial reaction center and the 32 and 34 kD peptides of the thylakoid photosystem II.

Hearst and colleagues were the first to note the sequence homology of the 32 kD peptide coded by a D-1 gene to both the L and M subunit of purple bacteria, shown to be the two subunits of the bacterial reaction center [24]. Later a D-2 gene coding for a 34 kD peptide has been sequenced and found to be homologous to the 32 kD peptide [25] and therefore also to the L and M subunit of the bacterial system. Hydropathy plots of the four peptides show a high homology of all four peptides. In particular, they have five transmembrane helical spans in common at exactly the same (equivalent) position in the sequence. This specifies which of the helices predicted by RAO et al. [26] for the 32 kD peptide are actually transmembrane,and which are parallel to the membrane. The homology suggests that the general folding of the two subunits of PS II (D-1/D-2) and of the bacterial system (L and M) through the membrane is the same. The bacterial reaction center from Rh. viridis has been crystallized [27] . The data of its X-ray structure, now available, show the exact location of the reaction center bacterio-chlorophyll, bacterio-pheophytin, Fe and the quinone binding site on the L and M subunit [28] . Also first data on the peptide folding are available, showing that indeed, as predicted from the hydropathy plot, both L and M subunit of the reaction center span the membrane in five helices [29] . DEISENHOFER et al. [28,29] , we [30] and others suggested, in view of the similarities in function,and in particular the homologies in peptides sequences,that the 32 kD (D-1) and the 34 kD (D-2) peptide is homologous to the L and M subunit,and therefore are the reaction center peptides of PS II. The few data available so far on amino acid residues involved in binding of prosthetic groups in the bacterial center from the X-ray studies [29,30] indicated that most of these principle amino acids are indeed conserved in the D-1 and D-2 peptides (like the histidines in transmembrane helices IV and V in Fe binding). Those amino acids identified so far in Q_B and herbicide binding in the 32 kD peptide [31,32] all appear in the right position to each other in transmembrane helix IV and the membrane parallel helix just before transmembrane helix V. This actually supports the suggestion that the folding of photosystem II peptides is like the bacterial photosystem. The genes for the 47 and 44 kD peptide have also been sequenced recently [31] (for review see [30]). Their sequence shows no hydropathy homology to the folding of the bacterial reaction center peptides. The 34 and 32 kD peptides on a gel do not carry chlorophyll, but the 47 and 44 kD peptides do. This is expected, as the cooperation of two subunits is required to bind chlorophyll in a reaction center, whereas larger core antenna peptides can easily keep chlorophyll attached,like those of the LHCP. Of course, it is important to get direct data on a reaction center complex between the 32 and 34 kD peptide of photosystem II that still does carry chlorophyll and is free of the core antenna peptides.

1 D.A. Berthold, G.T. Babcock and C.F. Yocum: FEBS Lett. 134, 231-234 (1981)
2 T.G. Dunahay, L.A. Staehelin, M. Seibert, P.D. Ogilvie and S.P. Berg: Biochim. Biophys. Acta 764, 179-193 (1984)
3 A.W. Rutherford: Biochem. Soc. Trans., in press (1985)
4 B.A. Diner, in Encyclopedia of Plant Physiol. (L.A. Staehelin and C.J. Arntzen, eds.), Springer-Verlag, Berlin, Heidelberg, in press (1985)
5 N. Murata and M. Miyao: Trends Biochem. Sci. 10, 122-124 (1985)
6 M. Ikeuchi, M. Yuasa and Y. Inoue: FEBS Lett. 185, 316-322 (1985)
7 T.M. Bricker, H.B. Pakrasi and L.A. Sherman: Arch. Biochem. Biophys. 237, 170-176 (1985)

8 K. Pfister, K.E. Steinback, G. Gardner and C.J. Arntzen: Proc. Natl. Acad. Sci. USA 78, 981-985 (1981)
9 A. Yamagishi and S. Katoh: Biochim. Biophys. Acta 807, 74-80 (1985)
10 J.G. Metz and M. Seibert: Plant Physiol. 76, 829-832 (1984)
11 J.G. Metz, T.M. Bricker and M. Seibert: FEBS Lett. 185, 191-196 (1985)
12 H.Y. Nakatani, in The Oxygen-Evolving System of Photosynthesis (Y. Inoue et al., eds.), pp. 49-54, Academic Press Japan, Tokyo (1983)
13 H.Y. Nakatani, B. Ke, E. Dolan and C.J. Arntzen: Biochim. Biophys. Acta 765, 347-352 (1984)
14 Y. Inoue, in The Oxygen-Evolving System of Photosynthesis (Y. Inoue et al., eds.), pp. 439-450, Academic Press Japan, Tokyo (1983)
15 E.L. Camm and B.R. Green: J. Cell. Biochem. 23, 171-179 (1984)
16 A. Yamagishi and S. Katoh: Biochim. Biophys. Acta 765, 118-124 (1984)
17 T. Omata, N. Murata and K. Satoh: Biochim. Biophys. Acta 765, 403-405 (1984)
18 C. de Vitry, F.A. Wollman and P. Delepelaire: Biochim. Biophys. Acta 767, 415-422 (1984)
19 M. Völker, T. Ono, Y. Inoue and G. Renger: Biochim. Biophys. Acta 806, 25-34 (1985)
20 A.Trebst: Ann. Rev. Plant Physiol. 25, 423-458 (1974)
21 L.P. Vernon and E.R. Shaw: Biochem. Biophys. Res. Commun. 36, 878-884 (1969)
22 G. Renger: Biochim. Biophys. Acta 440, 287-300 (1976)
23 K.E. Steinback, K, Pfister and C.J. Arntzen, in Biochemical Responses Induced by Herbicides. ACS Symposium Series 181 (D.E. Moreland, J.B. St.John and F.D. Hess, eds.), pp. 37-55, American Chemical Society, Washington (1982)
24 D.C. Youvan, E.J. Bylina, M. Alberti, H. Begusch and J.E. Hearst: Cell 37, 949-957 (1984)
25 J.-D. Rochaix, M. Droh, M. Rahire and P. Malnoe: Plant Molec. Biol. 3, 363-370 (1984)
26 J.K.M. Rao, P.A. Hargrave and P. Argos: FEBS Lett. 156, 165-169 (1983)
27 H. Michel: J. Mol. Biol. 158, 567-572 (1982)
28 J. Deisenhofer, O. Epp, K. Miki, R. Huber and H. Michel: J. Mol. Biol. 180, 385-398 (1984)
29 J. Deisenhofer, H. Michel and R. Huber: Trends Biochem. Sci. 10, 243-248 (1985)
30 W.A. Cramer, W.R. Widger, R.G. Herrmann and A. Trebst: Trends Biochem. Sci. 10, 125-129 (1985)
31 J. Hirschberg and L. McIntosh: Science 222, 1346-1349 (1983)
32 J.-M. Erickson, M. Rahire, J.-D. Rochaix and L. Mets: Science 228, 204-207 (1985)
33 J. Morris and R.G. Herrmann: Nucl. Acid Res. 12, 2837-2850 (1984)

Part III

Electron-Transfer: Theory and Model Systems

Application of Electron-Transfer Theory to Several Systems of Biological Interest

R.A. Marcus
Arthur Amos Noyes Laboratory of Chemical Physics*,
California Institute of Technology, Pasadena, CA 91125, USA

N. Sutin
Department of Chemistry, Brookhaven National Laboratory,
Upton, NY 11973, USA

Abstract. Electron-transfer reaction rates are compared with theoretically calculated values for several reactions in the bacterial photosynthetic reaction center. A second aspect of the theory, the cross-relation, is illustrated using protein-protein electron-transfers.

I. Introduction

Studies of electron-transfer reactions for a variety of inorganic and organic reactions have established various factors as influencing the reaction rates. In particular, quantities such as the standard free-energy of reaction, $\Delta G°$, the reorganization parameters for the vibrational (λ_i), and solvational (λ_o) changes, the separation distance r, and the work terms (e.g., coulombic interactions), affect the reaction rate. The experimental evidence and comparisons with theoretical relationships and calculations were presented in a recent review [1]. In addition, other factors, such as the mutual orientation of the reactants, when the latter are nonspherical, the nature of the material between the reactants, and the electronic energy levels of the donor and acceptor, are expected to influence the rate in the case of "nonadiabatic" [1] reactions (defined later). Various theoretical calculations on some or all of these other factors (many referred to in [1]) have been made, though the experimental data on them are, for the most part, presently scanty. Another factor influencing the electron-transfer rate is present in sufficiently viscous media, more precisely, in media with slow dielectric relaxation times. It has been found for reactions which are intrinsically very fast and for which, at the same time, the ratio λ_i/λ_o is sufficiently small [2]. (References to the data are cited in [1]. A detailed treatment, which includes the dependence on λ_i/λ_o is given in [2].)

The article mentioned earlier [1] included a review of various electron-transfer reactions of biological interest. In the present communication we shall describe some of the results for the rates of reactions in the photosynthetic bacterial reaction center, and for protein-protein electron-transfers. In particular, a comparison is made of predicted and experimental rate-constants. We consider, first, an example of the application of the theoretical cross-relation [3] to reactions of metalloproteins, in order to illustrate an aspect of the theory which has been frequently tested with purely inorganic or organic systems.

---

*Contribution No. 7248 from the A. A. Noyes Laboratory.

II. Application of the Cross-Relation to Protein-Protein Electron-Transfers

In the cross-relation [3] the rate-constant k_{12} for the reaction between two different redox species,

$$\mathrm{red}_1 + \mathrm{ox}_2 \xrightarrow{k_{12}} \mathrm{ox}_1 + \mathrm{red}_2 \tag{1}$$

is related to two self-exchange rate-constants k_{11} and k_{22}:

$$\mathrm{red}_1 + \mathrm{ox}_1 \xrightarrow{k_{11}} \mathrm{ox}_1 + \mathrm{red}_1 \tag{2}$$

$$\mathrm{red}_2 + \mathrm{ox}_2 \xrightarrow{k_{22}} \mathrm{ox}_2 + \mathrm{red}_2 \quad . \tag{3}$$

The relation is

$$k_{12} \approx (k_{11}k_{22}K_{12}f_{12})^{1/2} \tag{4}$$

where K_{12} is the equilibrium-constant for reaction (1), f_{12} is a known function of k_{11}, k_{22} and K_{12} given, for example, in [1] (frequently close to unity), and the k's appearing in (4) are corrected for the various 'work' terms, by a factor labelled W_{12} in [1]. In the case of metalloproteins, the self-exchange rate-constants k_{11} and k_{22} are typically determined by NMR or EPR line-broadening studies. The cross-relation, (4), is compared in Table I with data in the literature, taken from [1]. (Extensive references to the literature for such studies and tests of the cross-relation are also given in [1].)

Considering the various approximations made in this application of (4), and various uncertainties in the data themselves, discussed in [1], the agreement in Table I can be regarded as quite reasonable. In making such

TABLE I. Comparison of Observed and Calculated Rate-Constants for Protein-Protein Reactions

Oxidant	Reductant	k_{12},obsd $[M^{-1}\ s^{-1}]$	k_{12},calcd[a] $[M^{-1}\ s^{-1}]$
Cytochrome c	Azurin	1.6×10^3	3×10^4
Plastocyanin	Cytochrome c	1.0×10^6	5×10^5 [b]
Stellacyanin	Cytochrome c	3.5×10^2	4×10^3
Azurin	Cytochrome c-551	6.0×10^6	8×10^5 [c]
Plastocyanin	Cytochrome c-551	7.5×10^5	5×10^5 [c]
Cytochrome c	Cytochrome c-551	6.7×10^4	1×10^5 [c]

[a] Unless otherwise indicated, the calculated rate-constants do not contain the W_{12} correction. References for the results are given in [1].

[b] The W_{12} correction is considerably larger for this reaction than for the other systems considered, and was included in the calculation of k_{12}.

[c] Calculated using $1.2 \times 10^7\ M^{-1}\ s^{-1}$ for the self-exchange rate-constant and +0.28 V for the reduction potential of the cytochrome c-551 couple using data cited in [1], where it is also noted that the net charge on the protein is -2/-3.

an application, the rates of both the self-exchanges and the cross-reactions should be determined under pH conditions, where both redox forms of each protein have the same number of protons. (This may not have been the case for all of the reactions in Table I, and, if so, the experimental rate-constants in those cases should be corrected accordingly.)

As a preamble to the consideration of absolute rate constants for reactions between reactants fixed in a membrane, we consider first some available data on the dependence of the electron-transfer rate on the separation distance between the reactants.

III. Distance-Dependence of the Rate-Constant

A knowledge of the dependence of the rate-constant on the separation distance between fixed redox sites is needed for understanding their electron-transfer rate-constants. Information on this distance-dependence comes from several sources: (i) the unusual time-dependence of reaction rates when there is a distribution of separation distances of the reactants in frozen or other solid or highly viscous media, e.g., [4,5], (ii) the conductivity of fatty acid salt layers of varying chain length between two electrodes, e.g. [6,7], (iii) intramolecular electron-transfer rates in model compounds containing a pair of fixed sites separated by more or less rigid organic bridges, e.g., [8-11], and (iv) intramolecular transfers between the metal center of a metalloprotein and an inorganic redox complex attached to various sites in the protein [12,13]. Of these sources, there are fairly extensive published data on (i) and (ii), while data are beginning to appear on (iii) and (iv). The published data for the latter two are too few to estimate a distance-dependence, although this situation will undoubtedly be remedied soon.

Experimental data for the rate-constant or, in the case of method (ii), for the conductivity are usually represented by an exponential dependence on the separation-distance r, e.g., [4-7],

$$k(r) \propto \exp[-\beta(r-r_0)] \qquad (5)$$

TABLE II. Some Tentative Values of β in the Literature

System	β [$Å^{-1}$]	Comment
biphenyl-aromatics[a]	1.2	frozen media
biphenyl$^-$-naphthalene	1.2	frozen media
biphenyl$^-$-phenylethylene	1.0	frozen media
biphenyl$^-$-acridine	1.1	frozen media
pyrene$^+$-TMPD[b]	1.15	frozen media
biphenyl$^+$-TMPD[b]	1.15	frozen media
biphenyl$^-$-cinnamoyl grp.	(< 1.0)	steroid bridged
Al/C_n/Hg, Al or Au	1.5	conduction
Al/C_n/Mg	1.0	conduction

[a] Many aromatic acceptors were used, and gave an average value of β = 1.2 $Å^{-1}$.

[b] TMPD denotes tetramethylphenylenediamine.

Values of β are given in Table II. A typical value is seen to be in the range 1.0 to 1.2 $\AA^{-1}$, and we shall use a value of 1.2 $\AA^{-1}$ when, as in Table I, the intervening material between the two redox sites is "saturated", i.e., nonconjugated. When a portion of the intervening material is conjugated (e.g., aromatic) the value of β for that part of the separation-distance is expected to be lower.

A functional form for $k(r)$, which takes into account the various nodal properties of the electronic wave functions, as well as the three-dimensional geometry, is expected to be somewhat more complicated than that given by (5) and to include, for example, a pre-exponential term which depends on r, e.g., [14]. However, (5) has been frequently used in the literature, since the presently available data do not yet document the need to use a more complicated expression.

IV. Expressions for the Rate Constant

In applying (5), or an equation based on it, to the experimental data for reactions between fixed sites, some cognizance should be taken of the nonspherical nature of the reactants. A distance of particular interest is that between the closest atoms (of the two reactants) that are strongly coupled to their respective redox sites. This distance, denoted by d, is used in the following approximate expression for the reaction-rate constant k, an expression discussed more fully in [1]:

$$k_{cl} \simeq 10^{13}\, e^{-\beta d}\, e^{-\Delta G^*/RT}\ s^{-1} \quad . \tag{6}$$

When the motion of the nuclei is treated classically ($k = k_{cl}$), we have [1]:

$$\Delta G^* = \frac{\lambda}{4}\left[1 + \frac{\Delta G^\circ}{\lambda}\right]^2 \tag{7}$$

where ΔG^* and ΔG° are the free energy barrier and the "standard" free energy of reaction when the reactants are a distance r apart. Strictly speaking, the pre-exponential factor in (6) depends weakly on the temperature, being inversely proportional to $\sqrt{T}$ in the case of a "nonadiabatic" reaction (a reaction for which $\exp(-\beta d) \ll 1$), e.g., discussed in [1].

When the motion of the nuclei is treated, instead, quantum mechanically, the following expression for the ratio of the quantum to the classical rate-constants can be used [1]. It is based on now standard expressions in the literature for k_q and k_{cl} that are cited, for example, in [1]:

$$\frac{k_q}{k_{cl}} \simeq \frac{\sqrt{4\pi\lambda RT}}{h\nu}\, \frac{e^{vy - S\coth y}\, I_v(S\,\mathrm{cosech}\, y)}{e^{-\Delta G^*/RT}} \quad , \tag{8}$$

where ν is some mean vibration frequency for the relevant nuclear motion (a one-reactive-mode model has been used for the quantum rate-constant in (8)), v is $-\Delta G^\circ/h\nu$, S is $\lambda/h\nu$, y is $h\nu/2RT$, and I_v is a modified Bessel function. The ratio k_q/k_{cl} reduces to unity when $h\nu/2RT$ becomes small, and is of the order of unity even when $h\nu/2RT \simeq 1$. For example, when $-\Delta G^\circ/h\nu$ and $\lambda/h\nu$ are, respectively, (1,2), (3,6), (5,10) and (10,20), the values k_q/k_{cl} are 1.1, 1.1, 1.2 and 1.5 in the case of $y = 1$. For $\Delta G^\circ = 0$ and the above values of $\lambda/h\nu$, the corresponding values of

k_q/k_{cl} are 1.3, 1.4, 1.6 and 2.3, respectively. Finally, when $-\Delta G^\circ = \lambda$, the values of k_q/k_{cl} are all equal to 0.87 for the above values of $\lambda/h\nu$. (It can be shown that, when $v = S$ and provided that v is not too small, the ratio k_q/k_{cl} is equal to $\sqrt{(\tanh y)/y}$ which is equal to 0.87 for $y = 1$.)

It is characteristic of both the classical and the quantum expressions for the rate-constant that a maximum appears in k_{cl} and k_q as a function of increasing $-\Delta G^\circ$, when $-\Delta G^\circ \simeq \lambda$ [1]. Further, the activation energy for the classical expression and, in effect, for the quantum expression (given by (72) of [1]) is zero when $-\Delta G^\circ = \lambda$. We use this result later.

The distance d appearing in (6) was taken in [1] to be the actual separation distance of the centers of the two closest (in the case of aromatics) carbon atoms of the two reactants, minus some amount to allow for the extension of the π-electronic orbitals beyond the carbon nuclei. For concreteness a value of $d \simeq 3$ Å was used. Thereby, when the frequency of motion along the reaction coordinate is about 10^{13} s^{-1} the reaction is "adiabatic" when these aromatic carbons are a distance 3 Å or less apart.

V. Reaction Rates in the Bacterial Photosynthetic Reaction Center

The reaction step involving the electron-transfer from an electronically-excited bacteriochlorophyll dimer $*(BChl)_2$ to a neighboring monomer BChl and pheophytin BPh has a lifetime of about 5 ps [15] at room temperature, while the transfer from BPh^- to the quinone has a lifetime of about 200 ps (about 100 ps below 100 K) [15]. Both reactions are essentially temperature-independent over this broad temperature-range, and we shall suppose, therefore, that $-\Delta G^\circ \simeq \lambda$ for each. The values for d for the $*(BChl)_2$ + BChl and BPh^- + Q reactions are about 1 Å and 7 Å, respectively, obtained after subtraction of the 3 Å, in each case, from distances estimated from the crystal structure data (of Rps. viridis) [16]. Equation (6) then yields the lifetimes given in Table III.

The value for ΔG° appears to be quite small for the $*(BChl)_2$ + BChl and $BChl^-$ + BPh reactions, as is the λ for each reaction of these large relatively rigid molecules (data are summarized in [1]), perhaps 0.1-0.2 eV. The value of $-\Delta G^\circ$ for the BPh^- + Q reaction, on the other hand, is quite large (~ 0.61 eV for Rps. sphaeroides and ~ 0.47 eV for Rps. viridis), λ for a reaction involving Q is expected to be larger than λ's for reactions involving only the large molecules, partly because of the change in C-O bond length accompanying the $Q \rightarrow Q^-$ reaction and partly because of the smaller size of the Q molecule. (Cf Ref. [1] for factors influencing reaction rates, e.g., (2)-(6) there.) Calculated rates are given in Table III.

Variation of the Q caused ΔG° to vary by ~ 0.2 eV [17] and caused experimentally a variation in the lifetime for the BPh^- + Q reaction of a factor of two, with some randomness due to the somewhat different positions, shapes, or orientations of the different Q's. Again, when a second quinone was present (presumably as Q^-), creating a coulombic repulsion with the other Q^- newly formed in the BPh^- + Q reaction, the rate-constant for the latter reaction was smaller by almost a factor of two [18]. Theoretical calculations for this effect, taken from [1], are included in Table III.

The lifetime for the reduction of the $(BChl)_2^+$ by ferricytochrome c-558 is about 0.2 μs at room temperature [19]. The reduction of this $(BChl)_2^+$

TABLE III. Absolute Rates for Electron-Transfer Reactions in Photosynthetic Reaction Centers

Reaction[a]	ΔG° [b] [kcal mol^{-1}]	τ_{obsd}	τ_{calcd}
cyt c^{II} ~~ $BChl_2^+$	-3.9	0.2 μs	0.02 μs
$^*BChl_2$ ~~ BChl ~~ BPh	-5(?)	~ 1 to 4 ps[c]	>0.3 ps[d]
BPh^- ~~ Q	-12	100 to 200 ps[e]	140 ps
BPh^- ~~ Q_A, Q_B^- present	~ 0.2 eV[f]	increases factor of 2	increases factor of 2
BPh^- ~~ Q, Q varied	~ 0.2 eV[g]	changes factor of 2	changes factor of 2
*Porphyrin ~~ Q	$-\lambda$	τ_{min} ~ 4 ps	τ_{min} ~ 9 ps

[a] Asterisks indicate electronically excited states.

[b] The ΔG° values are approximate.

[c] Assignment of time-constants is uncertain.

[d] The value of τ_{calcd} is larger than this value if smaller λ's are used, i.e., τ_{calcd} is about 1 ps when λ = 1.2 kcal mol^{-1}. (See also Appendix II of Ref. [1].)

[e] Value depends on temperature.

[f] This is the approximate decrease in $|\Delta G^\circ|$ due to the presence of Q_B^-.

[g] This is the approximate variation in $|\Delta G^\circ|$ as Q is varied.

is presumably by the closest ferrocytochrome c molecule. In the calculations in Table III, an estimated edge-to-edge separation distance of $(BChl)_2^+$ to the closest heme ring of about 11 Å was used, estimated from the crystal structure data [16], and hence d ≈ 8 Å, using the 3 Å correction mentioned earlier. The value of $-\Delta G^\circ$ is about 0.17 eV, and a value of λ of about 13.5 kcal mol^{-1} was estimated† for this reaction, as discussed in [1]. The measured and calculated lifetimes for the reaction are given in Table III.

There is also a study of model porphyrin-quinone systems [20], for which a plot of rate-constant vs. $-\Delta G^\circ$ gave a maximum rate, and for which d could also be estimated. The details are described in [1] and the result is included in Table III.

The experimental and theoretical results in Table III are seen to be in quite reasonable agreement, considering the various approximations and uncertainties present. Calculations utilizing the quantum expression in (8), rather than just the classical one in (7), give values [1] not too

†This rough estimate of λ was made [1] using the λ for the cytochrome c self-exchange reaction, the barrierless result for the $*(BChl)_2$ - BChl - BPh reactions (with their λ then estimated from $\lambda \simeq -\Delta G^\circ$ for those steps) and an additivity rule for λ's [3].

different from those in Table III, at least for the approximate value of $h\nu$ assumed (~ 200 cm^{-1}). Recent calculations employing a frequency of 333 cm^{-1} (instead of the 200 cm^{-1} used in [1]) again show that the ratio of quantum to classical rate-constants at 25 °C is close to unity when $\lambda \lesssim 6$ kcal mol^{-1} (and $\Delta G° = -3$ kcal mol^{-1}) for the second entry in Table III. The classical rate-constant at 25 °C is typically a good approximation to the quantum rate-constant at that temperature for the frequencies used. The temperature coefficient of the reaction can therefore be found by considering the ratio k_{cl} (25 °C)/k_q (0 K), calculated from (14) and (71), respectively, in [1]. In the barrierless case ($-\Delta G° = \lambda$) this ratio is found to be $\sqrt{h\nu/2RT}$ when Stirling's approximation is introduced into (71) (T is the temperature of the classical measurement, 25°C in this case). For barrierless reactions the ratio k_{cl} (25 °C)/k_q (0 K) is typically close to unity for the vibrational frequencies used. The above considerations also show, we believe, the usefulness of the studies of the reaction rate-constant as a function of $-\Delta G°$. The latter permits one to locate the maximum in the plot, namely where $\Delta G° \simeq -\lambda$, and thereby the $\Delta G°$ where the reaction is barrierless over a broad temperature-range. It thereby provides one way, for example, of estimating λ for the reaction, and provides also a direct measure of the factor $10^{13} \exp(-\beta d)$ in (7).

In the calculations in Table III we have adopted a value of β of about 1.2 $Å^{-1}$ as representing to a fair degree the data in Table II. A value of 1.0 $Å^{-1}$ would have altered the calculated results in Table III relatively little. There is some indication that β will depend on the orbital energy of the donor and of the acceptor, e.g. [21], and, in particular, that it may depend on whether or not a reactant is in an excited electronic state, e.g., [22]. There is some indication, for example, from data given in [1] (Table IV there), that β may indeed be less for photoexcited systems. However, in the case of the present Table III only one reaction involves a photoexcited molecule, and the value for its d is so small that calculated lifetime would be hardly affected by using a somewhat smaller value for β. Some discussion of the dependence of β on the orbital energies is given in (65)-(68) of [1].

A second aspect of β for which more experimental data is desirable concerns the role of the intervening material. Because of the delocalized nature assumed for an electronic orbital in an aromatic reactant in our computation of the distance d, it would be consistent not to include in the separation distance any length occupied by a conjugated molecule, at least as a first approximation.

In conclusion, we would like to note that the recent major breakthrough and focal point of the present conference - the determination of the crystal structure of Rps. viridis [16] - has provided an essential and previously missing ingredient required for the theoretical calculation of the reaction rates in the bacteriochlorophyll reaction center.

Acknowledgement. Research at Brookhaven National Laboratory (N.S.) was performed under contract DE-AC02-76CH00016 with the U.S. Department of Energy and supported by its Division of Chemical Sciences, Office of Basic Energy Sciences, while that at Caltech (R.A.M.) was supported by the National Science Foundation and the Office of Naval Research.

References

1. R. A. Marcus and N. Sutin: Biochim. Biophys. Acta 811, 265 (1985).
2. H. Sumi and R. A. Marcus: submitted.
3. R. A. Marcus: Disc. Faraday Soc. 29, 21 (1960); J. Chem. Phys. 43, 679 (1965).
4. J. R. Miller: Science 189, 221 (1975); J. R. Miller and J. V. Beitz: J. Chem. Phys. 74, 6746 (1981).
5. I. V. Alexandrov, R. F. Khairutdinov and K. I. Zamaraev: Chem. Phys. 32, 123 (1978).
6. B. Mann and H. Kuhn: J. Appl. Phys. 42, 4398 (1971).
7. E. E. Polymeropoulos: J. Appl. Phys. 48, 2404 (1977).
8. L. T. Calcaterra, G. L. Closs and J. R. Miller: J. Am. Chem. Soc. 105, 670 (1983).
9. A. Joran, B. Leland, G. Geller, J. J. Hopfield and P. B. Dervan: J. Am. Chem. Soc. 106, 6090 (1984).
10. N. S. Hush, M. N. Paddon-Row, E. Cotsaris, H. Oevering, J. W. Verhoeven and M. Heppener: Chem. Phys. Lett. 117, 8 (1985).
11. H. Heitele and M. E. Michel-Beyerle: submitted.
12. D. G. Nocera, J. R. Winkler, K. M. Yokom, E. Bordignon and H. B. Gray: J. Am. Chem. Soc. 106, 5145 (1984).
13. S. S. Isied, C. Kuehn and G. Worosila: J. Am. Chem. Soc. 106, 1722 (1984).
14. P. Siders, R. J. Cave and R. A. Marcus: J. Chem. Phys. 81, 5613 (1984).
15. C. Kirmaier, D. Holten and W. W. Parson: private communication; D. Holten, C. Hoganson, M. W. Windsor, C. C. Schenck, W. W. Parson, A. Migus, R. L. Fork and C. V. Shank: Biochim. Biophys. Acta 592, 461 (1980); W. Zinth, M. A. Franz, M. C. Nuss, W. Kaiser and H. Michel: data presented at this conference.
16. J. Deisendorfer, O. Epp, K. Miki, R. Huber and H. Michel: J. Mol. Biol. 180, 385 (1984).
17. M. R. Gunner, Y. Liang, D. K. Nagus, R. M. Hochstrasser and P. L. Dutton: Biophys. J. 37, 226A (1982).
18. M. J. Pellin, C. A. Wraight and K. J. Kaufman: Biophys. J. 24, 361 (1978).
19. D. Holten, M. W. Windsor, W. W. Parson and J. P. Thornber: Biochim. Biophys. Acta 501, 112 (1978).
20. M. R. Wasielewski, M. P. Niemczyk, W. A. Svec and E. B. Pewitt: J. Am. Chem. Soc. 107, 1080 (1985).
21. J. R. Miller: data presented at this conference.
22. D. N. Beratan and J. J. Hopfield: J. Am. Chem. Soc. 106, 1584 (1984).

Effects of Distance, Energy and Molecular Structure on Long-Distance Electron-Transfer Between Molecules [1]

J.R. Miller

Chemistry Division, Argonne National Laboratory, Argonne, IL 60439, USA

The centerpiece of this meeting is the exciting measurements of the structure of the photosynthetic reaction center of R. viridis by MICHEL, DEISENHOFER and co-workers [1-3]. To understand the marriage of structure to function in this beautiful machine, we need to understand the basic principles of several processes, one of which is electron transfer (ET) between molecules. Things we must know are how electron transfer rates depend on distance, energy, spatial orientation, polarity of the medium and molecular structure of all the material involved. Our work is aimed at learning these basic principles through experimental measurements of electron transfer between molecules. The molecules we use are not components of photosynthesis, but are usually aromatic molecules. To learn how ET rates depend on distance, it has been necessary to hold electron donors and acceptors at fixed distances. It turns out that fixed distances are very helpful, perhaps even necessary, to answer the other questions as well.

Here I present a short summary of findings from both published work and work in progress, which may help in considering how reaction centers work. I will close with speculative suggestions.

1 Electron Transfer (ET) Rates Decrease Exponentially With Distance

Experimental measurements [4,5] of ET rate-constants, k, over nine orders of magnitude in time, were well described by exponential dependence on distance,

$$k(r) = \nu_o F \exp - \alpha(r-r_o) \tag{1}$$

where the frequency factor ν_o is about $10^{13.5}$ s^{-1}, F is a relative Franck-Condon factor (see below), α characterizes the decrease of electronic interaction with distance, and r_o is a contact distance. Figure 1 shows unpublished data of Krongauz [5] measuring rate vs. distance for ET from the radical anion of the polycyclic aromatic hydrocarbon, pyrene.

The rate as a function of distance is measured by freezing a solution of two molecules, D and A, in a glass-forming solvent such as MTHF (2-methyltetrahydrofuran). Pulse radiolysis is then used to attach electrons to a small fraction of the molecules and the progress of the reaction

$$D^{\bar{\cdot}} + A \longrightarrow D + A^{\bar{\cdot}} \tag{2}$$

[1] Work performed under the auspices of the Office of Basic Energy Sciences, Division of Chemical Science, US-DOE under contract number W-31-109-ENG-38.

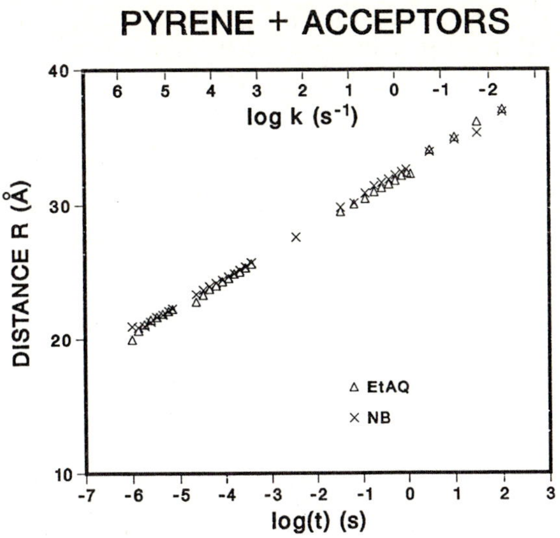

Figure 1 Kinetics of long distance electron transfer from pyrene radical anions to the stronger electron acceptors 2-ethylanthraquinone or nitrobenzene in MTHF glass at 77 K. The frozen solutions contained 0.3 M pyrene and 0.025 M of EtAQ or NB. Pulse radiolysis created pyrene$^{\bar{\cdot}}$ ions, which were observed by their strong absorption at 492 nm. The average distance from pyrene$^{\bar{\cdot}}$ to the nearest acceptor is 21 Å at [acceptor] = 0.025 M, but the distribution of distances gives the huge distribution of rates observed.

is followed by time-resolved optical absorption spectroscopy of the radical ions. The distributions of D-A distances gives a distribution of rates.

Figure 1 makes use of time-resolved measurements of the surviving fraction of Py$^{\bar{\cdot}}$ anions (Py ≡ pyrene) in the presence of electron acceptors to obtain a time-dependent reaction or tunneling distance, R, which is plotted in the figure. Because k(R), the rate constant for a D^-, A pair at distance R is just $(1.9\ t)^{-1}$ [6,7], where t is the time at which the tunneling distance equals R, Figure 1 can also be read to obtain the rate constant k(R). According to eq 1, a plot of log k(R) (top axis) vs. R (vertical axis) is linear with a slope $-\alpha \ln 10$. From the data in Figure 1, we obtained $\alpha = 1.13$ Å^{-1} [5].

In rigid solutions, e.g., glasses, it is possible to measure rate as a function of distance for many reactions. Two conditions are required [4,8,9]:

1. The distribution of D-A distances is known, which is possible when the distribution is random.
2. The distribution of rates observed (which typically spans many decades) is due predominantly to the distribution of distances.

It has been possible to devise tests which have strongly supported the methods used to obtain rates as a function of distance [8,9], but also point out that for weakly exothermic reactions which are very slow, the second condition breaks down, at least partially [9].

The first condition is often easily met. When D and A are neutral molecules, we need only determine that aggregation is negligible, a condition which makes many potentially interesting systems useless.

The second condition usually succeeds because the distribution of rates due to distance is so huge. However, the method yields only angle-averaged rate constants. It is possible to have large dependences of rates on angular orientation which have no effect on the spectrum of rates in a random distribution. As long as we accept angle-averaged rates, this is not a problem.

Distributions of redox energy levels [10] due to varied frozen-in reactant and/or solvent conformations can give large distributions in rates for weakly exothermic reactions. This problem has been shown to complicate measurements of rate vs. distance for very weakly exothermic ET reactions [9].

2 Distance Dependence and Tunneling Barrier Heights - Continuum or Superexchange

Knowing that ET rates are measured to decrease exponentially with distance, we then wish to ask about the damping coefficient α of eq 1. In the example above, ET from pyrene radical anion, α was about 1.1 $Å^{-1}$.

The simplest models of electron tunneling suggest that α depends fairly strongly on the binding energy, B, of the electron to the electron donor:

$$\alpha = 2\sqrt{2mB}/h \qquad (3)$$

B is the energy necessary to remove the electron from the electron donor and place it in a "free" conduction state between the donor and acceptor. Eq 3, which is based on the idea that the material between the donor and acceptor can be approximated by a region of relatively smooth potential energy, has been used, usually with notes of caution, to estimate α.

The early fast ET steps in bacterial reaction centers transfer electrons at redox potentials of about −0.5 to −0.1 eV (vs. NHE). We estimate that B is about 4 eV for these processes. Eq 3 would then estimate $\alpha \cong 2.1$ $Å^{-1}$. We have measured α for ET reactions occurring at a similar redox level, where the donor is the radical anion of 2-ethylanthraquinone [5]. The observed $\alpha = 1.43$ $Å^{-1}$ is much smaller than predicted by eq 3. Furthermore, reactions of several other donors, ranging to triptycene$^{\bar{\cdot}}$, which has a redox potential 2.2 v more negative than 2-ethylanthraquinone (so B is cut in half), α never becomes less than 1.1 $Å^{-1}$ [5]. It is clear that α is smaller than predicted by eq 3 and is much more weakly dependent on B than predicted by eq 3.

The value $\alpha = 1.43$ means the rate will decrease by a factor of 10 for each 1.61 Å increase in distance. For a 10 Å distance, the rate would decrease by a factor of 1.6×10^6. The prediction of eq 3 gives a much larger factor of 1.3×10^9. Such differences could be important in the operation of the reaction center.

The results are much better accommodated by a superexchange model, first used for electron-transfer by McConnel [11]. This model pictures the long distance electron exchange (or electron-transfer) interaction between donor and acceptor to be propagated through electronic states of the material in between them (see Figure 2). This leads to a different relation [12]:

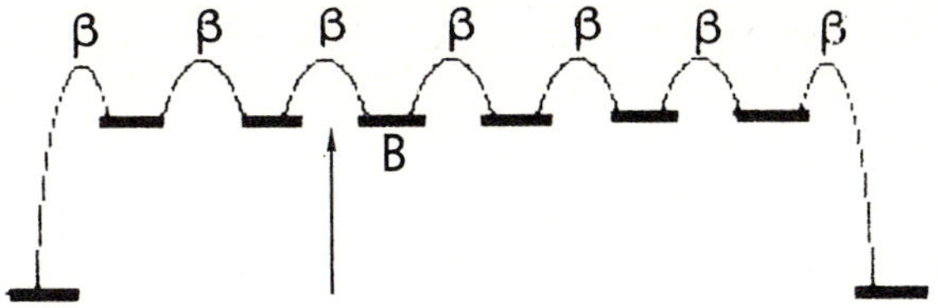

Figure 2 The superexchange model of electronic interactions for long-distance electron transfer: the interaction between the donor (D) and acceptor (A) occurs indirectly via interactions (β) with high energy states which place the electron in the matter between D and A.

$$\alpha = \frac{2}{d} \ell n(B/\beta) \qquad (4)$$

which is consistent with our measurements, obtained in MTHF glass at 77 K. In eq 4, d is the spacing between sites of the material in between the donor and acceptor. B is the energy needed to remove the electron from the donor and place it in one of these in-between sites, and β is the electron-transfer interaction between adjacent sites.

The failure of eq 3 and support of the superexchange model was also seen in reactions of radical cations [12], which occur at redox levels even more positive than the reaction between $(Bchl)_2^{+\bullet}$ and cytochrome c. These reactions appear to occur via interactions among filled, rather than empty, orbitals of the material between D and A. Thus, it may be appropriate to refer to them as "hole tunneling" rather than electron tunneling. The same mechanism is plausible in reaction centers. The superexchange picture received similar support from measurements on ET reactions in photoexcited metal complexes [13] or aromatic molecules [14], and in phenyl, diphenyl or naphthyl-bridged donor-acceptor molecules recently studied by HEITELE [15].

3 Rate vs. ΔG: Reorganization Energies and the "Inverted" Region

Figure 3 shows the dependence of ET rates on the thermodynamic driving force, ΔG° for the reactions of biphenyl$^{\bar{\bullet}}$ ions with several electron acceptors in MTHF glass at 77 K [4].

This curve and a similar one found when distance is fixed by separating D and A with a rigid molecular framework in molecules like that in Figure 4 [19], show that ET rates decrease for strongly exothermic reactions, as predicted by electron-transfer theories. This decrease could be a powerful way for the reaction center to prevent highly exothermic back reactions, which would dissipate stored redox energy.

A remarkable feature of Figure 3 is [4] how well the data conforms to very simple ET theories (solid line). When the electronic interaction is large, more complicated behavior may occur near the maximum, according to recent work of FISCHER [21], using soliton methods.

With the system in its relaxed initial state, $D^{\bar{\bullet}}$, A, sudden removal of the electron from D^- will be followed by relaxation of D and the surrounding solvent. Bond lengths and angles of D change to optimal values for the neutral molecule, and the solvent molecules, especially polar groups, will adjust to the loss of electric fields from $D^{\bar{\bullet}}$. Similar, but opposite reorganizational motions will occur at $A^{\bar{\bullet}}$ after addition of the electron

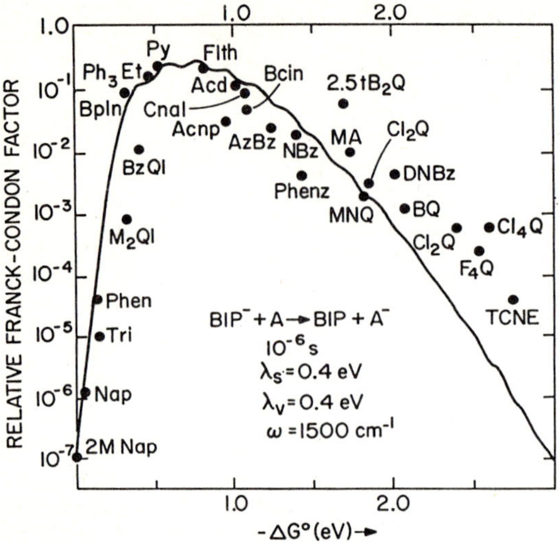

Figure 3 Measured dependence [4] ET rates, expressed as reactive Franck-Condon factors, for electron-transfer from the radical ion of biphenyl (BIP) to several different acceptors A in MTHF glass at 77 K. The rate is maximized when $-\Delta G^\circ$ equals the sum of the solvent reorganization energy, λ_s, and the reorganization energy due to changes in molecular vibrations of the donor and acceptor, λ_v. The curve drawn through the data is an interpretation [4], according to nonadiabatic ET theories [16-18].

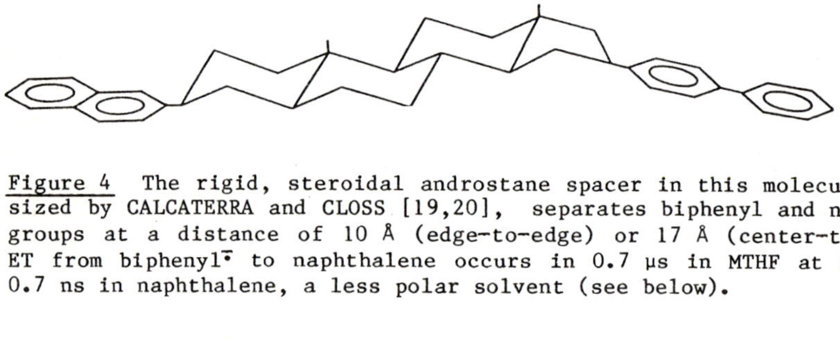

Figure 4 The rigid, steroidal androstane spacer in this molecule synthesized by CALCATERRA and CLOSS [19,20], separates biphenyl and naphthalene groups at a distance of 10 Å (edge-to-edge) or 17 Å (center-to-center). ET from biphenyl$^{\overline{\cdot}}$ to naphthalene occurs in 0.7 μs in MTHF at 295 K, but 0.7 ns in naphthalene, a less polar solvent (see below).

to dissipate a similar amount of energy. The ET rate in Figure 3 is maximized when the free energy change equals the total reorganization energy for the reaction, which is the sum of the energies dissipated by D, $A^{\overline{\cdot}}$ and the solvent.

The reorganization energy includes a large contribution, $\lambda_s = 0.4$ eV, from the solvent [4]. This seems remarkable. The solvent is a rigid glass in which diffusion is absent even on a time scale of hours or days, yet some substantial rotational reorientational motions apparently occur by 10^{-6}. It is possible that the large electric fields eminating from the charged species help to drive the solvent reorganization. At longer times, λ_s increases further, nearly doubling after another eight-decade change in time, according to the interpretation of reference [4].

4 ET Mechanisms and Reaction Center Function, Some Speculations

While the glasses are impressive at reorganizing around transient charged species despite their rigidity, bacterial photosynthetic reaction centers are impressively opposite: they allow the primary charge separation reactions to occur with very low reorganization energies. This is indicated by the observations that the first charge separation steps are very fast and are temperature-independent, despite their small exothermicities.

Small reorganization energy is a desirable feature for efficient operation of molecular charge separating devices, like photosynthetic reaction centers. When the reorganization energy is small, weakly exothermic reactions are fast, so forward reactions may be fast with little expenditure of energy, while highly exothermic back (charge recombination) reactions are slow.

The difunctional steroids provide an impressive example of how reorganization energy controls weakly and strongly exothermic ET rates [19]. For ET in these molecules, the reorganization energy is easily charged by changing solvent polarity, while ΔG remains almost the same [19]. Table I gives ET rates for ET from biphenyl$^{\cdot -}$ to either naphthalene ($\Delta G^\circ = -0.05$ eV) or benzoquinone ($\Delta G^\circ = -2.1$ eV) groups across the androstane spacers in moderately polar MTHF and non-polar isooctane.

TABLE I
Effect of Solvent Polarity on Rate Constants for Intramolecular ET from Biphenyl$^{\cdot -}$ to Acceptors Across a Rigid Androstane Spacer [19].

		$k(s^{-1})$	
Acceptor	ΔG°, eV	MTHF	Isooctane
naphthalene	−0.05	1.5×10^{6}	1.5×10^{9}
benzoquinone	−2.1	2.5×10^{8}	3.6×10^{6}

Upon lowering the solvent polarity, which makes the reorganization energy small, the weakly exothermic reaction speeds up by a factor of 1000, and the strongly exothermic reaction slows down by a factor of 70 [19].

It seems clear that small reorganization energies are a desirable feature of reaction centers. But how are reaction centers constructed to obtain these small reorganization energies?

One possible answer is suggested by analogy to the ET in glasses. The reaction center and the glasses are both rigid, but the glasses are not perfectly rigid, so reorganization does occur. But we saw in the glasses that the reorganization energy increased with time [4]. At short times (picoseconds) reorganization energies might be very small in both glasses and reaction centers.

This leads to the speculative suggestion that the close (almost touching) contact of $(BChl)_2$, BChl, BPheo and Q_1 in *R. viridis* reaction centers, and the resulting extremely fast ET rates have as one purpose, making ET so fast that reorganizational motions do not compete. These reorganizational motions may well be inevitable at long times, even in reaction centers. A larger spacing of the chromophores would, in this view, slow down the forward reactions, not only because of decreased electronic interactions, but also because much more reorganization of the medium would occur at each step. The relative importance of back reactions would then be increased. In this view, the reaction center is not necessarily very special among rigid environments. The special thing is that a series of very fast reactions are run in it between species which are close together and which may be oriented to maximize electronic interactions. A corollary is that slow ET reactions in proteins will have relatively large reorganization energies, and will not be fast unless their exothermicities are as large. This corollary is at least consistent with current measurements of ET in proteins [22-25]. This view would also suggest that the slow back ET from $Q_1^{\overline{\bullet}}$ to $BChl_2^{\overset{+}{\bullet}}$ would have a much larger reorganization energy than the fast forward reactions. This suggestion is consistent with the lack of temperature-dependence for this reaction, and the relative insensitivity of the rates to quinone substitution [26]. Whether the time-dependent relaxations I suggest here occur in reaction centers is not known. KIRMAIER, HOLTEN and PARSON [27] observe wavelength and temperature-dependence of ET in R. Sphaeroides reaction centers which might be due to such relaxations.

5 Concluding Remarks

Given that we have the structure of the viridis reaction center, and some knowledge of how ET rates depend on distance and energy, we might expect to quickly understand how function is related to structure. Here I have presented some data which will help to do this, but also a view that subtleties will make a real, quantitative understanding elusive for some time.

References

1. H. Michel: in this volume
2. J. Deisenhofer: in this volume
3. J. Deisenhofer, O. Epp, K. Miki, R. Huber, H. Michel: J. Mol. Biol. 180, 385 (1984)
4. J. R. Miller, J. V. Beitz, R. K. Huddleston: J. Am. Chem. Soc. 106, 5057 (1984)
5. V. Krongauz, R. K. Huddleston, J. R. Miller: to be published
6. R. K. Huddleston, J. R. Miller: J. Phys. Chem. 86 200 (1982)
7. A. B. Doktorov, R. F. Khairutdinov, K. I. Zamaraev; Chem. Phys. 61, 351 (1981)
8. R. K. Huddleston, J. R. Miller: Phys. Chem. 85, 2292 (1981)
9. R. K. Huddleston, J. R. Miller: J. Phys. Chem. 87, 4867 (1983); 86, 1347 (1982)
10. R. K. Huddleston, J. R. Miller: J. Chem. Phys. 79, 5337 (1983)
11. H. McConnel: J. Chem. Phys. 35, 508 (1961)
12. J. R. Miller, J. V. Beitz: J. Chem. Phys, 74 6746 (1981)
13. K. Simolo and G. L. McLendon, to be published
14. J. Miller, K. Hartman, S. Abrash: J. Am. Chem. Soc. 104, 4296 (1982)
15. H. Heitele, M.E. Michel-Beyerle: in this volume
16. J. Ulstrup and J. Jortner: J. Chem. Phys. 63, 4358 (1975)
17. R. P. Van Duyne, S. F. Fischer: Chem. Phys. 5, 183 (1974)
18. See many other references cited in [4]
19. J. R. Miller, L. T. Calcaterra, G. L. Closs: J. Am. Chem. Soc., 106, 3047 (1984)

20. L. T. Calcaterra, G. L. Closs, J. R. Miller: J. Am. Chem. Soc. 105, 670 (1983)
21. S. F. Fischer: in this volume
22. S. Peterson-Kennedy, J. McGourty, B. Hoffman: J. Am. Chem. Soc. 106, 5010 (1984)
23. G. McLendon, J. Miller: J. Am. Chem. Soc. in press
24. D. Nocera, J. Winkler, K. Yocum, E. Bordignon, H. Gray: J. Am. Chem. Soc. 106, 5145 (1984)
25. S. Isied, C. Kuehn, G. Worosilia: J. Am. Chem. Soc. 106 1722 (1984)
26. M. Gunnar, P. L. Dutton: to be published
27. C. Kirmaier, D. Holten, W. Parson: Biochim. Biophys. Acta, in press

Ultrafast Electron Transfer in Biomimetic Models of Photosynthetic Reaction Centers

M.R. Wasielewski, M.P. Niemczyk, W.A. Svec, and E.B. Pewitt

Chemistry Division, Argonne National Laboratory, Argonne, IL 60439, USA

1. Introduction

The primary electron-transfer event of photosynthesis involves oxidation of the lowest excited singlet state of a chlorophyll electron donor by a nearby electron acceptor [1]. The distance between the donor and the acceptor is restricted by the surrounding reaction center protein. The high quantum efficiency of photosynthetic charge-separation depends on favorable electron-transfer rates between electron donors and acceptors that are positioned in precise spatial relationships relative to one another, and that possess redox potentials which result in movement of an electron down a stepped potential gradient. We recently prepared a series of restricted distance porphyrin-quinone donor-acceptor molecules designed to study electron-transfer proceeding from the lowest excited singlet state of the porphyrin, Fig. 1 [2]. We have measured the dependence of photoinduced electron-transfer rates and subsequent dark charge recombination rates on the free energy of reaction in these molecules [3]. We find that the rate-constants for both radical ion pair formation and recombination in these molecules depend on the exothermicity of the respective electron-transfer reaction in the manner originally proposed by Marcus and later modified and extended by others [4].

Using this information, we have now synthesized a molecule for which the donor-acceptor electron-transfer rates are optimized to promote

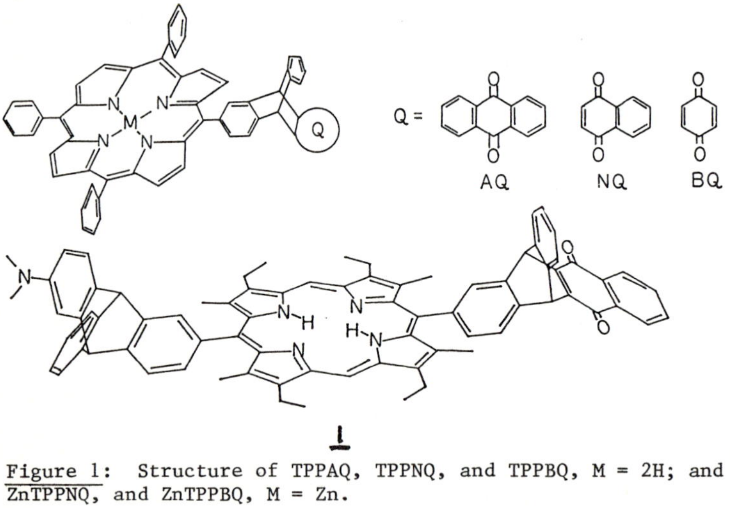

1

Figure 1: Structure of TPPAQ, TPPNQ, and TPPBQ, M = 2H; and ZnTPPAQ, ZnTPPNQ, and ZnTPPBQ, M = Zn.

efficient two-step charge separation over a known long distance. In addition to a primary porphyrin-quinone donor-acceptor pair, this molecule possesses a secondary N,N-dimethylaniline donor, 1. The donor-acceptor distances and orientations in 1 are strongly restricted by the polycyclic cage structures between them. The center-to-center distances [5] are: porphyrin-quinone, 10.5 Å; aniline-porphyrin, 10 Å; and aniline-quinone, 25 Å. Along with compound 1 was synthesized the analogous compound 2 which lacks the N,N-dimethylamino group and therefore lacks the secondary electron donor. Di-triptycycloporphyrin 3 which lacks both the N,N-dimethylamino and naphthoquinone groups was also prepared as a reference.

2. Experimental

The syntheses of the porphyrin-quinone molecules has been described [2]. The syntheses of compounds 1-3 will be described in a forthcoming publication [6].

The laser system that was used to obtain the data is as follows: The 600 nm, 1 ps, 0.5 nj output of a mode-locked Ar^+-synchronously pumped dye laser was amplified to 2.5 mj using a 4-stage dye amplifier pumped by a frequency-doubled Nd-YAG laser operating at 10 Hz. This pulse was split. A 2.0 mj pulse used to generate a picosecond continuum probe pulse, while a 0.5 mj pulse was used for sample excitation. Absorbance measurements were made with a double beam spectrometer which employed optical multichannel detection. Time delays between pump and probe pulses were accomplished with an optical delay line.

Fluorescence lifetimes were determined by the time-correlated photon counting method. Transient absorption measurements were carried out by exciting the porphyrin-quinone molecules with a 1 psec laser flash at 600 nm. The formation and decay of the porphyrin-quinone radical pair state was monitored at 460 nm and 650 nm with a 1 psec broadband probe light pulse. Measurements on the simple porphyrin-quinone molecules were performed at 21° in toluene (ε = 2.4) and in butyronitrile (ε = 20), while measurements on compounds 1-3 were performed in butyronitrile alone. Each transient was analyzed using the treatment of Provencher, and could be described by a single exponential rise or decay [7]. It is important to note that the radical ion pair state of each molecule decays cleanly to ground-state with no evidence of excited state formation [2].

Quantum yields of charge separation were determined by comparing the measured number of moles of excited molecules with the molarity of the resultant ions, as measured directly from the transient absorbance data as described previously [2]. Since the extinction coefficients of the ion are known, the quantum yield of ion formation is obtained directly.

3. Results and Discussion

The exothermicity of the charge separation, $-\Delta G_{cs}$ and that of the charge recombination, $-\Delta G_{cr}$ were obtained from Eqns. 1 and 2, respectively:

$$-\Delta G_{cs} = E(S_1) - E_D^{ox} + E_A^{red} + \Delta G(\varepsilon) \qquad (1)$$

$$-\Delta G_{cr} = E_D^{ox} - E_A^{red} - \Delta G(\varepsilon) \qquad (2)$$

where $E(S_1)$ is the energy of the lowest excited singlet state of the molecule, E_D^{ox} and E_A^{red} are the measured $E_{1/2}$ values for the one-electron oxidation and reduction of the porphyrin donor and quinone acceptor, respectively [2], ε is the solvent dielectric constant, and $\Delta G(\varepsilon)$ is a term which incorporates both the solvent dependent coulomb energy-change upon ion pair formation or recombination, and the free energy of solvation of the ions. $\Delta G(\varepsilon)$ was evaluated using the recent treatment of Weller, which is based upon application of the Born equation [8]. Referring to Eqn. 16 in Weller's paper, an ionic distance a = 10.5 A and an average ionic radius of 5.2 A was used to calculate $\Delta G(\varepsilon) = -0.45$ V for toluene and = -0.03 V for butyronitrile. The correction for toluene agrees well with the fact that we did not observe radical ion pair-state formation whenever $E(S_1) < E_D^{ox} - E_A^{red} + 0.4$ V [2,3].

Figure 2 is a plot of rate-constant vs reaction exothermicity for both radical ion pair formation from the lowest excited singlet state of the porphyrin and for radical ion pair-recombination leading back to the ground-states of the donor and the acceptor [9]. At the 10.5 ± 0.5 A donor-acceptor distance in these molecules the charge-separation rate-constants for $-\Delta G > 0.4$ V approach the 2-4 x 10^{11} sec $^{-1}$ rate-constants observed for the primary charge separation of photosynthesis [1].

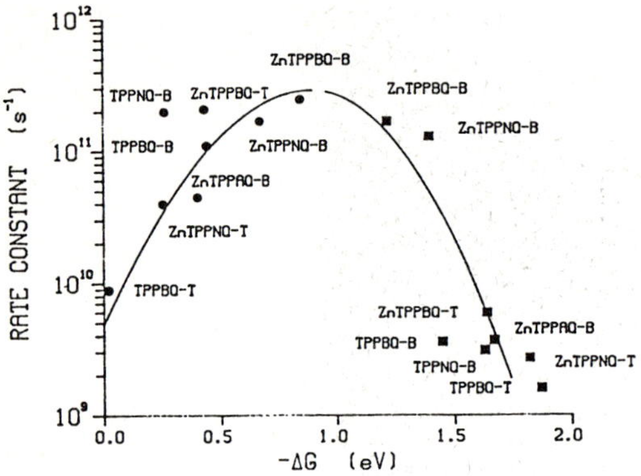

Figure 2: Plot of rate-constant vs exothermicity for the reaction $^{1*}P - Q \rightarrow P^+ - Q^-$, ; and for $P^+ - Q^- \rightarrow P - Q$, ; where P = porphyrin and Q = quinone. The -B and -T suffixes after the name of the compounds indicate data obtained in butyronitrile or in toluene, respectively. The maximum uncertainty in any given rate-constant is ± 20%.

Plotted on the same graph are the rate-constants for the subsequent radical ion pair-recombination reaction. Since each of these reactions proceeds directly to the ground electronic state without intermediate excited state formation, the charge recombination rate-constants can be measured at high exothermicities. The rate-constant for this process dramatically decreases over two orders of magnitude as the exothermicity of the recombination reaction increases.

This large decrease in rate is predicted by theory and results from increasingly unfavorable Franck-Condon factors between the radical ion pair state and the corresponding neutral ground-state [4]. Miller et al. recently observed solvent-dependent 30-500 fold decreases in rate for highly exothermic electron-transfer over a fixed 15 A distance between aromatic radical anions and neutral molecules [10]. The electron-transfer rate-constants observed in that work are about 1000 times smaller than those reported here, presumably because the anion-neutral molecule distance is longer than the porphyrin-quinone distance, 15 A vs 10.5 A, respectively. Both of these fundamentally different experiments demonstrate that a key prerequisite for clear observation of the so-called inverted region in the rate vs $-\Delta G$ dependence is the elimination of diffusion control of the electron-transfer kinetics [4j]. This is highlighted by the fact that several of our measured rate-constants are faster than the diffusion control limit.

Although the charge separation and recombination reactions involve different electronic states of the porphyrin donor, the dependencies of the rate-constants for each reaction on exothermicity shown in Figure 2 reach a common maximum at about $k = 2.5 \times 10^{11}$ sec^{-1} and $-\Delta G = 0.9$ eV. This is reasonable because the nuclear configuration, and hence the vibrational energy-level structure of the lowest excited π^* singlet states of rigid molecules with extended π systems, e.g., porphyrins [11], is often very similar to that of the ground singlet state. Using this assumption, the maximum of the curve at about $-\Delta G = 0.9$ V can be interpreted as the approximate total reorganization energy for both the charge-separation and recombination reactions [4].

The question remains as to what fraction of this energy-change is due to relaxation processes involving the solvent vs the porphyrin-quinone molecules. Interestingly, the measured reorganization energy is approximately solvent-independent. The fastest electron-transfers may be competing with dielectric relaxation of the solvent [12] and with vibrational relaxation [13] within the porphyrin-quinone molecules. These effects would tend to reduce the solvent-dependence. However, this is not a complete explanation. Comparisons between x-ray structural data of porphyrins and porphyrin radical cations show that oxidation results in little change in nuclear configuration [14]. Ir and raman spectroscopic evidence suggests that reduction of quinones to their radical anions principally results in changes in the C-O distance [15], so that nuclear reorganization is somewhat more important for the quinone than that for the porphyrin. Nevertheless, the solvent should contribute strongly to the total reorganization energy. Current work is focusing on these problems.

Given the data base of information on rate vs exothermicity for the electron-transfer reactions of our porphyrin-quinone molecules, we decided to prepare a two-step electron-transfer model that would optimize the exothermicity of each step in order to promote highly efficient unidirectional electron-transfer. The resultant radical ion pair would possess a long cation-anion distance after the second electron-transfer step. The molecule that we designed is compound 1.

One-electron redox potentials for compound 1 and related reference compounds are listed in Table I. The redox potentials of each donor and acceptor are not altered by their incorporation into 1 and 2. In high dielectric constant solvents the sum of the redox potentials for oxidation of the donor and reduction of the acceptor is a good estimate of the radical pair energy-level [8]. Thus, in compounds 1 and 2 the energy of P^+Q^- is 1.53 V above the ground-state, while in compound 1 the energy of D^+PQ^- is 1.39 V above the ground-state, where P = porphyrin, Q = quinone, and D = N,N-dimethylaniline. Since the lowest excited singlet state of these molecules is the 1.95 V singlet state of the porphyrin, the exothermicity of the reaction $^{1*}PQ \rightarrow P^+Q^-$ is 0.42 V and that of the reaction $DP^+Q^- \rightarrow D^+PQ^-$ is 0.14 V. Using the measured rate vs free energy data for

Table I. Energetics

Compound	$E^{-}_{1/2}$ [a]	$E^{+}_{1/2}$ [a]	$E^{+}_{1/2} - E^{-}_{1/2}$
1	-0.61; -1.42	0.78; 0.92	1.39; 1.53
3	-1.41	0.92	2.23
2-(N,N-dimethyl-amino)triptycene	-	0.78	-
triptycenenaphtho-quinone	-0.61; -1.42	-	-

[a] Redox potentials (volts vs SCE) were measured by cyclic voltametry at a Pt disc electrode. The measurements were performed in butyronitrile containing 0.1M tetra-n-butylammonium perchlorate at 21°.

porphyrin-quinone molecules in Fig. 2 we predict the following reaction rate-constants: $^{1*}PQ \rightarrow P^{+}Q^{-}$, 10^{11} sec^{-1}; $P^{+}Q^{-} \rightarrow PQ$, 8×10^{9} sec^{-1}; and $DP^{+}Q^{-} \rightarrow D^{+}PQ^{-}$, 2.5×10^{10} sec^{-1}. Thus, the secondary electron-transfer reaction $DP^{+}Q^{-} \rightarrow D^{+}PQ^{-}$ should be able to compete very effectively with the $DP^{+}Q^{-} \rightarrow DPQ$ charge recombination reaction in 1.

The ground-state optical absorption spectra of 1 and 2 are simply the sum of contributions from the individual chromophores, while the emission spectra are that of the unperturbed porphyrin alone. The fluorescence quantum yields of 1 and 2 are both 0.0002 and are highly quenched relative to the 0.06 fluorescence quantum yield of 3. The fluorescence lifetimes of both 1 and 2 are < 30 ps. These data indicate that a better than 99% efficient nonradiative pathway for decay of the singlet excitation energy in 1 and 2 exists.

The nature of this pathway was determined by transient absorbance measurements. Compounds 1 and 2 were excited with a 1 ps laser flash at 600 nm. The formation and decay of the porphyrin cation were monitored at 460 nm and at 650 nm [16], while the formation of the naphthoquinone anion and the N,N-dimethylanilinium cation were monitored at 480 nm [17] and 470 nm [18], respectively.

Excitation of either 1 or 2 results in formation of a transient spectrum due to $P^{+}Q^{-}$ with a rate-constant of 1.1×10^{11}, Fig. 3. In compound 2 the $P^{+}Q^{-}$ state decays cleanly back to ground-state with a rate-constant of 5.6×10^{9}, Fig. 4A. For compound 1 the transient due to $P^{+}Q^{-}$ decays more rapidly with a rate-constant of 1.4×10^{10} to a state that posseses absorbance in the blue region of the spectrum characteristic of D^{+} and Q^{-} with no persistent changes in the red, which would indicate the presence of the porphyrin cation, Fig. 4B. The $D^{+}Q^{-}$ state decays to the ground-state of 1 with a lifetime of 2.45 μsec, Fig. 5. A comparison between both the rate-constants and the absorbance change data for the reactions $P^{+}Q^{-} \rightarrow PQ$ and $DP^{+}Q^{-} \rightarrow D^{+}PQ^{-}$ both support the conclusion that $D^{+}PQ^{-}$ is formed from $DP^{+}Q^{-}$ in about 71% yield.

The measured electron-transfer rate-constants all agree within a factor of two of the values predicted by our rate vs free energy of reaction data. Thus, the appropriate choice of reaction exothermicities

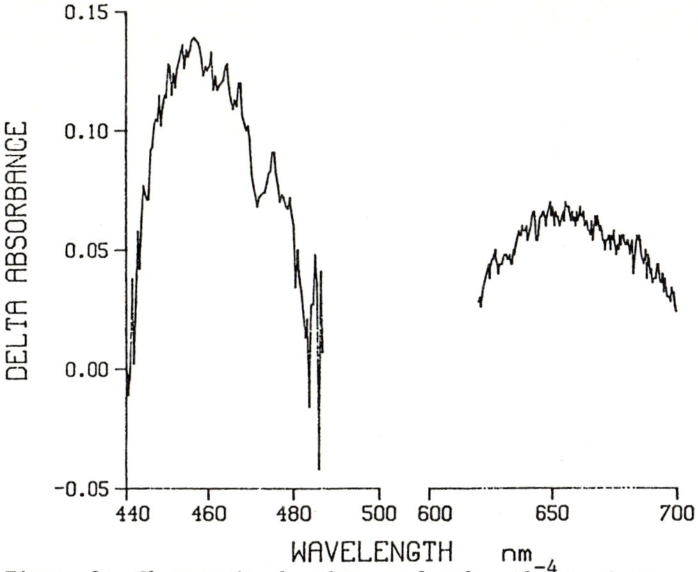

Figure 3: Changes in absorbance of a 2×10^{-4}M solution of 1 in butyronitrile at 20 ps following a 0.5 mJ, 1 ps, 600 nm laser flash. Cell pathlength, 2 mm. Filters that reject stray light cut out the 490-620 nm wavelength region, while the sharp cutoff at 440 nm is due to the intense absorption of the porphyrin Soret band at 421 nm.

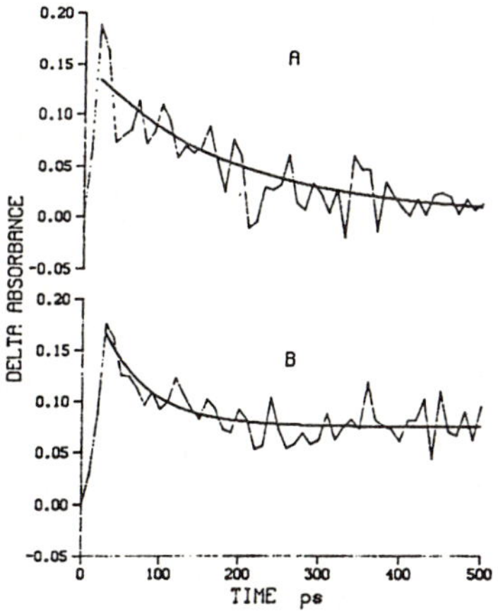

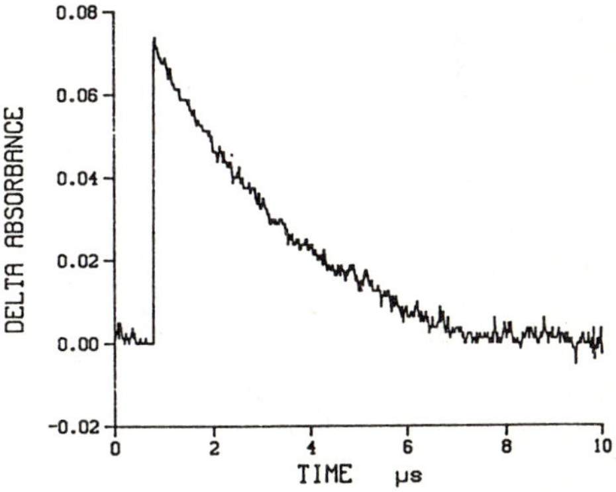

Figure 4: Transient absorbance decays measured at 460 nm of 2×10^{-4}M solutions of A) 2 and B) 1 in butyronitrile following a 0.5 mJ, 1 ps, 600 nm laser pulse. The superimposed smooth curves are the exponential fits to the data.

Figure 5: Transient absorbance decay of 1 in butyronitrile measured at 460 nm following a 0.5 mJ, 1 psec, 600 nm laser pulse.

results in a molecule that produces a charge-separated state in high quantum yield that is long lived and stores about 70% of the photon energy.

4. Conclusions

Two principal features of the simple porphyrin-quinone molecules allow us to clearly observe inverted rate vs $-\Delta G$ behavior at high exothermicities. First, the donor-acceptor distance is restricted, and second, the highly exothermic charge recombination reactions do not produce electronically excited states of the donor or acceptor.

Our results for compound 1 which undergoes sequential electron-transfer reactions show that it is possible to design molecules with very specific charge-separation capabilities, provided that three requirements are met. First, the distances between the various donors and acceptors must be highly restricted. Second, for a given donor-acceptor distance the dependence of the electron-transfer rates on the free energy of reaction must be known. This second requirement enables one to choose the appropriate redox potentials for the donors and acceptors to optimize the electron-transfer rates, and thereby maximize the quantum yield of charge-separation. Third, long-lived charge separation over a fixed distance requires more than a single optimized electron-transfer step to achieve long-lived radical ion pair states in a manner analogous to photosynthesis [19].

5. Acknowledgement

This work was supported by the Division of Chemical Sciences, Office of Basic Energy Sciences, Department of Energy under contract W-31-109-Eng-38.

6. References

1. D. Holten, C. Hoganson, M.W. Windsor, C.C. Schenck, W.W. Parson, A. Migus, R.L. Fork, C.V. Shank: Biochim. Biophys. Acta 592, 461 (1980).
2. M.R. Wasielewski, M.P. Niemczyk: J. Am. Chem. Soc. 106, 5043 (1984).
3. M.R. Wasielewski, M.P. Niemczyk, W.A. Svec, E.B. Pewitt: J. Am. Chem. Soc. 107, 1080 (1984).
4. a) R.A. Marcus: J. Chem. Phys. 24, 966 (1956). b) V.O. Levich: Adv. Electrochem. Eng. 4, 249 (1966). c) R.R. Dogonodze: in Reactions of Molecules at Electrodes, ed. N.S. Hush (Wiley-Interscience, New York, 1971). d) R.P. Van Duyne, S.F. Fischer: Chem. Phys. 5, 183 (1974). e) N.R. Kestner, J. Logan, J. Jortner: J. Phys. Chem. 78, 2148 (1974). f) J. Ulstrup, J. Jortner: J. Chem. Phys. 63, 4358 (1975). g) D. DeVault: Q. Rev. Biophys. 13, 387 (1980). h) M. Redi, J. Hopfield: J. Chem. Phys. 72, 6651 (1980). i) B.S. Brunschwig, J. Logan, M.D. Newton, N. Sutin: J. Am. Chem. Soc. 102, 5798 (1980); 77, 2387 (1982). j) R.A. Marcus, P. Siders: J. Phys. Chem. 86, 622 (1982).
5. The donor-acceptor distances were estimated using Corey-Pauling-Koltun molecular models.
6. M.R. Wasielewski, M.P. Niemczyk, W.A. Svec, E.B. Pewitt: J. Am. Chem. Soc. 107 (1985) in press.
7. S.W. Provencher: J. Chem. Phys. 64, 2772 (1976).
8. A. Weller: Z. Phys. Chem. (Wiesbaden) 133, 93 (1982).
9. The curves drawn in Figure 2 for both the charge-separation and the charge-recombination data are each the best fit to a quadratic function, see ref. 4A.

10. J.R. Miller, L.T. Calcaterra, G.L. Closs: J. Am. Chem. Soc. 106, 3047 (1984).
11. P.G. Seybold, M. Goutermann: J. Mol. Spectrosc. 31, 1 (1969).
12. D. Huppert, H. Kanety, E.M. Kosower: Faraday Discuss. Chem. Soc. 74, 161 (1982).
13. M. Bixon, J. Jortner: Faraday Discuss. Chem. Soc. 74, 17 (1982)
14. L.D. Spaulding, P.G. Eller, J.A. Bertrand, R.H. Felton: J. Am. Chem. Soc. 96, 982 (1974).
15. a) B.R. Clark, D.H. Evans: J. Electroanal. Chem. 69, 181 (1976).
    b) G.N.R. Tripathi: J. Chem. Phys. 74, 6044 (1981).
16. R. Bonnett, C. Lambert, E.J. Land, P.A. Scourides, R.S. Sinclair, T.G. Truscott: Photochem. Photobiol. 38, 1 (1983).
17. I.B. Afanas'ev, N.I. Polozova, G.I. Samohkvalov: Zhur. Org. Khim. 12, 2536 (1976).
18. T. Shida, W.H. Hamill: J. Chem. Phys. 44, 2369 (1966).
19. T.A. Moore, D. Gust, P. Mathis, J.-C. Mialocq, C. Chachaty, R.V. Bensasson, E.J. Land, D. Doizi, P.A. Liddell, W.R. Lehman, G.A. Nemeth, A.L. Moore: Nature 307, 630 (1984).

Electron Transfer Through Aromatic Spacers in Bridged Electron-Donor-Acceptor Molecules

H. Heitele and M.E. Michel-Beyerle

Institut für Physikalische und Theoretische Chemie,
Technische Universität München, D-8046 Garching, F.R.G.

1. Introduction

In many biological systems, such as the respiratory chain or photosynthetic reaction centers, electrons are transferred between redox centers over large distances with high efficiency at unusually high rates. One reason for fast rates and directionality could be that a particular protein structure enhances the electronic interaction between the redox centers. Therefore, it is interesting to understand how matter between redox components influences electron-transfer processes. Simple organic molecules, in which electron donors are linked to acceptors via molecular spacers, are suitable as model systems for an investigation of such a medium effect.

Within the formalism of nonadiabatic electron-transfer theory /1/ the dependence of the electron-transfer rate on the detailed electronic structure of the molecule has been described /2/. Among the factors determining the electronic properties are: the local electronic overlap between the electron donor and acceptor states and the states of the spacer, the energies of these states, and the length of the bridge.

2. Experimental Results

To test these theoretical considerations and to evaluate their possible implications in biological electron-transfer chains, several bridged electron-donor-acceptor systems with different spacers (1,2,3) have been

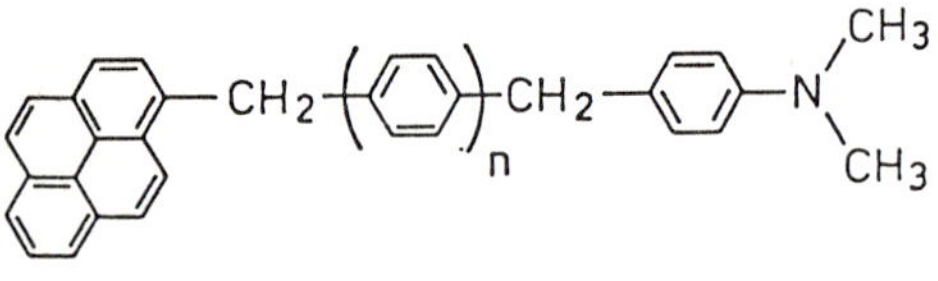

n=1 : 1
2 : 2

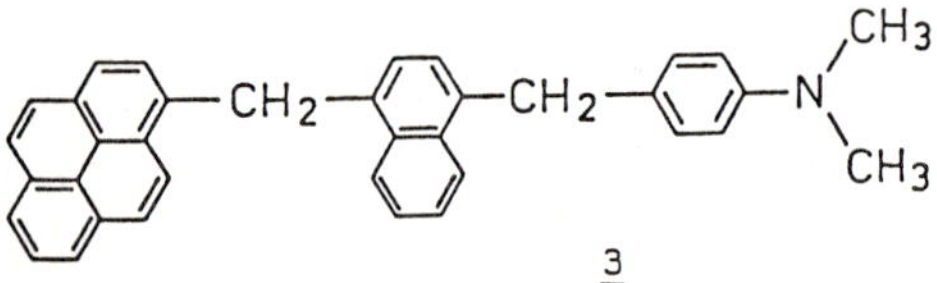

3

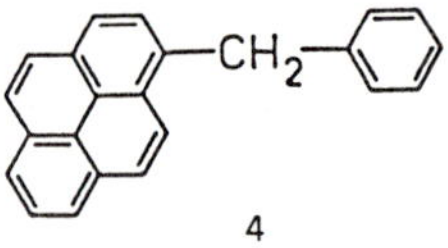

4

synthesized /3/. In these systems, pyrene serves as the electron acceptor, dimethylaniline as the donor and a phenyl-, biphenyl- or naphthyl-residue as spacer; 4 is used as reference compound. Compared to the phenyl-bridge the biphenyl-spacer increases the donor-acceptor-distance by 4Å. With the two spacers phenyl and naphthyl the distances are the same, but the electronic states of the spacers are different.

The rotations around the sigma-bonds at the methylene-groups in (1-4) allow a large variety of conformations. But, contrary to donor-acceptor compounds in which pyrene or anthracene is linked to dimethylaniline by a flexible methylene-chain /4/, intramolecular exciplex formation is not possible. Moreover, the possible conformations in the phenyl- and biphenyl-bridged compounds (1,2) are the same /5/. In the naphthyl-bridged system 3 the hydrogen atoms at the 5- and 8-positions of the naphthyl-residue somewhat restrict the rotational freedom of donor and acceptor. In the following this difference compared to the compounds 1,2 is neglected and it is assumed that comparing these compounds conformational effects cancel.

The fluorescence lifetime of the compounds 1-4 in hexane and acetonitrile solution were measured by time-resolved single photon counting (Table 1). The concentration of the compounds was $5\cdot10^{-6}$ - 10^{-4}M. Compared to the reference substance 4 the lifetimes of the donor-acceptor-systems in acetonitrile solution are considerably diminished, whereas in hexane they are almost unchanged. As steady state fluorescence measurements have shown, the decrease in lifetime is not accompanied by the occurrence of a charge-transfer emission at longer wavelengths.

Table 1

compound	1	2	3	4
τ_{Hex} (ns)	274±15	279±15		287±10
τ_{MeCN} (ns)	1.2±0.1	28.5±1.5	0.9±0.1	207±7
k_{ET} ($10^8 s^{-1}$)	8.3±0.7	0.30±0.02	11.0±1.3	

Table: Fluorescence lifetimes in hexane (τ_{Hex}) and acetonitrile (τ_{MeCN}) and intramolecular electron-transfer rates k_{ET}

In polar solution dimethylaniline quenches the excited singlet state of pyrene, thereby forming radical ions /6/. The free enthalpy change ΔG of this electron-transfer process is approximately -0.5eV in acetonitrile; in hexane the reaction is strongly endergonic because of insufficient stabilization of the ions by the solvent, i.e., the formation of ion pairs is not possible /6/. The presence of ion pairs has been explicitly proved for compound 1 in acetonitrile by monitoring the transient absorption spectrum between 410 nm and 510 nm 4ns after excitation with a nitrogen laser pulse (337 nm) /7/. The absorption bands of the pyrene anion around 490 nm and of the dimethylaniline cation around 460 nm could be clearly identified. The absorption of the ion pairs vanishes with a decay time of ca. 4 ns. The formation of pyrene triplets was not observed. In analogy to these results, the decrease of the fluorescence lifetimes of all the compounds investigated here is interpreted as intramolecular electron-transfer. The electron transfer-rate is given by $k_{ET}=1/\tau-1/\tau_0$, where τ denotes the fluorescence lifetime of the donor-acceptor-system and τ_0 the lifetime of the reference substance. The electron transfer-rates given in Table 1 show, that increasing the donor-acceptor-distance by 4 Å decreases the rate by a factor of about 27 (compare 1 and 2). Moreover, the substitution of the phenyl- by the naphthyl-spacer enhances the rate by a factor of about 1.4. Experiments in frozen matrices /8/ and with saturated hydrocarbon-spacers /9/ have shown a much more pronounced distance-dependence: an increase of the distance by 4 Å leads to a decrease of the rate by a factor of at least 100 up to 600.

3. Discussion

This unusual dependence of the electron transfer-rate on the distance as well as on the nature of the spacer indicates that the bridge actively participates in the electron-transfer process. Within nonadiabatic electron transfer theory the influence of the spacer is frequently discussed in terms of the so-called superexchange mechanism /2,10/. It can be assumed, that both, the solvent and the intramolecular reorganization during the electron-transfer are not seriously affected by the spacer. The Franck-Condon factors are then the same for all compounds, and the ratio of two transfer rates to be compared is given by the ratio of their electronic factors only. In its most elementary form, superexchange theory then predicts $k(\underline{1})/k(\underline{2})=|\Delta E/\beta|^2$ /10a/, where β is the resonance integral between the fragment orbitals localized on the two phenyl-rings of the biphenyl-spacer and ΔE corresponds to the (average) energy difference between a fragment orbital and the donor and acceptor states. This formula is only applicable when $|\beta| \ll \Delta E$. Using thermodynamic and spectroscopic data ΔE for the phenyl-bridged compound 1 was estimated to be 1-1.5 eV /11/. $k(\underline{1})/k(\underline{2}) \cong 27$ then yields $|\beta| \cong 0.2-0.3$ eV. Theoretically, β can be estimated with the Wolfsberg-Helmholtz formula $\beta=KS_{ij}(H_{ii}+H_{jj})/2$ /12/. Here S_{ij} is the overlap matrix element between the two fragment orbitals, K is a constant (K=1.75), H_{ii} and H_{jj} are the diagonal elements of the Hückel-matrix for the two fragment orbitals. For the matrix element between two benzene-like HOMOs, localized on the two phenyl-rings of the biphenyl-spacer, this relation gives $|\beta| \cong 1$ eV, i.e., $|\beta| \cong \Delta E$ and, therefore a more exact approach should be used.

An improved treatment has to take into account the whole manifold of states of the spacer as well as the local overlap between the spacer and donor and acceptor. On a semiquantitative level, this can be done using Extended-Hückel molecular orbital calculations /2,13/. The ability of a spacer to transfer an electron is intimately related to the energy-splitting Δ between the symmetric and the antisymmetric linear combination of a pair of (isoenergetic) orbitals which interact via the spacer. Δ corresponds to the splitting of the two (adiabatic) potential surfaces at the intersection. This splitting is due to the term in the total Hamiltonian of the system that gives rise to the electron-transfer. For the matrix element V_{if} of this term between the inital and ion pair states the relation $\Delta=2|V_{if}|$ holds /1a,b/. The nonadiabatic electron-transfer rate is proportional to $|V_{if}|^2$.

For a simulation of the molecules $\underline{1}$-$\underline{3}$ the fragments $\underline{1'}$-$\underline{3'}$ were used. The two p-orbitals at the terminal CH_2-groups correspond to the p-orbitals in the pyrene- and dimethylaniline-residues adjacent to the bridge. The splitting of this pair of orbitals has been calculated for various conformations.

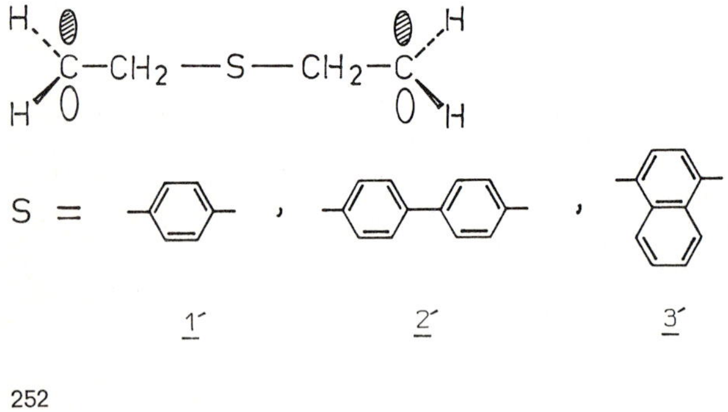

In these calculations the quantity which corresponds to ΔE as defined above is identified with the energy difference between the average of the symmetric and antisymmetric linear combination of the two p-orbitals and the HOMO of the spacer. To mimic the situation in the systems 1-3, this difference for the structure 1' is adjusted to the range 1-1.5 eV (as has been estimated for ΔE of compound 1). A slight change in the Coulomb integral of the p-orbitals of the two terminal C-atoms has the desired effect. The same Coulomb integral is then used for all the structures 1'-3'.

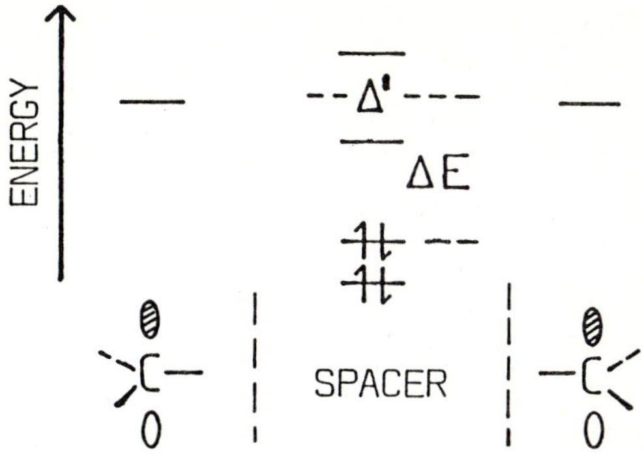

Energy level scheme

To obtain the matrix elements V_{if} of the experimental systems 1-3 it is now assumed that the following relation approximately holds: $\Delta = C_A \cdot C_D \cdot \Delta'$; here Δ, Δ' are the splittings of the systems 1-3 and 1'-3', and C_A, C_D are the coefficients of those p-orbitals in the donating and accepting states that correspond to the terminal p-orbitals in the model structures 1'-3'. These two orbitals provide the main contribution to the electronic overlap between the spacer and donor/acceptor. The proportionality to $C_A \cdot C_D$ is exact in a perturbation calculation, where Δ is proportional to $\eta_A \cdot \eta_D$ /2/; here η_D, η_A are the local resonance matrix elements between spacer- and donor- and acceptor-orbitals. It is assumed that a more exact calculation gives only minor corrections to this relation. Additionally, for the ratios of two splittings in structures with different spacers, a partial cancellation of these corrections can be expected.

Not unexpectedly, the splitting is strongly dependent on the geometry. For most conformations the interaction between the p-orbitals is small, whereas for the following orientation the splitting is significantly enhanced. Since

Conformation with maximum splitting

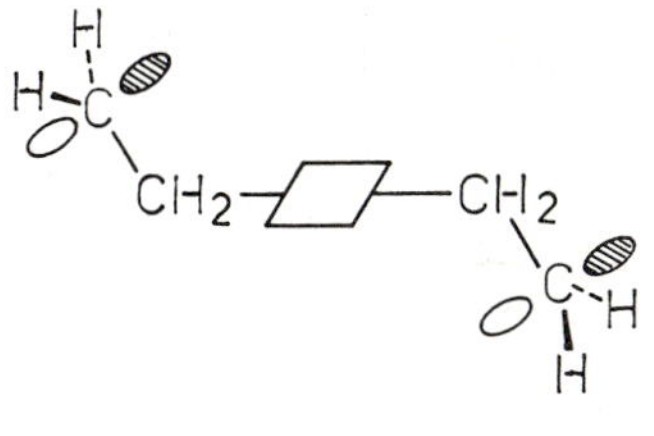

the rate is proportional to Δ^2, the importance of this and related conformations is even more pronounced. One should expect that this conformation occurs with similar probability for all the molecules investigated here. Therefore it can be assumed that averaging over all conformations leads to the same geometric factor in the rate for all compounds. The ratio of two transfer rates is then given by $k/k' = |V_{if}/V'_{if}|^2$, where V_{if}, V'_{if} are the maximum electronic matrix elements for the corresponding spacers. To give an example for the values that typically result from the calculations, the V_{if}'s are given for a situation where ΔE for the phenyl-spacer 1' has been adjusted to 1.1 eV. The ΔE values for 2' and 3' are calculated ones. The V'_{if} values have been corrected with the appropriate atomic coefficients and refer in the notation V_{if} to the experimental systems 1-3.

Table 2

compound	1'	2'	3'
ΔE (eV)	1.10	0.81	0.63
$\|V_{if}\|$ (eV)	0.0200	0.0052	0.0264

The direct interaction between the terminal p-orbitals is negligible, i.e., the participation of molecular orbitals localized on the spacers in electron-transfer is essential.

Due to the crudeness of the method, the absolute values for V_{if} must be considered as a first guess, whereas for the ratios k/k' extensive cancellation of errors can be expected. Variation of ΔE in the range 1-1.5 eV yields k(1)/k(2)=12-21 (27) and k(3)/k(1)=1.6-1.8 (1.4). A comparison with the experimental values in brackets (derived from Table 1) shows that the approach used can qualitatively and even semiquantitatively explain these ratios. In other words, the participation of molecular orbitals localized on the spacer in the electron-transfer provides a sufficient rationalization of the electron-transfer rates for the compounds investigated here.

Preparation of compounds with a differently substituted naphthyl-spacer (1,5- or 2,6- instead of 1,4-substitution) offers the possibility to study the effect of the local overlap between spacer and donor/acceptor on the transfer-rate. Using the same estimates as above, one should expect a ratio of 4:2:1 for the electron-transfer rate in 1,4-, 1,5-, and 2,6-substituted derivatives of the naphthyl-spacer. Experiments with these compounds are in progress.

ACKNOWLEDGEMENTS

We are highly indebted to Prof. Gerhard Wegner for his support of the synthetic work. Fluorescence decay measurements were possible through the generous help of Dr. J. Vogel and of Prof. S. Schneider in this institute, Dr. N. Ernsting, Prof. A. Müller and Prof. F. Schäfer, MPI Göttingen, and Dipl.-Phys. P. Finckh and Prof. M. Klingenberg, University of München.
With respect to computational problems, we profited a lot from the critical discussions with Prof. N. Rösch in this institute. We also thank Prof. S. Fischer and Prof. J. Jortner for their inspiring engagement in the interpretation of measured electron-transfer rates.
Financial support from the Alfried Krupp von Bohlen und Halbach-Stiftung, the Stiftung Volkswagenwerk and the Deutsche Forschungsgemeinschaft is gratefully acknowledged.

References:

/1/ a) J. Ulstrup "Charge Transfer Processes in Condensed Media", Springer Verlag, West Berlin 1979.
b) M.D. Newton and N. Sutin, Ann. Rev. Phys. Chem. 35, 437 (1984).
c) N.R. Kestner, J. Logan, and J. Jortner, J. Phys. Chem. 78, 2148 (1974)
d) R.P. Van Duyne and S.F. Fischer, Chem. Phys. 5, 183 (1974).
/2/ S. Larsson, J. Am. Chem. Soc. 103, 4034 (1981).
/3/ H. Heitele, M.E. Michel-Beyerle, and G. Wegner, to be published.
/4/ M.K. Crawford, Y. Wang, and K.B. Eisenthal, Chem. Phys. Lett. 79, 529 (1981).

M. Migita, M. Kawai, N. Mataga, Y. Sakata, and M. Soichi, Chem. Phys. Lett. 53, 67 (1978).
M. Schulz, Dissertation, University of Göttingen, 1974.
J. Fuhrmann, Dissertation, University of Göttingen, 1981.
/5/ For simplicity a fixed twist angle of 45° between the two phenyl-rings in biphenyl has been assumed in this article. Experimentally in solution an angle between 40° and 50° is found.
/6/ H. Knibbe, H. Staerk, and A. Weller, Ber. Bunsenges. Phys. Chem. 72, 257 (1968).
/7/ H. Heitele, A. Ogrodnik, N. Remy-Richter, M. Volk, and M.E. Michel-Beyerle, to be published.
/8/ S. Strauch, G. McLendon, M. McGuire, and T. Guarr, J. Phys. Chem. 87, 3579 (1983).
J. Miller and J. Beitz, J. Chem. Phys. 74, 6746 (1981).
J.R. Miller, J.A. Peeples, M.J. Schmitt, and G.L. Closs, J. Am. Chem. Soc. 104, 6488 (1982).
/9/ A.D. Joran, B.A. Leland, G.G. Geller, J.J. Hopfield, and P.B. Dervan, J. Am. Chem. Soc. 106, 6090 (1984).
/10/ a) H.M. McConnell, J. Chem. Phys. 35, 508 (1961).
b) A.M. Kuznetsov and J. Ulstrup, J. Chem. Phys. 75, 2047 (1981).
/11/ H. Heitele and M.E. Michel-Beyerle, J. Am. Chem. Soc., in press.
/12/ M. Wolfsberg and L. Helmholtz, J. Chem. Phys. 20, 837 (1952).
/13/ S. Larsson, Chem. Phys. Lett. 90, 136 (1982; J. Chem. Soc., Faraday Trans. 2 79, 1375 (1983). S. Larsson and J.M.O. Matos, J. Mol. Struc. 120, 35 (1985).

Electron Transfer in Rigidly Linked Donor-Acceptor Systems

S.F. Fischer, I. Nussbaum, and P.O.J. Scherer

Physik-Department, Technische Universität München,
D-8046 Garching, F. R. G.

1. Introduction

Electron-transfer processes have been a challenging topic to inorganic chemists and to physicists over the years /1,2/. As primary process in photosynthetic reaction centers the electron transfer gives us a number of questions to answer. The very first step of charge - separation takes place within approximately 5 ps in the bacterium Rhodopseudomonas viridis /3/. So far no other system has been synthesised which is equally fast and equally efficient in the sense that the rate of recombination of the charges is essentially negligible on that time-scale. Since we know the structure of the reaction center /4/ we can attempt to evaluate the rate for the electron transfer, learn about the biological function and find out in which way nature has optimized the process. More specifically, one would like to know:

(i) What is special about the pigments within the reaction center?

(ii) What is the role of the special pair? Is it needed for the charge separation?

(iii) Is there anything important about the protein environment that differs from other solutions?

(iv) Is the accessory monomer directly involved in the electron-transfer or does it act as a bridging element?

Even though we do not have answers to these questions,we can point at some properties which may play a role in this context.

(i) The bacteriochlorophyll molecules or more generally porphyrin ring systems show a relatively small Stokes shift in the absorption spectrum /5/. In addition, their ionisation energy is low and their electron affinity is relatively high.

(ii) A dimer can be more easily polarized in the excited state than a monomer. In particular the very special arrangement of the close neighbourhood of the acetyl group of one monomer and the magnesium of the other can effect the charge distribution and cause sensitive reorganization within the dimer during the transfer process.

(iii) The dielectric relaxation of the protein environment can be much faster than the corresponding relaxation of a very polar medium.

(iv) Since the accessory monomer is also a bacteriochlorophyll, there are more options for the transfer mechanism such as involvement of excitonic coupling or hole-transfer etc.

If some of these special features play a role in the electron-transfer process one would need a theory that is flexible enough to deal with them.

Essential to any theoretical model is the description of the initial and the final state as well as the interaction that couples these states.In this note we want to investigate these questions for a donor-acceptor system which is linked by a rigid bridge. We develop a theory for the initial and the final state as selfconsistently determined trapped states. That means the degree of their special localisation and the degree of internal as well as environmental reorganization is determined simultaneously. This approach leaves only nuclear fluctuations as coupling mechanism.

2. The Conventional Rate Expression

Before we introduce a detailed model and the methods to evaluate the transfer-rate let us discuss briefly the commonly accepted rate expression for a nonadiabatic electron-transfer process /1,2/. The system may consist of a donor with an energy for the transferable electron E_1, a spacer with the lowest unfilled orbital of energy E_s and an acceptor molecule with energy E_N. The molecules are coupled by a transfer term V,which makes the transition from the donor to the acceptor possible. The common procedure to project out the initial and final states is quite simple. It is assumed that the electron is in its initial state entirely localized at the donor site and finally at the acceptor site. The transfer is then induced by the pure electronic coupling which, in the case of a spacer with higher energy than both the donor and the acceptor takes the form of a super exchange interaction. For instance for a spacer consisting of N-2 elements all coupled by the same V it reads/6/

$$V_{ex} = V \left(V/(E_s - E_1) \right)^{N-2} \quad . \tag{1}$$

The rate expression is otherwise completely analogous to the rate expression for nonradiative transitions in molecules /7,8/. One only needs to incorporate the proper Franck-Condon factors for the internal degrees of freedom for the donor and the acceptor as well as the corresponding Stokes shift for the medium modes. The latter may be treated classically so that the rate takes the following form

$$k=\frac{2\pi}{\hbar}V_{ex}^2(4\pi kT\lambda)^{-1/2}\sum_{m_j}\prod_j (m_j!)^{-1} g_j^{2m_j}\, e^{-g_j^2} \exp\left[-\frac{(E_1-E_N+\lambda-\hbar\Omega_j m_j)^2}{4kT\lambda}\right] \tag{2}$$

Here g_j is the dimensionless equilibrium change of the normal mode R_j with frequency Ω_j for the donor as well as for the acceptor. The thermal population of the internal modes is neglected and the medium modes are treated in the classical limit where kT is larger than the medium mode energy.

The criticism we would make towards this approach is twofold. First it seems quite arbitrary to localize the electron entirely at the donor or the acceptor site. The Born-Oppenheimer approximation would be in conflict with this approximation since the electron spreads into the bridge for any given nuclear configuration. Secondly it seems in conflict with the transition state theory /9/because within the spirit of this theory or the Landau Zener theory /10/ for the nonadiabatic limit, the interaction should be taken for a nuclear configuration describing the activated complex or the crossing of the diabatic potential curves. At such a point the interaction can differ appreciably from the super exchange interaction given in Eq. (1). Finally, it seems desirable to have a theory which can describe the adiabatic as well as the nonadiabatic limit with the same conceptual tools. To achieve these goals we have to introduce some concepts related to the soliton theory in solids /11/.

3. The Rate Constant between Selftrapped States

We choose a model hamiltonian of the following form:

$$H= \sum_n E_n \,|n\rangle\langle n| + \sum_n^{N-1} V_{n,n+1}(|n\rangle\langle n+1| + |n+1\rangle\langle n|) + \sum_n \Lambda_n Q_n |n\rangle\langle n|$$

$$+ \sum_k(\lambda_{k1} q_k |1\rangle\langle 1| + \lambda_{kN} q_k |N\rangle\langle N|) + \frac{1}{2}\sum_n (P_n^2+\Omega^2 Q_n^2) + \frac{1}{2}\sum_k (p_k^2+\omega_k^2 q_k^2) \tag{3}$$

Here E_n is the energy of the transferable electron at site n, $V_{n,n+1}$ the transfer interaction and Λ_n is the coupling constant of an internal mode Q_n with frequency Ω. The medium modes q_k are coupled only to the donor site and to the acceptor site with coupling constants λ_{k1} and λ_{kN} , respectively. Taking the ground state expectation value with an electronic wavefunction $\psi=\sum_n \varphi_n |n\rangle$ one can determine φ_n and the equilibrium positions Q_n^0 or q_k^0 by a variational procedure which gives for example /12/

$$Q_n^0 = -\frac{\Lambda_n}{\Omega^2} |\varphi_n|^2 \quad . \tag{4}$$

By substituting the equilibrium values of the coordinates into the ground state expectation value of H , one arrives at a nonlinear functional in φ_n

$$H(\underline{\varphi}) = \sum_n E_n |\varphi_n|^2 + 2\sum_n^{N-1} V_{n,n+1}\varphi_n\varphi_{n+1} - \frac{1}{2}\sum_n \Omega^{-2}\Lambda_n^2 |\varphi_n|^4$$

$$- \frac{1}{2}\left(\sum \omega_k^{-2}\lambda_{k1}^2 |\varphi_1|^4 + \sum \omega_k^{-2}\lambda_{kN}^2 |\varphi_N|^4\right) \quad . \tag{5}$$

The equation $dH(\underline{\varphi})/d\varphi_n = 0$ can be solved for all φ_n. Let us assume that the acceptor has a lower energy than the donor. In this case one can always find a state localized at the acceptor site. Its energy is lowered with regard to E_n by the solvent and internal reorganization energies

$$\Delta E^N_{med} = \sum_k \lambda^2_{kN}/\ 2\omega^2_k \text{ and } \Delta E^N_{int} = \Lambda^2_N\ /\ 2\Omega^2 \quad \text{, respectively.}$$

It is also possible to find a self-trapped state at the donor site as long as the energy difference of the unrelaxed donor and acceptor state is smaller than twice the reorganization energy of the donor state

$$E_1 - E_N < 2\Delta E^1_{med} + 2\Delta E^1_{int}$$

For larger energy differences a state largely localized at the donor site can only be stabilized if it is an excited state, that means its wavefunction must have a node. We call this an excited charge transfer state. Between these two regions there exists a small regime where the excited state loses its character of charge localization.

It is possible to calculate the electron-transfer from either the self-trapped donor state or the excited charge transfer state ($\varphi_n{}^i$)to the relaxed ground state ($\varphi_n{}^f$).The interaction is given by

$$V^i = \sum_n \gamma_n \Lambda_n (Q_n - Q^0_n) + \sum_k \gamma_1 \lambda_{1k}(q_k - q^0_{k1}) + \sum_k \gamma_N \lambda_{Nk}(q_k - q^0_{kN}) \quad (6)$$

$$\text{with} \quad \gamma_n = \varphi^i_n\ \varphi^f_n - |\varphi^i_n|^2 \sum_m \varphi^i_m\ \varphi^f_m \quad (7)$$

The last term in Eq. 7 reflects the fact that the initial and the final electronic states are not orthogonal. For strong localisation we get $\gamma_1 \approx \gamma_N \approx \varphi_N{}^i$ and $\gamma_{n\neq 1,N} \ll \gamma_1$. The important result is that $\varphi_N{}^i$ shows a resonance like behaviour close to $E_i - E_f = 2\Delta E^1{}_{med} + 2\Delta E^1{}_{int}$. For the relaxed states $E'_i = E_i - \Delta E^1{}_{med} - \Delta E^1{}_{int}$ and $E'_f = E_f - \Delta E^N{}_{med} - \Delta E^N{}_{int}$ it reads $E'_i - E'_f = \Delta E = \Delta E^N{}_{med} + \Delta E^N{}_{int} + \Delta E^1{}_{med} + \Delta E^1{}_{int}$.

The transfer rate has the approximate form /13/

$$k_i = (4\pi kT\lambda)^{-1/2} \exp(-G^2_1 - G^2_N) \sum_{m_1 m_N} \frac{G_1^{2m_1} G_N^{2m_N}}{m_1!\ m_N!} \exp\left(-\frac{(\Delta E - \hbar\Omega(m_1+m_N))^2}{4kT\lambda}\right)$$

$$\left(\left(\gamma_1^2 \frac{(m_1 - G_1{}^2)^2}{G_1{}^2} + \gamma_N^2 \frac{(m_N - G_N{}^2)^2}{G_N{}^2}\right)\Delta E_{int}\Omega + (\gamma_1^2 \Delta E^1_{med} + \gamma_N^2 \Delta^N_{med})\frac{kT}{\hbar}\right) \quad (8)$$

$$\text{with } G^2_n = \frac{\Delta E_{int}}{\hbar\Omega} (\varphi^{i\,2}_n - \varphi^{f\,2}_n)^2 \text{ and } \lambda = \Delta E^1_{med} + \Delta E^N_{med} \quad (9)$$

λ is the solvent Stokes shift. Contributions originating from coupling to the bridging sites are neglected. The last term in brackets is the effective interaction V ,which consists of contributions describing transitions induced by the internal modes and those induced by medium modes. The latter become more important at higher temperatures.

4. Discussion of the Results

The theory presented here differs conceptually from the conventional description as described in chapter 2. The main differences are:

(i) The initial and the final state are both not entirely localized and the degree of delocalization depends on the solvent coupling as well as on the internal energies and coupling parameters.

(ii) The electron-transfer is due to fluctuations of the solvent or the internal vibrational degrees of freedom.

(iii) The theory applies not only to the nonadiabatic limit; it incorporates the adiabatic as well as the limit of internal conversion. The adiabatic limit is reached for large V, where the coupling constants γ_N and φ_N become independent on V. The internal conversion relates to the inverted regime, where the initial state has a node.

In principle it should be possible to test the predictions against experiments. In practice this is not so easily done. If the length of the spacer is changed by one /14,15/ the rates should change by a factor of $V^2/(E_1-E_s)^2$ if the formula (1) is used. Our result predicts in the limit of weak coupling essentially the same change, since we get in that limit $k \sim \gamma_N{}^2 \sim V^2/(E_1-E_s)^2$. The often observed weak temperature dependence at low temperature is commonly related to nuclear tunneling,and the increase at higher temperatures to an activated transfer. The two contributions we find in Eq. 8 predict a similar behaviour for different reasons. If the overall Stokes shift of the medium is larger than the one from the internal degrees of freedom,the contributions induced by the internal modes dominate at low temperatures,and the medium modes dominate at higher temperatures. The cross-over should be reached for $kT=\hbar\Omega\Delta E_{int}/\Delta E_{med}$.

A very different behaviour of our theory compared to the conventional one is predicted for the dependence of the rate upon the free energy change. Many data collected by Miller et al. /16/ as well as Wasielewski et al. /17/ seem to fall quite well on one curve, if the rate constants for different systems, are plotted against the free energy change. There is a smooth increase in the rate until the free energy change matches the Stokes shift and a even slower decrease for larger energy gaps, indicating unfavourable Franck-Condon transitions in the so called inverted regime /18,19/. For the interpretations of the experiments the electronic coupling has been approximated by a constant. We predict an almost resonance like increase and an even steeper exponential decrease in the inverted regime (see Fig. 1) which is due to variable electronic coupling around the resonance. Since the transfer in the inverted regime starts out from a state with a node (excited state), we interpret it as an internal conversion process with strong charge transfer

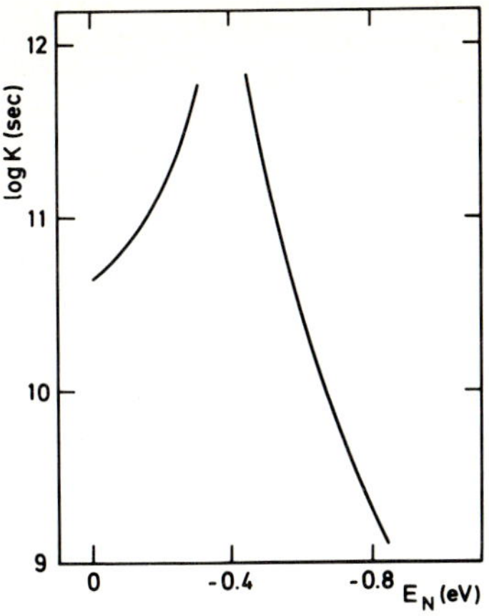

Fig.1 Transfer rate as a function of the energy gap E_1-E_N

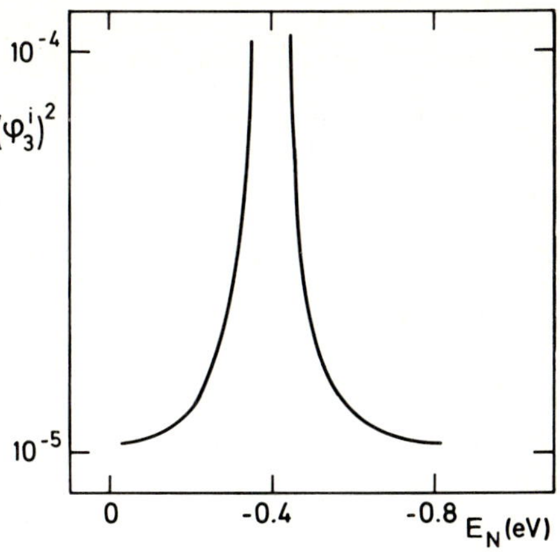

Fig. 2 Probability $(\varphi_3{}^i)^2$ of finding the initially prepared electron at the acceptor site as a function of the energy gap E_1-E_N

The parameters of both figures are N = 3, ΔE_{int} = 0.1 eV, $\Delta E_{med}{}^1$ = 0.05 eV, $\Delta E_{med}{}^N$ = 0.15 eV, $\hbar\Omega$ = 0.125 eV, V = 0.05 eV, T = 300K, E_1 = -0.1 eV.

character. Close to the resoncance the initial state starts to delocalize. Experimentally it may be difficult to be observed, since the dielectric relaxation may limit the transfer. In Fig. 2 it can be seen how the initial state starts to delocalize as the resonance is approached. The probability of finding the electron in the initial state at the site of the acceptor $(\varphi_N{}^i)^2$ is here plotted against the energy of the acceptor site E_N. The transfer is proportional to this probability and therefore the rate shows also a resonance like behaviour. In order to relate these results to the early transfer steps in the reaction center, we have chosen parameters, which mimic the situation for Rhodopseudomonas viridis. The spacer is the accessory monomer with energy E_s set up 0. The donor is the special pair with an energy E_1 =-0.1 eV and the acceptor is the pheophytine with energy E_N = - 0.3 eV. In addition we have introduced couplings to the internal modes with a Stokes shift of $\Delta E_{med}{}^N$ = 0.15 eV. The interaction between the dimer and the monomer as well as the one of the monomer to the pheophytine are put to V =- 0.05 eV. The choice of the parameters is based on preliminary estimates and experimental results /20/. We find a rate of $5\cdot10^{11}$sec^{-1} which corresponds closely to the observed rate /20/. From Fig. 1 it becomes clear that we get this fast rate only because we are already close to the resonance. The temperature dependence corresponding to these parameters is shown in Fig. 3. The process is essentially temperature independent. For comparison we have also drawn the temperature dependence for a process to an energetically lowered state (E_N = - 0.5 eV) which falls into the inverted regime and has an even weaker temperature dependence.

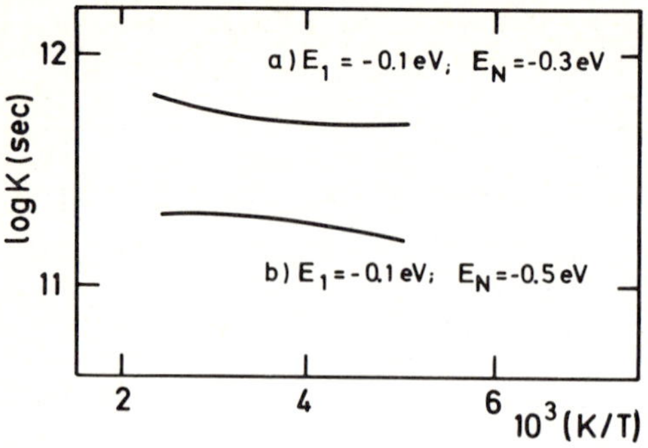

Fig. 3 Transfer rate as a function of temperature
a) normal regime b) inverted regime
Parameters: $N=3, \Delta E_{int}=0.1eV, \Delta E_{med}^{1}=0.05eV$
$\Delta E_{med}^{N}=0.15eV, \hbar\Omega=0.125eV, V=-0.05eV$

These results demonstrate, that the early fast electron transfer step in Rhodopseudomonas viridis from the dimer to the pheophytine might be explained as a single transfer, which involves the accessory monomer only as an intermediate leading to an effective coupling via a super exchange mechanism. The dynamic treatment of this interaction together with the medium relaxation may be essential to understand the fast transition.

Acknowledgement

The authors like to thank Prof. M. E. Michel-Beyerle for many discussions. The work has been supported by the DFG via SFB 143.

References

1. J. Meyer, Progr. Inorg. Chem. 30 389 (1983)
2. J. Jortner, Biochim. Biophys. Acta 594 193 (1980)
3. D. Holten, M. W. Windsor, W. W. Parson, J. P. Thornber, Biochim. Biophys. Acta 501 112 (1978)
4. J. Deisenhofer, O. Epp, K. Miki, R. Huber, and H. Michel, J. Mol. Biol. 180 385 (1984)
5. R. J. Platenkamp, H. J. Den Blanken, and A. J. Hoff, Chem. Phys. Lett. 76 1 (1980)
6. H.M. McConnel, J. Chem. Phys. 35 508 (1961)
7. R. Englman, J. Jortner, Mol. Phys. 18 145 (1970)
   S. F. Fischer, J. Chem. Phys. 53 3195 (1970)
8. E.W. Schlag, S. Schneider, and S. F. Fischer, Ann. Rev. Phys. Chem. 12 465 (1971)
9. Unimolecular Reactions: P. J. Robinson and K. A. Molbrook, Wiley Interscience (1978)
10. C. Zener, Proc. Roy. Soc. A137 696 (1972)
11. A. S. Davydov, Physica 3D 1 (1980)
12. G. Venzl and S. F. Fischer, Phys. Rev. in press
13. G. Venzl and S. F. Fischer: Photoreaktive Festkörper Wahl Verlag, Karlsruhe 1984 ed. H. Sixl pp. 723

14. H. Heitele and M. E. Michel-Beyerle, J. Am. Chem. Soc. in press
15. A. D. Joran, B. A. Leland, G. G. Geller, J. J. Hopfield, and P. B. Dervan, J. Am. Chem. Soc. 106 6090 (1984)
16. J. R. Miller, L. T. Calcaterra, and G. L. Closs, J. Am. Chem. Soc. 106 3047 (1984)
17. M. R. Wasielewski, M. P. Niemcyk, W. A. Svec, and E. B. Pewitt, J. Am. Chem. Soc. 107 1080 (1985)
18. R. A. Marcus, Ann. Rev. Phys. Chem. 15 155 (1964)
19. R. Van Duyne and S. F. Fischer, Chem. Phys. 5 183 (1974)
20. D. Holten and W. Windsor, Ann. Rev. Biophys. Bioeng. 7 189 (1978)

Electron Conduction Along Aliphatic Chains

R. Bittl, H. Treutlein, and K. Schulten

Physik-Department, Technische Universität München,
D-8046 Garching, F. R. G.

1. Introduction

The efficiency of photoinduced charge separation in the photosynthetic reaction center is achieved by a choice of suitable electron-transfer rates which are fast enough in the forward and slow enough in the backward direction. The separated charges seem to be paying a minimum price in terms of irreversible energy losses. The cardinal question is why the reaction center functions so efficiently. An answer to this question is tantamount to explaining which properties of the reaction center protein material are responsible for the realization of its transfer rates.

The rate of electron transfer by organic molecules in the simplest description,which assumes a separation of the motion of electron and nuclear degrees of freedom (Born-Oppenheimer approximation),is given by expressions of the type

$$\text{rate of electron-transfer} = V^2(r)\ f(T) \qquad (1)$$

Here f(T) is the Frank-Condon weighted density and describes the temperature- dependent matching of the vibrational modes of freedom of the electron acceptors and donors and their surrounding. This factor has received most attention in theoretical studies. Its classical description dates back to the 1956 work of MARCUS [1], the quantum mechanical description is due to the work of Pekar and others covered in a 1959 review [2]. A most recent review is Ref. 3.

The first factor in (1) describes the tunneling of the electron between donor and acceptor without accounting for a possible participation of vibrational degrees of freedom along its path. It originates from an overlap of the frontier electron orbitals between donor and acceptor, where the overlap can involve intermediate virtual orbitals either of the solvent (protein) or of intramolecular groups (e.g. the chains of the reaction center chromophores). In this study we want to investigate this first factor of the electron-transfer rate, which despite its wide neglect [4] in the literature is of obvious importance, too.

Reasons for the neglect of the electron-transfer matrix element $V^2(r)$ are at least two: (1) the calculation of this element from first principles is very difficult because its magnitude is actually small compared to the natural energies of electronic states in molecules. (2) The factor cannot be altered easily by the experimentalist, the opposite being true for the Franck-Condon weighted density which varies with the temperature. Because of the poor state of knowledge on $V^2(r)$ one should not expound new theories on this term but rather contribute to its measurement. An interesting attempt has been made in this respect by POTASEK and HOPFIELD [5] who obtained information on the size of $V^2(r)$ through measurements on charge-transfer absorption.

Our approach to the electron-transfer element is through the exchange interaction $J(r)$ between paramagnetic groups. The exchange interaction of two doublet molecules measures the energy separation of the singlet and the triplet states in which such pairs can exist. This exchange energy is related to the electron transfer matrix element by the expression

$$J(r) = V^2(r)\ g \qquad (2)$$

where the factor g depends only on intramolecular electronic properties. r is the distance between the paramagnetic groups. Measuring the distance-dependence of $J(r)$ yields information on $V^2(r)$.

The most simple model system which yields a distance-dependent exchange interaction are two hydrogen atoms separated in the vacuum beyond a distance r of 5Å, say. HERRING and FLICKER [6] determined

$$J(r) = -7.71\cdot 10^9 \text{Gauss}\ r^{5/2}\ \exp(-2r/a_o) \qquad (3)$$

where r is in Ångstrom. In case of the electronic interaction in the reaction center the protein should modify (3). If one likens the interior of the reaction center to an organic solvent, one may replace (3) by an expression suggested by deKANTER et al. [7] on the basis of CIDNP experiments involving two radicals chained together by polymethylene

$$J(r) = 9.46 \cdot 10^9 \text{Gauss} \exp(-2.136 r/\text{Å}) \qquad (4)$$

We have pointed out above in discussing the second factor in equation (1) that electron-transfer rates are strongly influenced by nuclear motions. The low-frequency part of this factor stems from the stochastic motion of electron acceptor and donor as well as solvent molecules. The effect of this stochastic motion can also be investigated through the exchange interaction, as demonstrated in [7].

In order to obtain information on J(r) we will analyze recent experiments by WELLER, STAERK and TREICHEL [8] which demonstrated the existence of magnetic field effects on reactions between the paramagnetic end groups of a polymer in an organic solvent. The experiment and the underlying spin dynamics will be explained in Section 2. In Sections 3 we provide a description of the folding motion of polymers. In Sections 4,5 we present an analysis of the experiments in Ref. 8. We show that the observed magnetic field effects can be explained through the folding motion of the polymer. This folding stochastically alters the exchange interaction between the paramagnetic end groups, and thereby affects their spin dynamics. The analysis provides evidence in favour of (4). In Section 6 we discuss this result [9].

2. Spin Pair Dynamics of Biradical Systems

An example of a reaction which is influenced by weak magnetic fields is the light-induced electron-transfer reaction [10,11,12] involving molecules ^1A (electron acceptor, here pyrene) and ^1D (electron donor, here dimethylaniline) which yield triplet molecules $^3A^*$ according to the scheme

$$^1A + {}^1D + \text{photon} \rightarrow {}^1A^* + {}^1D \rightarrow {}^1({}^2A^- + {}^2D^+) \leftrightarrow {}^3({}^2A^- + {}^2D^+) \rightarrow {}^3A^* + {}^1D \qquad (5)$$

This route to triplets involves the intermediate doublet pair

$^2A^- + ^2D^+$ which is born in an overall singlet state denoted by $^1(^2A^- + ^2D^=)$. This pure spin state is perturbed by the interactions collected in the Hamiltonian

$$H = H_1 + H_2 + V(J) \tag{6}$$

$$H_i = g\mu\underline{B}\cdot\underline{S}_i + \sum_{k=1}^{N_i} a_{ik}\underline{I}_{ik}\cdot\underline{S}_i \quad ; \qquad i=1,2 \tag{7}$$

$$V(J) = J(r)[\tfrac{1}{2} + 2\underline{S}_1\cdot\underline{S}_2] \tag{8}$$

H_i describes the interaction of the electron spin $\underline{S}_i$ of ^{2}pyrene$^-$ (i=1) and ^{2}dimethylaniline$^+$ (i=2) with the external magnetic field $\underline{B}$ (Zeeman interaction) and with the N_i nuclear spins $\underline{I}_{ik}$ (hyperfine interaction). The hyperfine coupling constants assumed for the compounds ^{2}pyrene$^-$ and ^{2}dimethylaniline$^+$ are given in Ref. 9. V(J) represents the exchange interaction. The coupling constant J depends on the distance between the paramagnetic molecule as described by (4).

For an observation [10,11] the molecules ^1A and ^1D are solvated in a polar organic liquid, for example methanol. ^1A is excited electronically by a ns laser flash and the formation of $^3A^*$ is monitored by a second light beam. The transient concentration of $^3A^*$ is found to depend strongly on the strength of the external magnetic field $\underline{B}$. The magnetic field effect can be explained by means of the probability $p_T(t,B)$ that a singlet pair $^1(^2A^- + ^2D^+)$ assumes triplet character at time t, i.e. reaches the state $^3(^2A^- + ^2D^+)$. This probability is given by

$$p_T(t,B) = \frac{1}{Z}\, \mathrm{Tr}\,[\, Q_T\, \mathbf{U}(t)\, Q_S \,] \quad \text{where} \tag{9}$$

$$Q_S = \frac{1}{4} - \underline{S}_1\cdot\underline{S}_2 = \rho(0) \tag{10}$$

$$Q_T = \frac{3}{4} + \underline{S}_1\cdot\underline{S}_2 \tag{11}$$

are the projection operators on the singlet and triplet states and $\mathbf{U}(t)$ describes the propagator of the density matrix $\rho(t)$ of the system. The trace involves all electron - nuclear spin states and $Z=\mathrm{Tr}Q_S$ is the number of all nuclear spin states. The yield of triplet products ${}^3A^*$ is then given by

$$\phi_T(B) = \int_0^\infty dt \, \exp(-t/\tau_o) \, p_T(t,B) \tag{12}$$

where τ_o is the life time of the doublet pair ${}^2A^- + {}^2D^+$. This observable is shown in Fig. 1a,b for situations with <u>fixed</u> exchange interactions $J(r) = 0$ and $J(r) = 150$ Gauss.

In the case $J = 0$ the triplet yield (12) decreases monotonically with increasing magnetic fields and reaches a constant value for fields $B \gg \Sigma_{ik} a_{ik}$. This behaviour is well understood [13]. In the case of a non-vanishing exchange interaction,the singlet and triplet levels are energetically shifted with respect to each other by 2J. For $J \gg \Sigma_{ik} a_{ik}$ singlet and triplet transitions cannot be induced by the hyperfine coupling except at fields $B \approx 2J$ when (for $J>0$) the triplet state T_{-1} becomes degenerate with the singlet state S_o. This degeneracy causes the maximum of the triplet yield $\phi_T(B)$ in Fig. 1b.

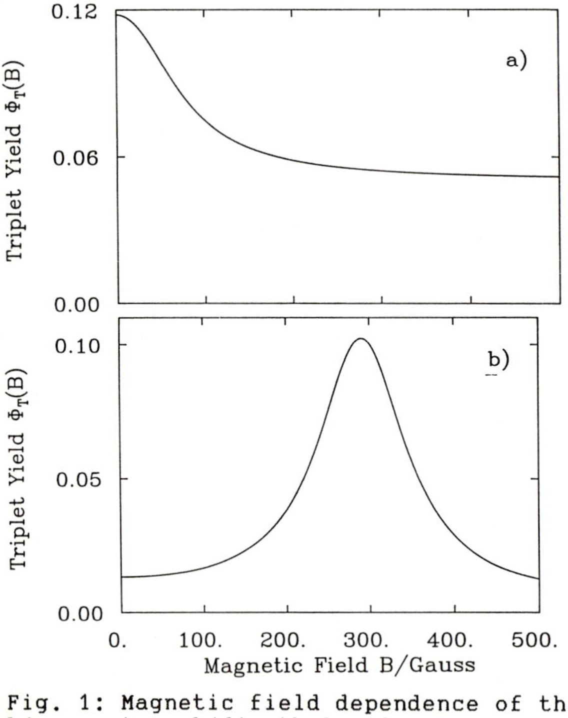

Fig. 1: Magnetic field dependence of the triplet yield $\phi_T(B)$ of a ^{2}(pyrene)$^-$ + ^{2}(dimethylaniline)$^+$ pair with a life time τ_o = 1 ns; (a) J = 0 Gauss, (b) 2J = 300 Gauss.

The yield in Fig. 1a for a vanishing exchange interaction agrees well with the observations when ^1A and ^1D are separate molecules which diffuse freely (except for the Coulomb attraction of the doublet pair) in the solvent. The magnetic field dependence observed in Ref. 8 on polymeric diradical systems exhibits a maximum at intermediate fields and resembles more the yield in Fig. 1b, i.e. is indicative of a non-vanishing exchange interaction.

However, the observations in Ref. 8 also show important differences compared to the yield as shown in Fig. 1b. First, the maximum value of the yield $\phi_T(B_{max})$ is only about 20 percent larger than the yield $\phi_T(B=0)$ at zero field. Second, the observed maximum is much broader. These features point to the possibility that the doublet pair $^2A^- -(CH_2)_n- ^2D^+$ formed as an intermediate experiences a distribution of exchange interactions. Here we want to demonstrate that this is, in fact, the case. However, the distribution of exchange interactions which explains the observation does not reflect the static folding pattern of the polymer, but rather originates from the dynamics of the folding polymer. The exchange interaction (2) has to be considered a stochastic variable for the doublet pair $^2A^- -(CH_2)_n- ^2D^+$.

3. Computer Simulation of the Polymer Folding

The origin of the stochastic variation of the exchange energy between the paramagnetic end groups of $^2A^- -(CH_2)_n- ^2D^+$ is the folding of its polymer moiety $-(CH_2)_n-$. The relevant stochastic variable is the distance r(t) between the two polymer ends. In a 'mean field' description the time-dependent distribution of end-end-distances p(r,t) satisfies the Fokker-Planck equation

$$\partial_t p(r,t) = L(r)\, p(r,t) \tag{13}$$

$$L(r) = D\, \partial_r p_o(r)\, \partial_r [p_o(r)]^{-1} \quad . \tag{14}$$

Here $p_o(r)$ denotes the equilibrium distribution function of the end-end-distances and D is the effective diffusion coefficient, the value of which should be somewhat smaller than the sum of the diffusion coefficients of the separate groups $^2A^-$ and $^2D^+$. The static distribution $p_o(r)$ entails information on the sterical interactions of the polymer chain $-(CH_2)_n-$.

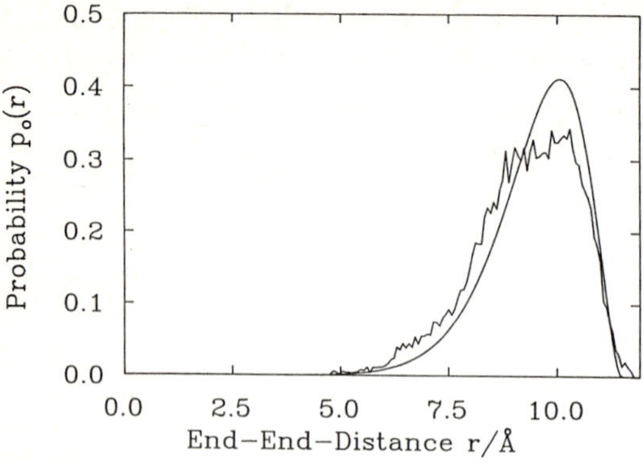

Fig. 2: Distribution of end-end distances for the polymer $CH_3-(CH_2)_9-CH_3$ resulting from a molecular dynamics calculation involving the computer program of KARPLUS et al. [14]. The hydrogen atoms in this compound were not explicitly described, but only included through an increase of the van der Waals radii of the carbon atoms. These atoms were subjected to random forces and friction. For actual calculations it is convenient to employ a smooth equilibrium distribution $p_o(r)$. Therefore, we approximated the molecular dynamics result by the smooth curve (——).

In lieu of available information on $p_o(r)$ we have generated this distribution by a computer simulation which models the force field between all atoms of the system and integrates the Newtonian equations of motion for all degrees of freedom. The simulation program employed has been developed by KARPLUS and coworkers [14] for the description of the dynamics of biological macromolecules. In our simulation we added random forces and friction to all atoms in order to describe the effect of a solvent. The distribution resulting from sampling the r(t) values of a long-time trajectory of the polymer is presented in Fig. 2.

4. Computer Simulation of the Spin Dynamics

In order to evaluate the propagator $\mathbf{U}(t)$ for the density operator in the representation of the electron-nuclear spin states, we employ the von Neumann equation

$$\partial_t \rho = -i\ \mathbf{H}[J(t)]\ \rho \tag{15}$$

where $\mathbf{H}[J(t)]=[\ H[J(t)],\ \ldots\ ]$. Here $\mathbf{H}[J(t)]$ denotes a time-dependent superoperator, $H[J(t)]$ the Hamiltonian (6) and $[\ ,\ ]$ the commutator. The propagator can be formally written

$$\mathbf{U}(t) = \langle \exp_{-}\{\ -i \int_o^t dt'\ \mathbf{H}[J(t')]\ \}\ \rangle \tag{16}$$

where $\langle\ \rangle$ represents the ensemble average over all histories of

J[r(t)]-values when the sytem assumes initially the distribution of end-end-distances $p_o(r)$. The exponential operator in (16) is time-ordered and is defined by the limit $\tau \to 0$ of

$$\exp\{-i\tau \mathbf{H}[J(N\tau)]\} \cdot \cdot \cdot \exp\{-i\tau \mathbf{H}[J(2\tau)]\}\exp\{-i\tau \mathbf{H}[J(\tau)]\} \qquad (17)$$

with $N\tau$ constant.

The magnetic field dependence of the triplet yield $\phi_T(B)$ resulting [9] for a $(pyrene)^- - (CH_2)_n - (dimethylaniline)^+$ pair of paramagnetic groups with an end-end equilibrium distribution of Fig. 2 is presented in Fig. 3. The yield shown closely resembles the observation of Ref. 8 in that the maximum of $\phi_T(B)$ in Fig. 3 lies at the field B_{max}=300 Gauss ,which compares well with the observation of B_{max}=285 Gauss. The ratio $\phi_T(B_{max})/\phi_T(B=0)$ is found theoretically to be 1.1 and experimentally to be 1.3. These latter values have to be compared with a ratio of 5 obtained if a fixed exchange interaction of 2J = 300 Gauss is assumed (see Fig. 1b). The good agreement with the observation shows that one can explain the magnetic field dependence ot the triplet yield $\phi_T(B)$ in polymethylene-linked donor acceptor systems with the stochastic modulation of the distance - dependent exchange interaction (4). This finding argues against an influence of a 'through bond exchange' along the aliphatic chain (which should not be modulated stochastically) and settles the dispute [15] about the origin of the exchange interaction in this polymeric system in favour of an exchange straight through the space (solvent) between the end groups.

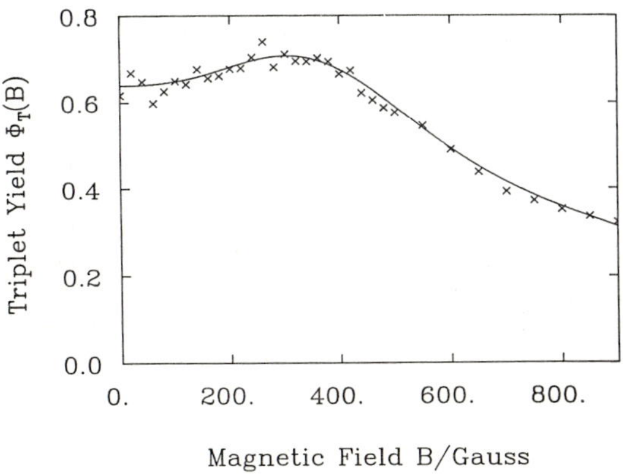

Fig. 3: Magnetic field dependence of the triplet yield $\phi_T(B)$ of the polymeric compound $^2(pyrene)^- - (CH_2)_9 -$ $-^2(dimethylaniline)^+$ resulting from the computer simulation of Ref. 9 (+ + +) and resulting from the approximation (19) presented in Sect. 5 (——).

The computer simulation which yielded the results in Fig. 3 is extremely time consuming, and demanded a CPU time of about 100 hrs on a VAX-11/750. This vast amount of computer time is due to the Monte Carlo character of the algorithm [9] which requires large enough samples to produce significant results. The simulation does not yield many insights about the behaviour of the stochastic spin system, e.g. does not provide an answer about the question why the maximum of $\phi_T(B)$ is observed at about 300 Gauss. The computer time of the theoretical description can be drastically reduced and insight can be gained if one employs an approximate description presented in the next Section.

5. Spectrum of the Stochastic Exchange Interaction

In this Section we seek an approximation of the propagator (16). The necessary ensemble average for the fluctuating exchange interaction J(t) can be evaluated formally by means of the stochastic Liouville equation [16]

$$\partial_t \mathbf{U}(t) = \{ -i\mathbf{H}[J(r)] + \mathbf{L} \} \mathbf{U}(t) \tag{18}$$

where $\mathbf{L}$ denotes the Fokker-Planck operator (14). The approximation introduced by us can be cast into the form

$$\mathbf{U}(t) \approx \int_{-\infty}^{\infty} dj\ q(j) \exp[-it\mathbf{H}(j)] \tag{19}$$

where the propagator $\mathbf{U}(t)$ with a time-dependent (stochastic) superoperator $\mathbf{H}[J(t)]$ is replaced by a superposition of propagators $\exp[-it\mathbf{H}(j)]$ with time-independent superoperators $\mathbf{H}(j)$. Here $\mathbf{H}(j)$ denotes the same superoperator as $\mathbf{H}[J(t)]$ except that the stochastic exchange interaction J(t) is replaced by the constant j. q(j) represents the spectrum of the stochastic exchange interactions and is defined by

$$q(j) = \frac{1}{\pi} \mathrm{Re}\langle [i(2j-2\mathbf{J}) - \mathbf{L}]^{-1} \rangle \tag{20}$$

Equation (19) implies that the spectrum q(j) can be interpreted as a dynamical distribution of exchange interactions j of the polymer system. This distribution is shown in Fig. 4 for the four cases $D = 0$, $4\cdot10^{-6}cm^2s^{-1}$, $4\cdot10^{-5}cm^2s^{-1}$, $4\cdot10^{-4}cm^2s^{-1}$. The

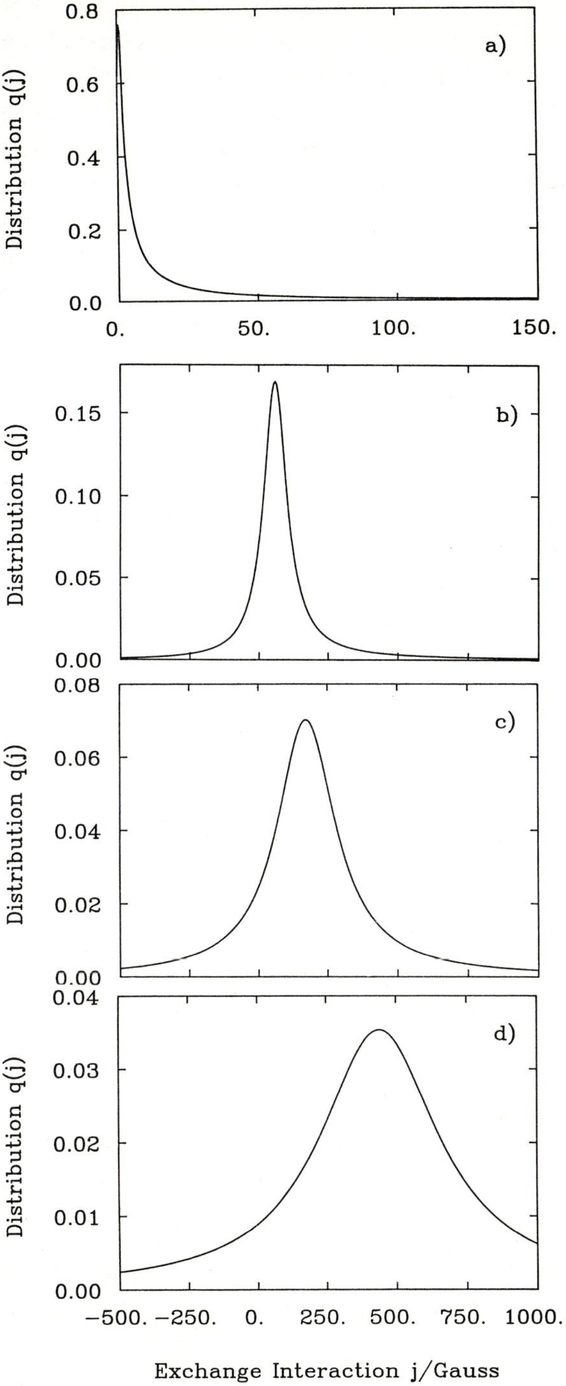

Fig. 4: Spectrum q(j) of the exchange interaction as defined in Eq. (20); (a) for $D = 0$; (b) $D = 4 \cdot 10^{-6}\ cm^2\ s^{-1}$; (c) $D = 4 \cdot 10^{-5}\ cm^2\ s^{-1}$; (d) $D = 4 \cdot 10^{-4}\ cm^2\ s^{-1}$.

different distributions in Fig. 4 show that the folding dynamics shift the spectrum of exchange interactions from small values for D = 0 to larger values for increasing D.

The approximation (19) holds exactly for short times, the time scale being defined by the strength of the hyperfine coupling, not the exchange and Zeeman interactions. One can expect that the suggested approximation results in exact triplet yields $\phi_T(B)$ as long as the life time τ_0 is only a few nanoseconds. However, the approximation provides a qualitatively correct magnetic field dependence of $\phi_T(B)$ also for longer life times [28]. The approximation (19) holds exactly for long times in case of either very slow or for very fast stochastic motion of r(t).

Figure 3 compares the magnetic field dependence of the triplet yield $\phi_T(B)$ obtained by means of the approximation (19) with the results from a simulation discussed in Sect. 3. The agreement found is satisfactory. This implies that an interpretation of the observed triplet yield (which also agrees with the curve in Fig. 3) should refer to the spectrum of exchange interactions q(j). The maximum of $\phi_T(B)$ does indeed occur approximately at the field $B_{max} = 2\, j_{max}$ which corresponds to the maximum $q(j_{max})$ of q(j) as can be seen from a comparison of Fig. 3 and Fig. 4c. Therefore, the spectrum q(j) relates the magnetic field effect of $\phi_T(B)$ to the folding dynamics of the polymer. The approximation (19) can be employed to study the dependence of $\phi_T(B)$ on the polymer length. The results of such calculations compare well with the available observations [17].

6. Summary

The analysis of the experiments in Ref. 8 show that the spin pair dynamics of polymeric biradicals is influenced by a stochastic folding motion. This folding explores the exchange interaction J(r) between the radical ends. Therefore, the observable, the magnetic field-dependent triplet yield $\phi_T(B)$, is very sensitive to the strength and the distance-dependence of the exchange interaction. The analysis in Sections 4 and 5 revealed that the interaction (4) of de Kanter et. al. describes the observations accurately. One has to conclude that the exchange interactions and electron-transfer rates in the photosynthetic reaction center, the interior of which should resemble an organic solvent, is governed by a $V^2(r)$ similar to that which underlies (4). The decay constant of (4) is slightly out of the range of 1-2Å^{-1} as given in Ref. 6.

The discussion above shows very clearly the dominant role which the stochastic dynamics of molecules play on electronic interactions. This should also be the case for the electron transfer in the reaction center. However, in the latter case the description of the stochastic motion is simpler than in case of the folding polymer. The reason is that the polymer folding alters the end-end-distance very drastically, whereas the stochastic motion of the protein atoms extends only a few tenths of an Ångstrom from their equilibrium positions [18]. As a result the coupling to a stochastic matrix element $V^2[r(t)]$ can be assumed in the 'motional narrowing' limit.

7. Acknowledgements

The authors like to thank A. Weller, H. Staerk and R. Treichel for several illuminating discussions and for making their data available prior to publication. The authors are grateful to A. Brünger and M. Karplus for the CHARMM program for the evaluation of the equilibrium distribution $p_o(r)$. This work has been supported by the Deutsche Forschungsgemeinschaft (SFB 143/C 1).

8. References

1 R.A.Marcus: J.Chem.Phys. 24, 966 (1956)

2 J.J.Markham: Rev.Mod.Phys. 31, 956 (1959)

3 R.A.Marcus and N.Sutin: Biochim.Biophys.Acta 811, 265 (1985)

4 Exceptions are, for example, the contribution of J.R.Miller to this conference; a review on organic conductors is H.Meier: Current Chemistry 61, 85 (1976)

5 M.J.Potasek and J.J.Hopfield: Proc.Natl.Acad.Sci. USA 74, 229 (1977)

6 C.Herring and M.Flicker: Phys.Rev. 134, A362 (1964)

7 F.J.J.de Kanter, J.A.den Hollander, A.H.Huizer and R.Kaptein: Mol.Phys. 34, 857 (1977); F.J.J.de Kanter, R.Z.Sagdeev and R.Kaptein: Chem.Phys.Lett. 58, 33 (1978);

F.J.J.de Kanter and R. Kaptein: Chem.Phys.Lett. 58, 340 (1978); F.J.J.de Kanter and R.Kaptein: J.Am.Chem.Soc. 104, 4759 (1982)

8 A.Weller, H.Staerk and R.Treichel: Faraday Discuss. 78, (1984); R.Treichel: Thesis, Göttingen (1985)

9 K.Schulten and R.Bittl: submitted to J.Chem.Phys.

10 K.Schulten, H.Staerk, A.Weller, H.-J.Werner and B.Nickel: Z.Phys.Chem. NF101, 371 (1976)

11 M.E.Michel-Beyerle, R.Haberkorn, W.Bube, E.Steffens, H.Schröder, H.J.Neusser and E.W.Schlag: Chem.Phys. 17, 139 (1976)

12 Actually this reaction has also been employed to study the photosynthetic reaction center; a review of this work is A.J.Hoff: Quart.Rev.Biophys.14, 599 (1981); see also Ref. 13

13 K. Schulten in "Festkörperprobleme (Advances in Solid State Physics), Vol. XXII, J.Teusch, ed., pp. 61 (Vieweg, Braunschweig, 1982)

14 B.R.Brooks, R.E.Bruccoleri, B.D.Olafson, D.J.States, S.Swaminathan and M.Karplus: J.Comp.Chem. 4, 187 (1983)

15 R.Hoffmann: Acc.Chem.Res. 4, 1 (1971); M.N.Paddon-Row, Acc.Chem.Res. 15, 245 (1982)

16 C.S.Johnson, Jr.: J.Chem.Phys. 41, 3277 (1964); M.Blume: Phys.Rev. 174, 351 (1968); S.Dattagupta and M.Blume: Phys.Rev. B 10, 4540 (1974)

17 R.Bittl, H.Treutlein and K.Schulten: submitted to Chem.Phys.Lett.

18 H.Frauenfelder, G.A.Petsko, and D.Tsernoglu: Nature 280, 558 (1979)

Part IV

Reaction Centers: Structure and Dynamics

Kinetics and Mechanisms of Initial Electron-Transfer Reactions in *Rhodopseudomonas sphaeroides* Reaction Centers

W.W. Parson, N.W.T. Woodbury, and M. Becker

Dept. of Biochemistry, Univ. of Washington, Seattle, WA 98195, USA

C. Kirmaier and D. Holten

Dept. of Chemistry, Washington Univ., St. Louis, MO 63130, USA

1. Introduction

The reaction center (RC) of the photosynthetic bacterium Rhodopseudomonas sphaeroides consists of three small polypeptides that hold a cluster of four molecules of bacteriochlorophyll-a (BChl), two molecules of bacteriopheophytin-a (BPh), one nonheme iron atom, and two molecules of ubiquinone [1,2]. The RCs can be purified readily, and they recently have been crystallized [3]. Although the crystal structure has not yet been solved, RCs from Rps. sphaeroides are functionally very similar to those from Rps. viridis whose structure has been solved to 3 Å resolution [4]. In both species, two of the four BChls form a closely interacting pair (P) that undergoes oxidation to a radical cation (P^+) when the RC is excited with light. The electron is transferred to a BPh, and from there to a quinone (Q) that is close to, but probably not bound to the Fe. (In Rps. viridis, Q is menaquinone instead of ubiquinone.) In Rps. sphaeroides RCs, the photochemical electron-transfer reaction from P to Q has a quantum yield of essentially 100% [5]. The kinetics of electron-transfer from the intermediate BPh anion radical (BPh^-) to Q have been measured under a variety of conditions, and are similar in the two species, in spite of the difference between the quinones [6-10]. It has been suggested that one of the two BChls that are not part of P mediates electron-transfer from the excited singlet state of P (P*) to the BPh [10-12]. However, the evidence for the formation of a P^+BChl^- radical pair prior to the formation of P^+BPh^- is inconclusive [13].

The absorption spectra of RCs from the two species also have many similarities, although the absorption bands occur at longer wavelengths in Rps. viridis because of the substitution of BChl-b and BPh-b for BChl-a and BPh-a. These functional and spectroscopic similarities suggest that the essential structural features of the two RCs will prove to be much the same. One significant difference is that RCs isolated from Rps. sphaeroides do not contain the bound c-type cytochromes that provide electrons to P^+ in the Rps. viridis complex; instead, P^+ reacts with the dissociable, water-soluble cytochrome c2.

We have used picosecond and nanosecond spectrophotometric techniques to investigate the transient excited states and radical-pairs that are created when Rps. sphaeroides RCs are excited with light. To study the thermodynamic properties of the P^+BPh^- and P^+Q^- radical pairs, we have measured the "delayed" fluorescence that is emitted during the lifetimes of the transient states. Since the fluorescence comes from P^*, its intensity is proportional to the amount of the excited singlet state that is in equilibrium with the radical pair, and the intensity thus depends on the effective standard free energy difference between the two states [14-16]. Measurements of the fluorescence decay kinetics can be used to follow relaxations that may not cause detectable changes in the absorption spectrum, such as movements of the amino acids near the electron carriers. This paper summarizes some of our recent results on the electron-transfer reaction from BPh^- to Q, with an emphasis on the temperature-dependence of the reaction. It then turns to the mechanism and kinetics of the formation of P^+BPh^-, and focuses on the question of whether a BChl molecule acts as the initial electron acceptor from P^*. The work will be described in greater detail elsewhere [13,17-18].

2. Materials and Methods

Reaction centers were isolated from *Rps. sphaeroides* as described previously [15]. For studies at low temperature, they were incorporated into films of polyvinyl alcohol [19]. Two types of apparatus were used for the picosecond spectrophotometric studies. The first [20] employed a mode-locked Nd-YAG laser with frequency doubling and Raman shifting to generate 5-Hz excitation pulses about 30 ps in width at either 600 or 867 nm. Probe pulses of broadband light with the same width were created by self-phase modulation of the Nd-YAG fundamental pulses, split into two parts that passed through excited and unexcited regions of the sample and then through a spectrometer, and then detected with a 2-dimensional optical multichannel analyzer. A 150-nm spectrum of the absorbance changes caused by the excitation was recorded on each flash and was sent to a computer for averaging over 300 to 1000 flashes. The second apparatus used a 600-nm rhodamine-6G dye laser that was pumped synchronously at 82 MHz by a mode-locked Ar^+ laser. The 2-ps dye pulses passed through a multi-stage dye amplifier system, which was pumped by a frequency-doubled Nd-YAG laser at 10 Hz, and then were used to excite the sample and to generate broadband probe pulses. The probe pulses were split and treated as above, except that absorbance changes at individual wavelengths were detected with Si photodiodes and sample/hold amplifiers. Delayed fluorescence was measured by single-photon counting, with 30-ps, 600-nm excitation at 0.8 MHz [16].

3. Results and Discussion

When *Rps. sphaeroides* RCs are excited with light, their optical absorption spectrum changes dramatically. Among the changes are the loss of the long-wavelength absorption band of P at 865 nm, and a blue-shift of an absorption band at 800 nm. The bleaching at 865 nm reflects the oxidation of P to P^+ [1,2]. The blue-shift probably is due mainly to the Stark effect of the P^+ cation on an absorption band of the other two BChls, and possibly to the development of a a new band near 785 nm due to P^+. At relatively long times after the excitation, there also are absorption changes indica-

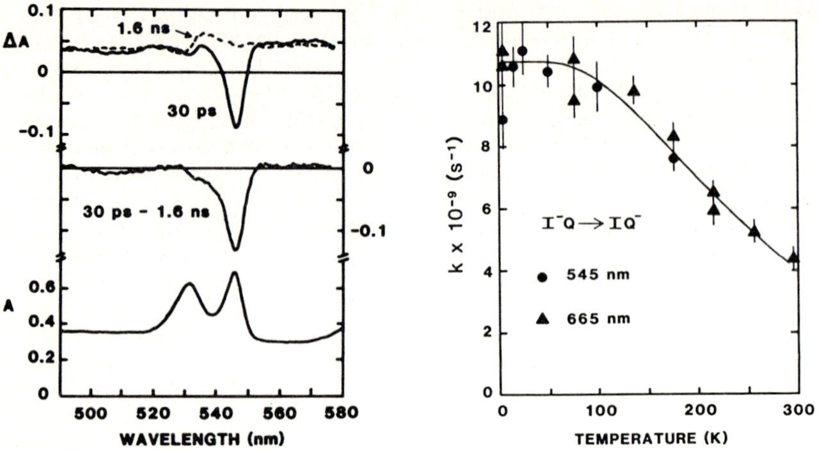

Figure 1 A: Absorption spectrum of Rps. sphaeroides RCs at 5K (lower trace); absorption changes caused by excitation with a 30-ps, 600-nm flash as measured at 30 ps and 1.6 ns after the flash (upper traces); and difference between measurements at 30 ps and 1.6 ns (middle trace). B: Temperature-dependence of the rate-constant for electron-transfer from BPh^- to Q, measured from absorbance changes at 545 nm (●) or 665 nm (▲) as in A.

tive of the reduction of Q to the anionic semiquinone (Q^-) [1]. These are much smaller than the absorption changes associated with P^+, and they are difficult to measure unless a reductant is added to reduce P^+ back to P. At short times after the excitation, one sees absorption changes that appear to reflect the transient reduction of a BPh. These are illustrated in Fig. 1A. The lower trace shows the absorption spectrum of unexcited RCs at 5K, in the region of the Q_x bands of the BPhs. There are two, well resolved bands at 530 and 545 nm, which probably are due to the two different BPhs. The top traces show the changes in absorbance at 5K, as measured at 30 ps and 1.6 ns after the RCs are excited with a flash lasting about 30 ps. At the earlier time (solid curve), the 545-nm band is bleached, suggesting that the BPh that absorbs here is reduced. The absorption band of the other BPh is hardly affected. The bleaching at 545 nm has decayed by 1.6 ns (dashed curve), leaving only a small red-shift of the 530-nm band. It is curious that the red-shift is restricted to the BPh that is not reduced initially. The middle trace in Fig. 1A shows the difference between the spectra measured at 30 ps and 1.6 ns. The decay of the 545-nm bleaching in this time-interval evidently reflects the transfer of an electron from BPh^- to Q, because it is blocked if Q is extracted from the RCs or is reduced chemically before the excitation flash [22]. P^+BPh^- is still formed on excitation under these conditions, but it lives for about 12 ns, decaying by several types of back reactions [2,15,16,21,22].

Along with the absorption changes in the 545-nm region, the transient spectrum includes a bleaching of the BPh's Q_y band at 765 nm, the formation of a broad band at 665 nm, and partial bleaching and shifting of the BChl band at 800 nm [2]. The 665-nm band is characteristic of the BPh radical anion (BPh^-), although it also could be due to $BChl^-$ [23]. The absorbance changes at 800 nm will be discussed below. The kinetics of electron-transfer from BPh^- to Q can be measured by following the decay of the transient absorbance changes at 545, 665 or 765 nm, but measurements at 665 nm are easiest to interpret, because they probably are not complicated by shifts of the absorption bands of the second BPh [17]. Figure 1B shows the rate-constant of the electron-transfer, as measured at 665 nm over a wide temperature-range, and at 545 nm at temperatures low enough so that the bands of the two BPhs are well resolved. The rate-constant increases with decreasing temperature down to about 100K, and then becomes essentially independent of temperature. The 1/e time-constant of the reaction is about 200 ps at 295K, and about 100 ps at temperatures below 100K. The temperature-dependence found here differs somewhat from that obtained in earlier, less extensive studies by PETERS et al. [9] and SCHENCK et al. [15].

Current theories of nonadiabatic electron-transfer emphasize the importance of nuclear overlap between the reactants and products [24-27]. The increase in the rate-constant with decreasing temperature can be understood if one assumes that the reaction occurs most favorably from the zero-point vibrational states of the reactants. This will be the case if the potential energy surfaces for the reactants and the products intersect near the minimum in the reactants' surface. The theory that fits the data most readily is one developed by KAKITANI and KAKITANI [26], which includes as a parameter a change in the frequency of low-frequency vibrational modes of the electron carriers or the surrounding protein. The solid curve in Fig. 1B was generated by this theory [17]. In addition to parameters describing the nuclear vibrations, the theory requires an estimate of the standard free energy-change in the reaction. The free energy-change (about 0.7 eV at 295K) was obtained from measurements of the delayed fluorescence associated with P^+BPh^- and P^+Q^-, as explained in the Introduction. At 295K, P^+BPh^- appears to lie about 0.16 eV below P^* in free energy, and P^+Q^- about 0.86 eV below P^* [14-16].

The free energy-gap between P^* and P^+BPh^- appears to decrease with decreasing temperature, indicating that the initial charge separation involves a significantly entropy increase [16]. On a simple model, one might have expected an entropy change in the opposite direction, because of the increase in vibrational frequencies that should occur in the protein around the charged products [27]. However, we still have very little understanding of the nuclear relaxations that probably follow the electron transfer reaction. When electron-transfer from BPh^- to Q is blocked, the intensity of the delayed fluorescence decays in several steps, perhaps because of the occurrence of such relaxations [16].

The interpretation of the transient absorption changes in the 800-nm band is unclear. We found that they decay significantly more rapidly than the changes at 545 or 665 nm [17]. The decay time measured at 795 nm is about 150 ps at 295K, and 70 ps at temperatures below 100K. It has been suggested that the bleaching near 800 nm reflects the presence of a small amount of P^+BChl^- in equilibrium with P^+BPh^- [2,10-12,28]. On this model, however, one would expect to find a component of the 665-nm absorption band with decay kinetics matching those at 795 nm, since either $BChl^-$ or BPh^- should absorb at 665 nm. There is no indication of such a component [17]. An alternative interpretation is that the absorbance changes near 800 nm are due to a nuclear relaxation that follows the formation of P^+BPh^-. This model, however, has difficulty explaining why the relaxation increases in rate with decreasing temperature, in much the same way that the electron-transfer from BPh^- to Q does. Also, it predicts that the relaxation might be seen in the absorbance changes near 800 nm even in RCs that have Q removed or reduced before the excitation. Attempts to detect a relaxation under these conditions are in progress.

The kinetics of the initial reduction of the BPh to BPh^- can be measured from the rise of the absorbance changes at 545 or 665 nm. Similar results are obtained at either wavelength: the rise is not instantaneous, but lags after the excitation by about 5 ps [11,12,20,29]. It has been reported that there is a transient bleaching at 800 nm prior to the development of the bleaching at 545 nm, suggesting that the formation of a P^+BChl^- radical pair preceeds that of P^+BPh^- [10-12]. However, the measurements that have been made of the initial absorption changes have not had sufficient time-resolution to be convincing, and the interpretations of the data have been open to criticism [13]. We have attempted to resolve the initial absorption changes more cleanly by using weak excitation flashes with a width of about 2 ps. Measurements at several wavelengths are shown in Fig. 2.

As shown in Fig. 2A, the bleaching of P's absorption band at 865 nm occurs within the time-resolution of the measurements. Because the 600-nm excitation flash is absorbed by all four BChls of the RC, this means that

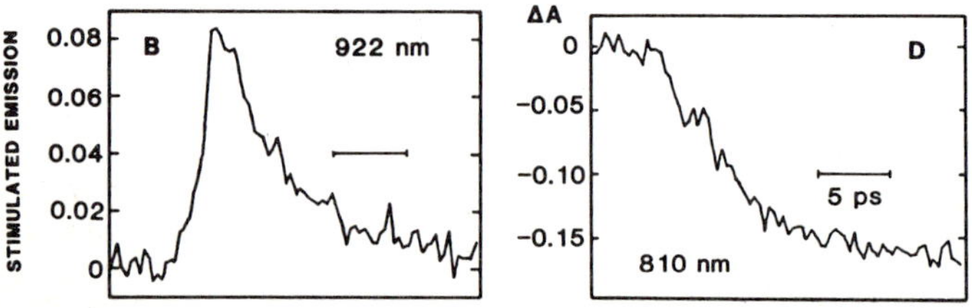

Figure 2 A, C, and D: Time courses of absorbance changes at 865, 785 and 810 nm caused by excitation with subsaturating 2-ps, 600-nm flashes at 295K. B: Time-course of stimulated emission at 922 nm, expressed as $\log(I/I_0)$ where I_0 and I are the relative intensities of probe pulses that pass through unexcited and excited regions of the sample.

the excitation must be localized on P^* in less than 2 ps. (The experiments of Fig. 2 were done at 295K; transfer of the energy to P appears to be slower at low temperatures, when the absorption bands are narrower [18].) As expected, the bleaching at 865 nm remains more or less constant during the period from 3 to 10 ps after the flash, when an electron is moving to the BPh (Fig. 2A). The absorption band is expected to bleach, no matter whether P is in the form P^* or P^+. When P is first excited, however, one also sees the development of a transient signal at 922 nm that we believe represents stimulated emission (fluorescence) from P^* (Fig. 2B). Operationally, what is measured is an increase in the 922-nm probe light passing through the sample. This could result either from a decrease in the absorbance of the sample at 922 nm, or from a transfer of energy from the sample to the probe beam, i.e. stimulated emission. The latter interpretation seems likely to be the correct one, because if the signal is interpreted as an absorbance decrease, the initial change is calculated to be larger than the measured absorbance of the RC sample at this wavelength. (922 nm is close to the emission maximum of the spontaneous fluorescence by P^* [16,30], and the absorbance here is small.) The oxidation of P to P^+ causes no net absorbance change at this wavelength. The decay of the stimulated emission should parallel the disappearance of P^*, which presumably occurs simultaneously with the formation of P^+. If our interpretation of the measurements is correct, the time-constant of the initial electron-transfer reaction is approximately 5 ps (Fig.2B). This is close to the lifetime of 7 ps that ZANKEL et al. [30] calculated for the spontaneous fluorescence from measurements of the fluorescence yield. It also is the same as the time-constant that has been obtained for the formation of BPh^-, as indicated by the bleaching at 545 nm [11,12,20,29].

Absorbance changes that appear with the same lag after the excitation can be seen at numerous places in the near infra-red [12,13]. Figures 2C and D show measurements at 785 and 815 nm. Although there probably are multiple contributions to the signals at these wavelengths (see below), the largest contribution seems likely to come from the blue-shift of the 800-nm absorption band. As mentioned above, the blue-shift is probably due mainly to a Stark effect of P^+ on the neighboring BChls, along with the formation of an absorption band of P^+ itself, so one might expect it to occur as P^+ is formed.

Figure 3 shows spectra of the absorbance changes measured at -0.7, -0.3, 0.9, 3.3 and 15.3 ps with respect to the center of the 2-ps excitation flash. The spectrum measured at the longest time is essentially identical to those measured previously for P^+BPh^- at 20 to 30 ps after excitation, with flashes lasting from 7 to 30 ps [6,7,10-13,17,20]. The spectra measured at the earliest times, when the RCs evidently are mainly in the excited singlet state P^*, differ in having a broad, weak absorbance increase from 750 to 795 nm, and a relatively small bleaching centered near 815 nm. These features are qualitatively similar to the spectral changes associated with the formation of an excited triplet state of P (state PR) [2,21,28], and to those observed following excitation of RCs in which electron-transfer to the BPh is blocked [31]. The shoulder in the 920-nm region in the early spectra is the component that is interpreted above as stimulated emission. None of the spectra shows the pronounced

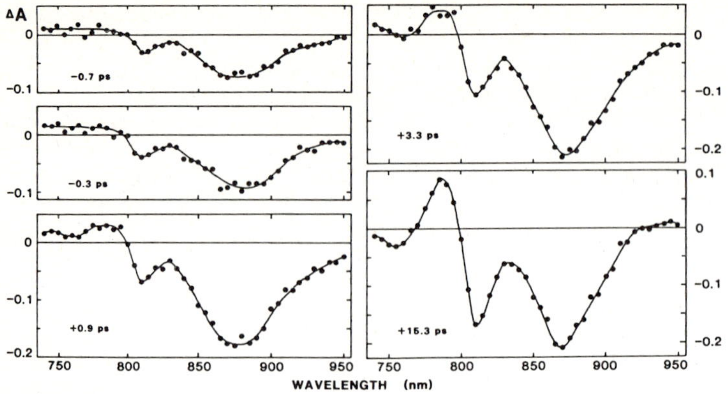

Figure 3 Spectra of absorbance changes measured as in Fig. 2 at various times relative to the center of the excitation flash.

bleaching at 800 nm that one might expect to see if there were a significant population of P^+BChl^- prior to the formation of P^+BPh^-.

The measurements shown in Figs. 2 and 3 favor the conclusion that, in *Rps. sphaeroides* RCs at 295K, P^* passes an electron directly to the BPh with a time-constant of about 5 ps. There is no indication of the formation of P^+BChl^- as a distinct intermediate. This is a surprising result, because in the x-ray structure of *Rps. viridis* RCs, the additional BChl lies closer to P than the BPh does [4]. It is, however, consistent with recent molecular orbital calculations suggesting that P^+BChl^- is likely to lie above P^* in energy [32]. If the BChl does not act as the initial electron acceptor, what does it do? One possible view is that the excited states of the BChl mix with those of P so that they expand the region of space occupied by P^* [32]. The excited molecular orbital in P^* thus is not confined to P, but to some extent spreads out over the other BChls to overlap with the vacant orbitals of the BPh. An increase in the distance over which the initial charge-separation occurs would help to overcome the coulombic attraction between the ions in the radical-pair product. Another viewpoint might be that, even if the energy of P^+BChl^- is above that of P^* and P^+BPh^-, it is far below the ionization energy of P. The presence of BChl in the region between P and the BPh thus lowers the potential energy barrier through which an electron must tunnel to reach BPh. This is basically a superexchange mechanism. A more refined theoretical analysis of the electron-transfer should be possible when the structure of the RC is understood in greater detail.

4. Acknowledgements

This work was supported by grants from the National Science Foundation (PCM-8312371, PCM-8316161 and PCM-8302477) and from the Competitive Research Grants Office of the U.S. Dept. of Agriculture (84-CRCR-1-1455). We thank D. Middendorf, A. Scherz, A. Warshel, and M. Wasielewski for helpful comments and suggestions, and P. Linnemeyer for purifying the RCs.

5. References

1. M.Y. Okamura, G. Feher, N. Nelson: in Photosynthesis: Energy Conversion by Plants and Bacteria Vol. 1 ed. Govindjee (Academic Press, New York, 1982), pp. 195-272
2. W.W. Parson, B. Ke: in Photosynthesis: Energy Conversion by Plants and Bacteria Vol. 1, ed. Govindjee (Academic Press, New York, 1982), pp. 331-385
3. J.P. Allen, G. Feher: Proc. Natl. Acad. Sci. U.S.A. 81, 4795 (1984)
4. J. Deisenhofer, O. Epp, K. Miki, R. Huber, H. Michel: J. Mol. Biol. 180, 385 (1984)
5. C.A. Wraight, R.K. Clayton: Biochim. Biophys. Acta 333, 246 (1974)
6. M.G. Rockley, M.W. Windsor, R.J. Cogdell, W.W. Parson: Proc. Natl. Acad. Sci. U.S.A. 72, 2251 (1975)
7. K.J. Kaufmann, P.L. Dutton, T.L. Netzel, J.S. Leigh, P.M. Rentzepis: Science 188, 1301 (1975)
8. D. Holten, M.W. Windsor, W.W. Parson, J.P. Thornber: Biochim. Biophys. Acta 501, 112 (1978)
9. K. Peters, P. Avouris, P.M. Rentzepis: Biophys. J. 23, 207 (1978)
10. V.A. Shuvalov, A.V. Klevanik, A.V. Sharkov, J.A. Matveetz, P.G. Kryukov: FEBS Lett. 81, 135 (1978)
11. S.A. Akhamanov, A.Y. Borisov, R.V. Danielius, R.A. Gadonas, V.S. Kazlowski, A.S. Piskarkas, V.A. Shuvalov: FEBS Lett. 114. 149 (1980)
12. V.A. Shuvalov, A.Y. Klevanik: FEBS Lett. 160, 51 (1983)
13. C. Kirmaier, D. Holten, W.W. Parson: FEBS Lett. in press (1985)
14. H. Arata, W.W. Parson: Biochim. Biophys. Acta 638, 201 (1981)
15. C.C. Schenck, R.E. Blankenship, W.W. Parson: Biochim. Biophys. Acta 680, 44 (1982)
16. N.W.T. Woodbury, W.W. Parson: Biochim. Biophys. Acta 767, 345 (1984)
17. C. Kirmaier, D. Holten, W.W. Parson: submitted to Biochim. Biophys. Acta (1985)
18. N.W.T. Woodbury, M. Becker, W.W. Parson: to be submitted
19. C.C. Schenck, W.W. Parson, D. Holten, M.W. Windsor, A. Sarai: Biophys. J. 36, 479 (1981)
20. C. Kirmaier, D. Holten, W.W. Parson: Biochim. Biophys. Acta 725, 190 (1983)
21. W.W. Parson, R.K. Clayton, R.J. Cogdell: Biochim. Biophys. Acta 387, 268 (1975)
22. K.J. Kaufmann, K.M. Petty, P.L. Dutton, P.M. Rentzepis: Biochem. Biophys. Res. Commun. 70, 839 (1976)
23. J. Fajer, D.C. Brune, M.S. Davis, A. Forman, L.D. Spaulding: Proc. Natl. Acad. Sci. U.S.A. 72, 4956 (1975)
24. J. Jortner: J. Chem. Phys. 64, 4860 (1976)
25. A. Sarai: Biochim. Biophys. Acta 589, 71 (1980)
26. T. Kakitani, H. Kakitani: Biochim. Biophys. Acta 635, 498 (1981)
27. T. Kakitani, N. Mataga: Chem. Phys. 93, 381 (1985)
28. V.A. Shuvalov, W.W. Parson: Proc. Natl. Acad. Sci. U.S.A. 78. 957 (1981)
29. D. Holten, C. Hoganson, M.W. Windsor, C.C. Schenck, W.W. Parson, A. Migus, R.L. Fork, C.V. Shank: Biochim. Biophys. Acta 592, 461 (1980)
30. K.L. Zankel, D.W. Reed, R.K. Clayton: Proc. Natl. Acad. Sci. U.S.A. 61, 1243 (1968)
31. C.C. Schenck, W.W. Parson, D. Holten, M.W. Windsor: Biochim. Biophys. Acta 635, 383 (1981)
32. W.W. Parson, A. Scherz, A. Warshel: this volume

Femtosecond Studies of the Reaction Center of *Rhodopseudomonas viridis*: The Very First Dynamics of the Electron-Transfer Processes

W. Zinth, M.C. Nuss, M.A. Franz, and W. Kaiser

Physik-Department, Technische Universität München, D-8000 München, F.R.G.

H. Michel

Max-Planck-Institut für Biochemie, D-8033 Martinsried, F.R.G.

The X-ray structure analysis of crystals made up of reaction centers (RC) of Rhodopseudomonas viridis provides us with the information on the location and orientation of the various pigments in the protein matrix /1,2/. After many years of speculation one is now in the position to predict unequivocally the path of the electron in the RC following the absorption of a photon by the special pair in the RC. This note is concerned with the time-dependence of the very first events.

We present here results from ultrafast time-resolved experiments. The RC were excited by a first ultrashort light pulse which triggered the photochemical reactions. A second, properly delayed probing pulse monitored absorbance changes induced by the various transient intermediates. The measurements with highest time resolution (1×10^{-13} s) were made with exciting and probing pulses of 150 fs (1.5×10^{-13} s) duration at a wavelength of 620 nm. These pulses were generated in a cw dye-laser system operating in the colliding pulse mode (CPM) /3/. The exposure of the sample to light was held on a low level by the following two techniques: First, an electro-optical modulator operating at 100 KHz cuts short pulse trains of five individual pulses from the 100 MHz repetition rate emission of the CPM laser. Second rotation of the sample cell with 25 Hz ensured that each short pulse train illuminated a new portion of RC /4/. The light intensity was kept so low that in the excited-sample volume less than 10^{-2} of the reaction centers were excited. The changes of absorption were monitored by delayed probe pulses at the same wavelength. To supplement our femtosecond data, excite and probe measurements using single picosecond pulses from a Nd-glass laser system were made at different probe frequencies. Tuning of the probe pulse was achieved by frequency converters based on the stimulated Raman process. Excitation wavelength was 620 nm. In the picosecond experiments less than 15% of the RCs were excited.

The reaction centers of Rhodopseudomonas viridis studied here were prepared according to the procedure given in /1/. We used reaction center preparations where 20 mM ascorbate was added in order to reduce the quinones. In that way it was guaranteed that oxidized special pairs did not accumulate. The absorption spectra of the RC were studied prior to and after each experimental run. An absorbance change indicating a decomposition of

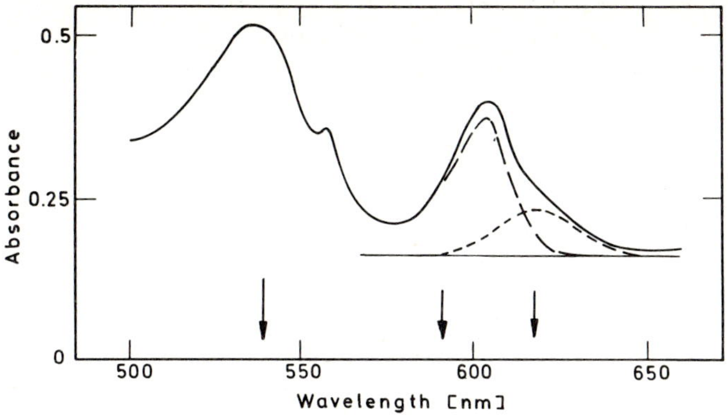

Fig. 1 Absorbance spectrum of a suspension of reaction center preparation of Rhodopseudomonas viridis. The broken lines represent a decomposition of the band around 605 nm into the contributions from the Q_x transition of the accessory $BChl_A$ ($\lambda_{max} \sim$ 605 nm) and of the special pair ($\lambda_{max} \sim$ 618 nm). (Sample: RC without ascorbate).

the RC during the picosecond and femtosecond experiments was not observed.

In Fig.1 part of the absorption spectrum of the reaction center in solution is depicted. The band at 610 nm corresponds to the Q_x transition of the two accessory bacteriochlorophyll b molecules ($BChl_A$) and the broad shoulder around 620 nm belongs to the Q_x transition of the special pair (P), the bacteriochlorophyll dimer. The position of the Q_x transition of P is well established from bleaching experiments /5/. Photooxidation of P reduces the absorption band at 960 nm, the Q_y transition of P, and simultaneously the absorption around 618 nm. The broad absorption at 530 nm in Fig.1 is made up of absorption bands of the two bacteriopheophytin b molecules (BPh) of the reaction center and of the four cytochrome units attached to the RC.

With light pulses of 620 nm we excited predominantly the special pair; approximately 20% of the incident radiation was absorbed by the neighboring $BChl_A$. In Fig.2 the absorption changes, initiated by the femtosecond excitation pulse and monitored by the delayed probe pulse, are presented as a function of delay time between the two pulses. Four successive processes are readily seen in the figure: During the passage of the excitation pulse of 150 fs the absorption of the sample decreases strongly. At the end of the excitation the absorption increases very rapidly with a time constant shorter than the pulse duration. The enhanced absorption decreases for approximately 1 ps and recovers with a time constant of 5 ps.

The experimental data suggest that four transient species are seen during the first 10 ps after excitation of the special pair: A first state having reduced absorbance (at 620 nm) lives shorter than the pulse duration; a second intermediate of enhanced absorbance lasts for 1 ps, a third one lives 5 ps, and a fourth one is stable during our subsequent observation time of 100 ps.

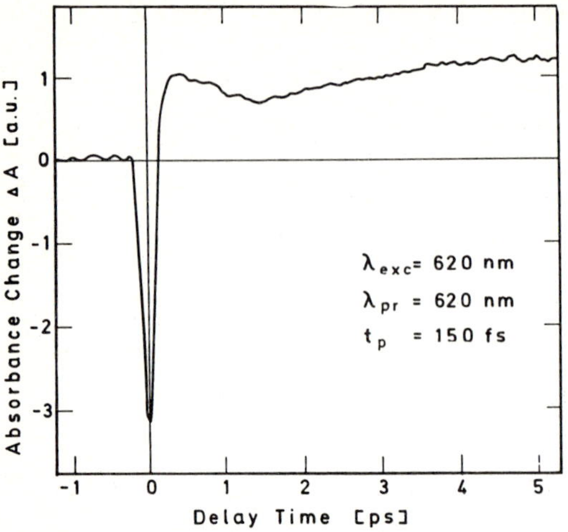

Fig. 2 Absorbance change induced by 150 fs pulses at λ = 620 nm measured as a function of time delay at the same wavelength. The transient absorbance changes indicate the existence of four intermediate states formed after optical excitation.

The assignment of the different absorption processes of Fig.2 to molecular states is supported by several sources of information. (i) As pointed out above, the x-ray work tells us the location of the different pigments within the protein, and thus gives strong indication on the course of the electron after excitation of the special pair. (ii) Our picosecond data (discussed below) at the frequency position of the BChl and BPh absorption bands help us to interpret our data for times exceeding 1 ps. (iii) Information on the absorption properties of the oxidized special pair P^+ may be deduced from the known spectra of the state P^+Q^-. Assuming that the negative charge at the quinone does not influence the visible spectrum, we deduce that P^+ has a broad absorption extending from a peak around 1.3 μm throughout the visible. Additional spectroscopic data /6/ give information on the absorbance changes induced by reducing the pigments BChl and BPh. In Fig.3 the absorption changes are depicted which occur when bacteriochlorophyll b and bacteriopheophytine b are reduced chemically to form $BChl^-$ and BPh^-, respectively. The curves in Fig.3 (redrawn from /6/) were taken in solutions of dimethylformamide ($BChlb^-$) and CH_2Cl_2 (BPh^-). While small shifts of the band are likely when going into the protein surrounding, the broad features of the spectra should basically remain unchanged, allowing the following discussion.

According to our present tentative picture the time-sequence of events is as follows: (i) During the excitation the special pair is promoted to the excited state; the reduced number of special pairs in the ground state leads to a decrease in absorption at the very beginning. The observed absorbance decrease is enhanced by the so-called coherence artifact, which exists only during the time duration of the exciting pulse. (ii) Immediately following the excitation, i.e. even during the excitation pulse of 150 fs, a second, strongly absorbing species is formed. To our

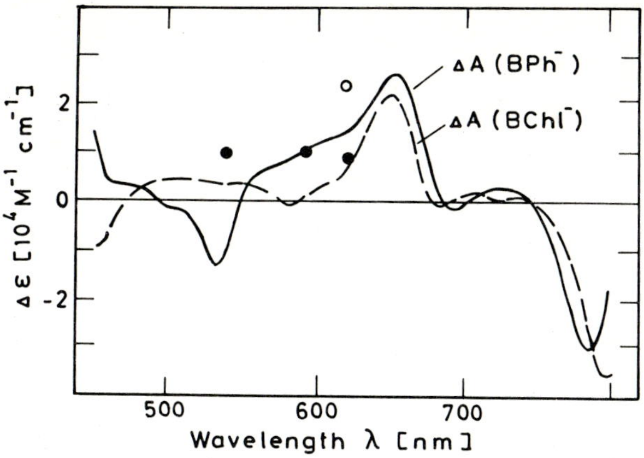

Fig. 3 Absorbance change induced by reduction of BChlb in dimethylformamide (solid curve) and of BPhb in CH_2Cl_2 (broken curve, after /6/). Absorption changes deduced from our time-resolved data are shown for the state P^{+-} ($\Delta A(P^{+-})$, open circle) and for the oxidized special pair ($\Delta A(P^+)$, full circle).

interpretation, fast charge separation in the excited special pair forms the P^{+-} state within 150 fs. (iii) The following decrease in absorption within 1 ps suggests to us the rapid passage of the electron at the neighboring $BChl_A$ forming a transient $P^+BChl_A^-$ state. The electron continues its path to the pheophytine within 5 ps; we see a new state P^+BPh^-. In our samples, where the quinones are chemically reduced, back-reactions return the RC to its initial state within less than 10^{-4} s.

Some comments should be added concerning the absolute values of the absorbance changes induced by the various intermediates. At late delay times one observes state P^+BPh^-. Its absorbance change ΔA is induced, in a first-order approximation, by oxidation of the special pair (giving $\Delta A(P^+)$) and reduction of a bacteriopheophytine ($\Delta A(BPh^-)$): $\Delta A(P^+BPh^-) \simeq \Delta A(P^+) + \Delta A(BPh^-)$. In a similar way we obtain $\Delta A(P^+BChl_A^-) \simeq \Delta A(P^{+-}) + \Delta A(BChl_A^-)$. Knowing $\Delta A(BChl^-)$ and $\Delta A(BPh^-)$ from Fig.3 we can deduce from Fig.2 the values of $\Delta A(P^+)$ (full circles in Fig.3) and $\Delta A(P^{+-})$ (open circle in Fig.3).

Additional information supporting the presented interpretation and yielding the absorbance changes $\Delta A(P^+)$ at other wavelengths can be deduced from the picosecond measurements: In Fig.4 the change of absorption at 592 nm is presented after picosecond excitation at 623 nm. As pointed out in Fig.1, at 592 nm the Q_x transition of $BChl_A$ absorbs stronger than the special pair. After excitation of the special pair at 623 nm the absorption at 592 nm rises during the pulse duration of 4 ps and remains constant for longer times. A more careful analysis of the data shows, however, that the build-up of the absorption is slowed down. The broken curve in the figure represents the measured integral over the autocorrelation curve. The difference between the instantaneous response and the experimental data are presented on an enlarged

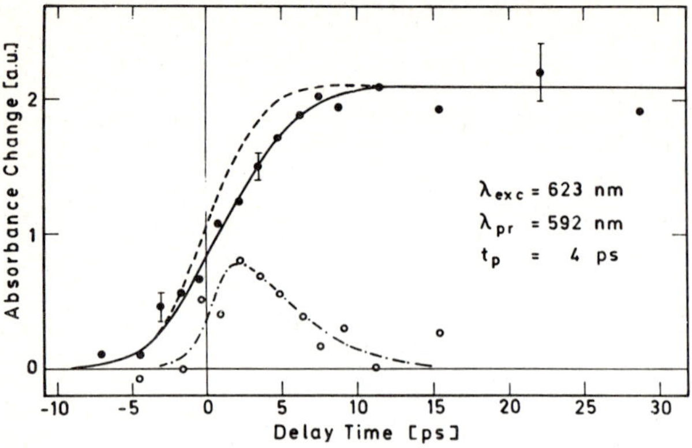

Fig. 4 Absorbance change induced by pulses at 620 nm with 4 ps duration measured at λ = 592 nm (full circles). The broken curve shows the integrated cross-correlation curve. Differences between the solid and the broken curve are drawn in an enlarged scale (x(-3)) as the open circles.

scale by the dash-dotted curve. One may explain the data of Fig.4 as follows: The excitation of the special pair at 623 nm leads to an enhanced absorption at 592 nm due to the rapid generation of P^{+-} and the slower formation of the final state P^+BPh^- within 5 ps. The delay in the build-up is an indication of a short-lived intermediate state, which absorbs weaker than the final state P^+BPH^-. We believe that this intermediate state is $P^+BChl_A^-$. Indeed, according to the data of Fig.3, the absorbance change $\Delta A(BChl^-)$ is considerably smaller than $\Delta A(BPh)$ at 592 nm.

The situation at the probing frequency at 540 nm is quite different. In Fig.5 the change of absorption after picosecond excitation at 620 nm is depicted. We find a rapidly rising (within

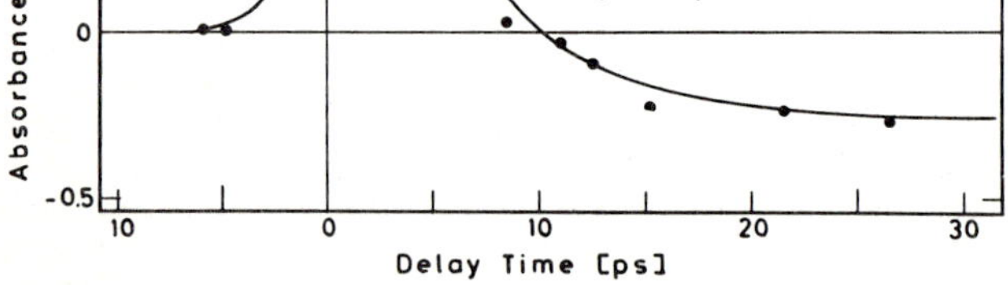

Fig. 5 Absorbance change induced by pulses at λ = 620 nm with 4 ps duration probed at λ=540 nm. At this wavelength the bacteriopheophytines give a negative absorbance change ΔA. The positive ΔA at earlier times is due to the oxidized special pair P^+.

the time resolution of 1 ps) enhanced absorption, followed by a negative absorption change. The short positive absorption change is believed to be due to a strong absorption of the P^{+-} and $P^{+}BChl^{-}$ states at 540 nm. These short-lived states are not resolved with pulses of a few picosecond duration. The negative absorption change for the transition BPh → BPh^{-} (see Fig.3) changes the sign of ΔA in Fig.5. In fact, the negative values of ΔA build up with a time-constant of 5 ps, as expected for the formation of the radical pairs $P^{+}BPh^{-}$. The final absorbance change is due to $\Delta A(P^{+})$ and $\Delta A(BPh^{-})$. From a numerical fit of the measured absorbance changes we estimate $\Delta A(P^{+})$ to be 80% of ΔA(BPh). The value $\Delta A(P^{+})$ determined by using $\Delta A(BPh^{-})$ is shown as full circle in Fig.3. It is interesting to note that the absorbance change induced by oxidation of the special pair $\Delta A(P^{+})$ is nearly constant in the observed spectral range.

A comment should be made concerning the intramolecular energy relaxation between the Q_x state and the lower lying Q_y state of the special pair. There exists strong evidence in the literature that in large polyatomic molecules intramolecular energy relaxation in the electronic excited state proceeds very fast, within 10^{-13} seconds /7/. It is very likely that the intramolecular energy relaxation in the special pair occurs during the first hundred femtoseconds prior to the formation of the P^{+-} state.

In summary we wish to say that experiments with the improved time resolution of 100 fs give new information not visible in previous picosecond experiments /8/. The proposed time sequence of charge separation and electron transfer

$$P \xrightarrow{h\nu} P^{*} \xrightarrow{<0.1\ ps} P^{+-} \xrightarrow{0.1\ ps} P^{+}BChl_A^{-} \xrightarrow{5\ ps} P^{+}BPh^{-}$$

is consistent with the present knowledge of the reaction center and with our femtosecond and picosecond experiments.

References

1 H. Michel, J. Mol. Biol. 158, 567 (1982)
2 J. Deisenhofer, O. Epp, K. Miki, R. Huber, H. Michel, J. Mol. Biol. 180, 385 (1984)
3 R.L. Fork, B.I. Greene, C.V. Shank, Appl. Phys. Lett. 38, 671 (1981)
4 For details on the experimental set-up see: M.C. Nuss, W. Zinth, W. Kaiser, E. Kölling, D. Oesterhelt, Chem. Phys. Lett. 117, 1 (1985)
5 W. Zinth, M. Sander, J. Dobler, W. Kaiser, H. Michel, same proceedings
6 M.S. Davis, A. Forman, T.K. Hanson, J.P. Thornber, J. Fajer, J. Phys. Chem. 83, 3325 (1979)
7 F. Graf, A. Penzkofer, Optics and Quant. Electron. 17,53 (1985)
8 D. Holten, C. Hoganson, M.W. Windsor, C.C. Schenck, W.W. Parson, A. Migus, R.L. Fork, C.V. Shank, Biochim. Biophys. Acta 592, 461 (1980)

Analysis of Time-resolved Fluorescence of *Rhodopseudomonas viridis* Reaction Centers

J.K.H. Hörber, W. Göbel, A. Ogrodnik, and M.E. Michel-Beyerle

Institut für Physikalische und Theoretische Chemie,
Technische Universität München, D-8046 Garching, F. R. G.

E.W. Knapp

Physik-Department, Technische Universität München,
D-8046 Garching, F. R. G.

1. Introduction

Knowledge on structure and dynamics of photosynthetic reaction centers (RCs) is crucial to understand their function i.e. the light-driven charge separation process. The recently obtained structure of Rhodopseudomonas viridis /1,2/ is an important step towards this goal. It has been shown that steady state optical spectra are mainly determined by the static structural arrangement of the pigments /3-7/. Time-resolved spectra are related more to the dynamical aspects of RCs. However, structural information on RCs is also useful to understand time-dependent spectra.

In this paper we analyse the fluorescence decay pattern /8/ of the Q_y band of the two bacteriochlorophylls forming the special pair (BC_P) in Rhodopseudomonas viridis RCs which emit at room temperature around 1020 nm. The lifetime of this fluorescent state is limited to less than 10 ps /9,10/ by the fast initial step of charge separation. As a consequence we observe prompt fluorescence with a low quantum yield. In reduced RCs the negative charge at the only quinone (Q) present blocks the electron transport chain at the bacteriopheophytin (BP). Now in addition to the prompt a delayed fluorescence signal occurs due to the recombination of the radical pair state $BC_P^+BP^-$. This interpretation is supported by two features: i) the lifetime of this delayed fluorescence component is identical to the one of the radical pair (15 ns) /9/ and ii) this lifetime increases by about 10% in a weak magnetic field. However, time-resolved fluorescence measurements of reduced RCs exhibit not only one, but two delayed components. Moreover, also for untreated RCs a delayed component of the fluorescence has been observed which has the same lifetime (2,5 ns) as the faster one of the two delayed components of reduced RCs.
A similar, multiexponential fluorescence decay pattern has been reported for RCs of Rhodopseudomonas sphaeroides /11,12/. To understand this emission phenomenon of the RCs of purple bacteria, we put different kinetic models to test /13/. As a result a special branching kinetics is required to describe the experimental observations, as shown in this paper.

2. Experimental Results /10/

In the RC preparation of Rhodopseudomonas viridis used /14/, only one quinone (menaquinone) was present whereas ubiquinone has been lost. Untreated RCs are kept in the dark, reduced RCs are obtained by illumination of the same sample with a CW-light source (about 20 photons/sRC) after addition of ascorbate (10 mM). Time-resolved fluorescence measurements are performed at

5° C using a Varian VP-152 photomultiplier operating at 190 K. Data processing is made by a Tektronix R7912 transient digitizer and a PDP 11-34 computer. The detecting system is only triggered when the excitation pulse from a dye-laser system is in an intensity regime where about 0.5 % of the RCs are excited. The excitation wavelength is 590 nm i.e. in tne Q_x band of the accessory bacteriochlorophyll monomers (BC_A). The fluorescence is probed in the Q_y band of the special pair at 1020 nm using an interference filter. The time resolution of the detecting device is about 1 ns.

Since the time-resolved fluorescence signals of untreated and reduced RCs are obtained from the identical sample without changing the geometry, the experimental data are normalized with respect to each other. The time-dependent fluorescence intensity $\phi(t)$ is assumed to obey a multiexponential decay law:

$$\phi(t) = \sum_n I_n e^{-t/\tau_n} \tag{1}$$

and is analysed using the Marquardt algorithm /15/. Comparison with the measured fluorescence intensity I(t) is made by convoluting the expression (1) with the excitation pulse shape f(t):

$$I(t) = \int_{-\infty}^{+\infty} dt' \, \phi(t-t')f(t') \tag{2}$$

A satisfying description of the measured fluorescence signals requires two exponential decay terms for untreated RC's and three exponential decay terms for reduced RCs. The relative fluorescence yields $y_n = I_n \tau_n$ and the lifetimes τ_n of the individual components are listed in Table 1.

Table 1 Measured fluorescence yields and lifetimes

	RC with MQ		RC with MQ⁻	
n	y	τ(ns)	y	τ(ns)
1	1	<<1	1.9	<<1
2	0.17	2.5	0.59	2.5
3	-	-	0.44	15

From Table 1 it is evident that reduction of menaquinone leads to i) an increase of the prompt fluorescence component 1 and ii) a slow delayed component 3 with the lifetime of the radical pair state $BC_P^+ BP^-$.

It is rather unexpected that untreated RCs also exhibit delayed fluorescence (component 2) with the characteristic lifetime of 2.5 ns. Such a component is also observed in reduced RCs, and most interestingly its amplitude increases dramatically by reduction of menaquinone.

3. The Model /13/

The structure of the RC exhibits two almost symmetrical branches (M and L) of pigments which emerge from the special pair BC molecules /1/. The sequence of pigments along the X(=M,L) branch consists of a BC_{XP}, an accessory BC_{XA} and a BP_X molecule. The center-to-center distance of nearest neighbour pigments in each branch is ~11 Å, the distance of the two BP molecules in the different branches is ~23 Å. In the RC preparations used, only the menaquinone of the L branch was present. The almost symmetrical pigment arrangement in the two branches leads us to the assumption that the first steps of charge separation may occur in either of the two branches. However, due to asymmetries of the protein environment at equivalent pigments, the two pigment branches behave differently with respect to the electron-transfer process, highly favouring the L branch. The electron transfer of the M branch can be observed in absorbance measurements, where after reduction of both Q and BP_L, the absorption band of BP_M is bleached /16/.

The resulting model system involves three states, one fluorescent state (P) corresponding to the BC_P and two non-fluorescent states M and L accounting for the participation of the pigments in either of the two branches.

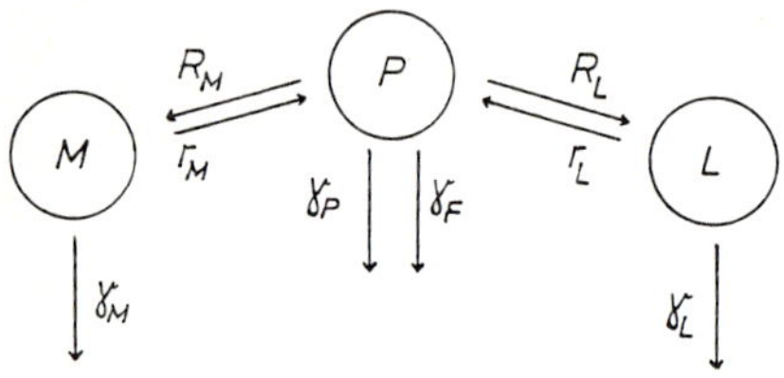

Fig. 1

The coupling scheme of the model.

The coupling scheme of the model is given in Fig. 1 together with the transfer rates of charge separation (R), recombination (r), and the rates of the irreversible processes of non-radiative (γ_P, γ_M, γ_L) and radiative decay (γ_F). The high efficiency of the charge separation in RCs requires that the following inequalities are valid:

$$R_L >> r_L \,,\; R_M >> r_M \quad \text{and} \quad \gamma_L >> r_L \tag{3}$$

The large value of R_L from the latter inequality accounts for the fast electron-transfer from the BP_L molecule to the quinone in untreated RCs. For reduced RCs we can assume that:

$$\gamma_M = \gamma_L = \gamma << R_L,\, R_M \tag{4}$$

Under the above conditions all quantities of the model can be solved in closed form /13/. The decay law of the fluorescent special pair state P reads for reduced and untreated RCs:

$$\phi_P^{red}(t) = e^{-Rt} + \frac{R_M R_L}{R^2}\,\frac{(r_M - r_L)^2}{Rr}\,e^{-(\gamma+r)t} + \frac{r_L r_M}{Rr}\,e^{-\gamma t} \tag{5}$$

and

$$\phi_P^{unt}(t) = e^{-Rt} + \frac{R_L r_L}{(R-\gamma_L)^2}\,e^{-\gamma_L t} + \frac{R_M r_M}{R^2}\,e^{-(\gamma_M + R_L r_M / R)t} \tag{6}$$

In the Eqs. (5) and (6), the following abbreviations are used:

$$R = R_L + R_M \tag{7}$$

$$r = (r_M R_L + r_L R_M)/R \tag{8}$$

The yield y_n of an individual component of the fluorescence is obtained from Eqs. (5) and (6) by multiplication of the corresponding preexponential factor I_n with the ratio of the radiative decay rate γ_F versus the total decay rate $1/\tau_n$ as follows:

$$y_n = I_n \gamma_F \tau_n \tag{9}$$

Also of interest are the probabilities to meet the excitation in the state L or M of the corresponding pigment branch. For reduced RCs the time dependent probabilities are given by:

$$\phi_M^{red}(t) = R_M/R\ (-e^{-Rt} + \frac{r_M - r_L}{r}\frac{R_L}{R}\, e^{-(\gamma + r)t} + \frac{r_L}{r}\, e^{-\gamma_M t}) \tag{10}$$

The corresponding probability for the state L is obtained from Eq. (10) by replacing the indices L by M and vice versa. For untreated RC's the probabilities are given by:

$$\phi_M^{unt}(t) = R_M/R\ (-e^{-Rt} + e^{(-\gamma_M + r_M R_L/R)t}) \tag{11}$$

and

$$\phi_L^{unt}(t) = \frac{R_L}{R - \gamma_L}\ (-e^{-Rt} + e^{-\gamma_L t}) \tag{12}$$

4. Discussion

The model is used to interpret the fluorescence data of RCs. For reduced and untreated RCs we employed the following procedure:

i) With the amplitudes and lifetimes of the three fluorescence components of reduced RCs we calculate from Eq.(5) the relative rates of the model scheme. After normalization with the fast forward rate R_L = 1/5 ps /9/ we obtain the absolute rates (Table 2).

ii) Assuming that the electrostatic effect of the excess charge at the menaquinone is restricted to the L branch, the values R_M, r_M and γ_M obtained from reduced RCs are also used for untreated RCs. Together with the experimental value of γ_L = 1/230 ps /9/ Eq.(6) yields the forward rate R_L = 1/450 ps.

The calculated rate parameters of the model given in Table 2 show the kinetic requirements which have to be met for the M branch to participate in the first steps of charge separation. It also clearly indicates that the predominant pathway is along the L branch.
For reduced RCs the forward rate on the L branch exceeds the one on the M branch by a factor of 2.5 whereas the recombination rate r_M on the M branch is faster than the corresponding one on the L branch by a factor 45.
For untreated RCs the branching ratio R_L/R_M is now 5.6 favouring even more the L branch than in reduced RCs. Eq. (6) also provides a delayed fluorescence component with the lifetime of 2.2 ns and a yield of 15 % in

Table 2 Calculated rate constants

Rates (ns)	R_M	r_M	γ_M	R_L	r_L	γ_L
untreated RC	80	0.45	0.067	450	-	-
reduced RC	80	0.45	0.067	200	0.01	0.067

relation to a prompt fluorescence of 100 %. This is in good agreement with the experimental lifetime of 2.5 ns and a fluorescence yield of 17% from untreated RCs (Table 1).

The large value of γ_L /9/ in untreated RCs implies a low intensity of the fluorescence component 3. Therefore recombination on the L branch competing with forward electron-transfer to the menaquinone does not contribute in a measurable extent to delayed fluorescence. This is the reason why r_L in untreated RCs can not be determined. With the absolute fluorescence yield of y = 3 (±1.5) 10^{-4} /10/ the radiative decay rate γ_F = 0.16 (±0.08) ns^{-1} of the state P can be obtained.

The model predicts also the time-dependent probability to meet the excitation in the state M, Eqs. (9) and (10). For untreated RCs the calculated participation of the pigments on the M branch does not exceed 15 % and decays very fast (τ_2 = 2.5 ns) (Fig. 2). For reduced RCs this participation increases (28 %) since the negative charge at the menaquinone slows down the forward electron-transfer rate to the BP_L (Tab. 2) as reflected by the increase of the prompt fluorescence upon reduction of Q by a factor of 1.9.

The maximum of ϕ_L predicted for the L branch in Fig. 3 is caused by the highly efficient delayed population of the P state during recombination on the M branch. It should lead to a small corresponding delayed component in the formation of radical pairs on the L branch. This participation of the M branch should hardly affect the quantum yield of the overall charge separation.

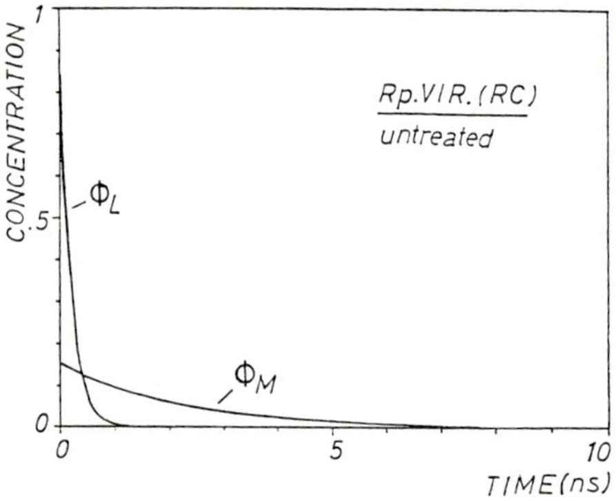

Fig. 2

Time-dependent concentration of the excitation in the M and L branch calculated for untreated RC's using Eqs. (11) and (12).

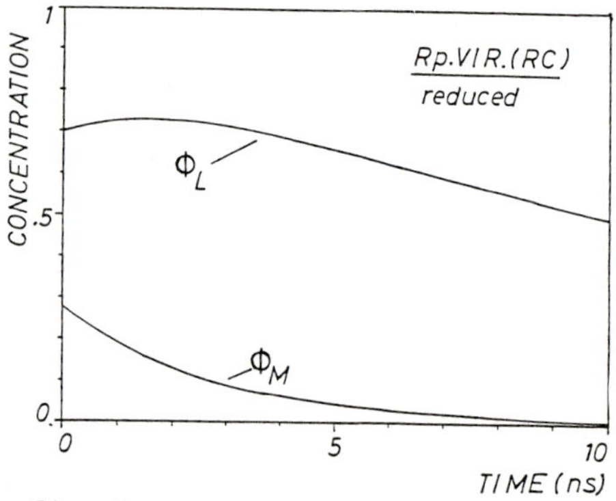

Fig. 3

Time-dependent concentration of the excitation in the M and L branch calculated for reduced RC's using Eq. (10).

The phenomenology of the multiexponential decay pattern described here is also found for RCs of Rp. sphaeroides /17/. Application of the kinetic model to RCs of Rp. sphaer. is leading to a more pronounced participation of the M branch.

In order to test this interpretation of multiexponential fluorescence decay we propose picosecond absorption measurements in the Q_X (or even Q_Y) bands of BP_M and BP_L in reduced and untreated RCs (Fig.2,3). However, electron transfer to the M branch may not lead to a complete bleaching of the M pheophytin. The nature of the non-fluorescent state in the M branch may differ considerably from the one in the L branch. The participation of the BP_M in the state M may for instance be much lower than the corresponding participation of the BP_L in the state L. This leads to a reduction of the bleaching signal from BP_M.

Whatever the origin of the two delayed fluorescence components is, a detailed analysis shows /13/ that only branching kinetics of the type depicted in Fig.1 is consistent with the experimental data. A sequential kinetics involving only pigments on the L branch cannot explain our finding of the 2.5 ns delayed fluorescence component,which is slow compared to the forward rate from the BP_M to Q.

Acknowledgements

We like to thank Prof. S.F. Fischer for many stimulating and critical discussions. This work is supported by the Deutsche Forschungsgemeinschaft under the project "Primary processes in bacterial photosynthesis", SFB 143.

References

/1/ J. Deisenhofer, O. Epp, K. Miki, R. Huber, and H. Michel: J. Mol. Biol. 180, 385 (1984).
/2/ J. Deisenhofer and H. Michel: This volume.
/3/ V.A. Shuvalov and A.A. Asadov: Biochim. Biophys. Acta 545, 296 (1979).
/4/ A. Vermeglio and G. Paillotin: Biochim. Biophys. Acta 681, 32 (1982).
/5/ W. Zinth, E.W. Knapp, S.F. Fischer, W. Kaiser, J. Deisenhofer, and H. Michel: Chem. Phys. Letters (1985) in press.
/6/ E.W. Knapp, W. Zinth, M. Sander, S.F. Fischer, W. Kaiser, J. Deisenhofer, and H. Michel: Proc.Natl.Acad.Sci. US (1985) in press.
/7/ E.W. Knapp and S.F. Fischer: This volume.
/8/ D. Holten, M.W. Windsor, W.W. Parson, and J.P. Thornber: Biochim. Biophys. Acta 501, 112 (1978).
/9/ W. Zinth, M.C. Nuss, M.A. Franz, W. Kaiser, and H. Michel: This volume.
/10/ J.K.H. Hörber, W. Göbel, A. Ogrodnik, M.E. Michel-Beyerle, and R.J. Cogdell: to be published.
/11/ N.W.T. Woodbury and W.W. Parson: Biochim. Biophys. Acta 767, 345 (1984).
/12/ P. Sebban and J.G. Barbet: FEBS Letters 165, 107 (1984).
/13/ J.K.H. Hörber, M.E. Michel-Beyerle, and E.W. Knapp: to be published.
/14/ J.P. Thornber, R.J. Cogdell, R.E.B. Seftor, and G.D. Webster: Biochim. Biophys. Acta 593, 60 (1980).
/15/ D.W. Marquardt: J. Soc. Indust. Appl. Math. 11, 431 (1963)
/16/ J.P.Thornber, R.E.B. Seftor, and R.J. Cogdell: FEBS Letters 134, 235 (1981)
/17/ J.K.H. Hörber, W. Göbel, M.E. Michel-Beyerle, and R.J. Cogdell: to be published

The Characterization of the Q_A Binding Site of the Reaction Center of *Rhodopseudomonas sphaeroides*

M.R. Gunner, B.S. Braun, J.M. Bruce[†]*, and P.L. Dutton*

University of Pennsylvania, Department of Biochemistry and Biophysics, Philadelphia, PA, 19104, USA

1 Introduction

Quinones are important participants in respiratory and photosynthetic electron-transfer chains. Some are found tightly associated with individual redox proteins, while others are considered to be free in the bilayer membrane, where, as a pool, they serve as electron or hydrogen carriers between specific sites on different proteins within the membrane.

The reaction center protein from photosynthetic bacteria contains a very tightly bound quinone that does not exchange with the pool. It is designated Q_A, and plays an obligatory functional role in the light-activated electron-transfer reactions in the reaction center protein. Cogdell et. al. (1) and Okamura et. al. (2) showed by extraction/reconstitution experiments that Q_A in the reaction center protein from Rhodopseudomonas sphaeroides is ubiquinone-10 (UQ_{10}). They also demonstrated that other quinones could bind to the site and reconstitute activity. Furthermore, Okamura et. al. (2) demonstrated that several quinones added as replacements for the UQ in the Q_A site could, in fact, function at cryogenic temperatures. The study of the affinity and functionality of different molecules at the Q_A binding site of this protein can therefore provide an opportunity to determine what aspects of a quinone govern its binding to the protein, its properties in the binding site, and its function.

A systematic study of the affinity of various molecules for the Q_A site has been carried out. This has yielded a quantitative view of the features of the quinone molecule that contribute to the energy of binding, and has permitted the identification of structural elements of the quinone that fix the orientation in the site. Concurrent with this work the three-dimensional structure of the reaction center protein from Rp. viridis (3) and the primary sequences of the reaction center proteins from several other sources are becoming available. As the details of the structure of the protein emerge, it should be possible to combine this information with that obtained from the biochemical binding studies of the kind presented here, to identify the specific contacts between the important regions of the quinone and the protein. This report briefly summarizes some of the

---

[†]Department of Chemistry, University of Manchester, Manchester, U.K.

conclusions of our studies, aimed at describing the factors that govern the binding of simple quinones and related compounds to the Q_A site.

2 Materials and Methods

The reaction center protein from Rps. sphaeroides was isolated using well-established procedures (4). The UQ_{10} at the Q_A site was removed from the protein using the procedure of Okamura et. al. (2) with the modification that the temperature of the high detergent extraction step is raised to 28°C. Under these conditions the quinone is removed in two hours or less. Typically 90-98% of the UQ_{10} was removed by this procedure.

Quinones were obtained from the Aldrich Chemical Company and the Eastman Kodak Company, while others were synthesized. The partition coefficients and purity were determined using an HPLC technique (5).

All of the quinones discussed in this paper fully reconstitute the Q_A-dependent flash-activated electron-transfer reactions in the isolated protein. The quinone-reaction center dissociation constants (K_d) were therefore determined by the measurement of activity as a function of the concentration of the quinone. The best non-linear least squares fit of the data to the complete quadratic form of the equation for the binding of one ligand to a single site was used to determine the K_d.

The non-quinone compounds examined do not reconstitute Q_A function. The affinity of each of these compounds for the Q_A binding site was determined by its potency as a competitive inhibitor of the binding of tetramethyl-1,4-benzoquinone, which is an active Q_A. The concentration of the non-quinone was adjusted to be at least ten times greater than that of the protein, so that the amount of inhibitor bound was always a small fraction of the total added. This permits the assumption that the free concentration of inhibitor is equal to the total added, i.e. $[I]_T=[I]_F$ to be made. Under these conditions the shape of the binding curve for the increase in activity with the addition of an active quinone is not affected by the presence of the inhibitor, so the data was fit using the procedure developed for the simple one-component titration. However, the apparent K_d (K_d^{app}) is related to the K_d in the absence of inhibitor by the relationship $K_d^{app} = K_d(K_i + [I]_T)/K_i$. This was then used to obtain K_i, the equilibrium dissociation constant of the inhibitor.

The assay for Q_A activity utilizes the enormous stabilization of the light-initiated separation of charge in reaction center protein when it contains a functional Q_A. In brief, the method relies on all protein molecules that absorb a photon reaching the state $(BChl)_2^{+}BPh^{-}$, but when forward electron-transfer from BPh^{-} to Q_A cannot take place, the $(BChl)_2^{+}BPh^{-}$ state undergoes charge recombination to reform the ground state at a rate of about 10^8 s^{-1}. In contrast, if

Q_A can be reduced by BPh^-, the rate of charge recombination of protein in the resultant state, $(BChl)_2^{+}BPhQ_A^-$, is very much slower, and it is now dependent on which quinone is serving as Q_A. If UQ_{10} is Q_A, the rate of charge recombination is $7s^{-1}$, but with other quinones it has been found to range from less than $1s^{-1}$ to greater than $10^4 s^{-1}$ (6). The change in the redox state of $(BChl)_2$ which can be determined by the $\Delta A_{605-540}$ is the most convenient chromophore to measure. The initial concentration of $(BChl)_2^{+}$ measured at times greater than 10µs after the flash is equal to the concentration of Q_A^- and so provides an assay of the protein with a functional Q_A bound at each quinone concentration. A more complete discussion of electron-transfer reactions in isolated reaction center protein can be found in the recent review by Parson and Ke (7).

Binding titrations were performed at 22°C using 250 to 350 nM reaction protein in 10 mM Tris-HCl pH 8.0 containing $5x10^{-4}$ to 10^{-3}% LDAO. We have found that although the absolute measured K_d values increase with increasing detergent, the relative affinities of different quinones do not appear to be very sensitive to the concentration of detergent until the detergent concentrations appproach 1%. Since the comparison of affinities, rather than their absolute values, are important for the conclusions drawn from this study, it should be sufficient to carry out all titrations at one, well-defined detergent concentration.

3 Results and Discussion

Photosynthetic bacteria are, depending on the species, found to utilize one of several different chemical forms of quinone at the Q_A site; the most common are ubiquinone (UQ), and menaquinone (MK) (8). Rhodopseudomonas sphaeroides, the species used in the present study, uses UQ_{10} as the quinone in the Q_A site. The head groups of these quinones are quite different. For UQ it is 2,3-dimethoxy-5-methyl-1,4-benzoquinone (UQ_0) whereas 2-methyl-1,4-naphthoquinone (MK_0) is the head group of MK. We have measured the affinity of the methyl derivatives of these molecules to evaluate the contribution of the head group to binding. The methyl derivative of UQ (6-methyl UQ_0) binds rather weakly to the Q_A site of the Rps. sphaeroides reaction center protein; it has a K_d of 10 µM. The MK derivative (3-methyl-MK_0) binds much more tightly with a K_d of 40 nM. This simple difference in affinity may explain why in those bacterial species, such as Rp. viridis, and Chr. vinosum, that contain both MK and UQ, the occupant of the Q_A site is found to be a menaquinone (9). It can also be argued that the chemical nature of the quinone occupant of the Q_A site in the reaction center protein from these different bacteria is simply an expression of the availability of a tighter binding quinone species (MK) rather than of any particularly significant differences in the architecture of the site itself. This provides a basis for the use of the information derived from the crystal structure of the protein from Rp. viridis in the analysis of the requirements for binding to the protein from Rp. sphaeroides.

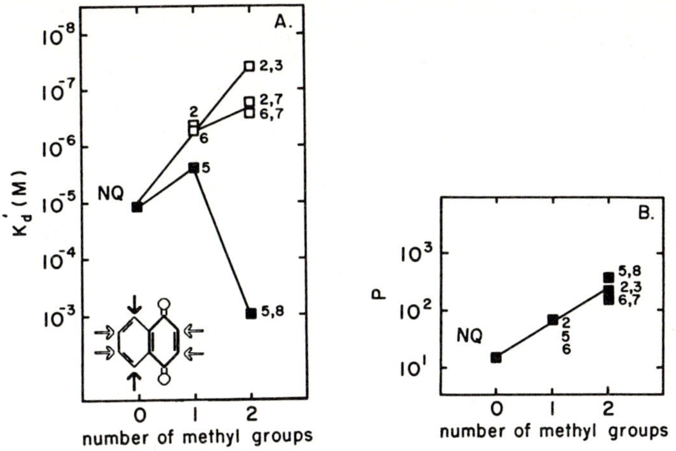

Figure 1: a) Dissociation constants and b) cyclohexane/water partition coefficients of methyl substituted 1,4-naphthoquinones

Although 6-methyl UQ_0 binds weakly to the protein, the affinity of UQ_{10} is extremely high. This difference demonstrates the significant contribution of the isoprene tail to the binding energy. Each of the first two isoprene units add at least 10^3 (4 kcal) to the affinity and UQ-2 has a K_d no greater than 100 pM, which represents the current limit for K_d determination.

Some details of how the head group binds to the site can be seen in Figure 1a which plots the K_d values for naphthoquinones (NQs) with methyl groups at the ortho (2 and 3), beta (6 and 7), and alpha (5 and 8) positions. Figure 1b shows the cyclohexane/water partition coefficients of the same compounds. A comparison of the two plots shows that although all isomers display almost identical hydrophobic character, there can be very large differences in their affinity for the Q_A site. This cannot be explained by a model in which the association of quinone with the protein is viewed simply as a partitioning of the relatively hydrophobic molecule out of the solution into the site. It supports a view in which the quinone binds to the site in a structurally orchestrated manner, so that different parts of the quinone molecule are associated with specific regions of the site and therefore make different contributions to the affinity.

The open symbols in figure 1a show the effects of methyl additions at the ortho and beta positions. These extend the molecule in a direction that is generally in the plane of the ring and orthogonal to the carbonyl axis. Each addition at an ortho position increases the affinity by 17 fold (1.7 kcal). Methyl additions to the ortho positions of benzoquinone add the same amount of binding energy (10), which suggest that they interact with equivalent positions in the protein. In contrast, the second beta addition on NQ makes a smaller contribution to the affinity. This is similar to the effect of additions to the beta positions on anthraquinone, which also increase the affinity by less than a factor of two (10). In general, the contribution of a methyl group appears to depend more on its position relative to the carbonyl axis (i.e. whether it is in an ortho or a beta position) than on the size of the quinone ring system. This suggests that the carbonyl plays an important role in fixing the position of the quinone in the site.

The filled points in Figure 1a describe the effects of substitution at the alpha positions, which are on the ring adjacent to the one with the quinone carbonyls; these substituents are parallel to the axis of the carbonyls. Addition of a methyl group to an alpha position changes the K_d of the unsubstituted quinone by an amount that is significantly less than that found with addition at either the ortho or beta positions. However, the addition of a second methyl in the other alpha position, para to the first, causes a dramatic decrease in the affinity: the K_d of 5,8-dimethyl NQ is at least 500 times larger than that of 5-methyl NQ. This pattern suggests i) that a close contact between one carbonyl and the protein is required for binding, and that this interaction can be disrupted by the addition to an adjacent, alpha position and ii) that the side of the quinone molecule containing the other carbonyl interacts only weakly with the protein, so that it contributes little or nothing to the binding energy.

The change in K_d with the addition at the alpha positions of methoxy groups to NQ, and methyl, methoxy and amino groups to AQ shows the same remarkable pattern in which the first alpha addition increases the affinity by a small amount and the second addition para to the first strongly weakens the association (10). This provides further evidence that hydrophobic forces do not play a major role in the interaction of the alpha position of the quinone with the protein, since all of these modifications have the same effect on the affinity, but they have very different influences on the partition coefficient. Additions of amino groups and alpha methoxy groups make the molecule more hydrophilic, while additions of methyl groups increase the hydrophobicity (5).

Although the presence of two carbonyls characterizes a molecule as a quinone and are vital for redox function, for the purpose of examining the contributions made by different parts of the quinone molecule to the association with the Q_A site, the carbonyls can be considered as simply another functional group. Their role in binding to the site can be determined by the measurement of the affinity of quinone analogues where one or both carbonyls have been removed. Anthrone, with only one carbonyl group, has an affinity of only a factor of 3 (0.6 kcal) smaller than AQ; but anthracene, where both carbonyls have been removed, binds at least 10^3 (more than 4 kcal) less tightly (10). The same pattern is seen in the affinity of NQ analogues with less than two carbonyls. This demonstrates conclusively that one carbonyl

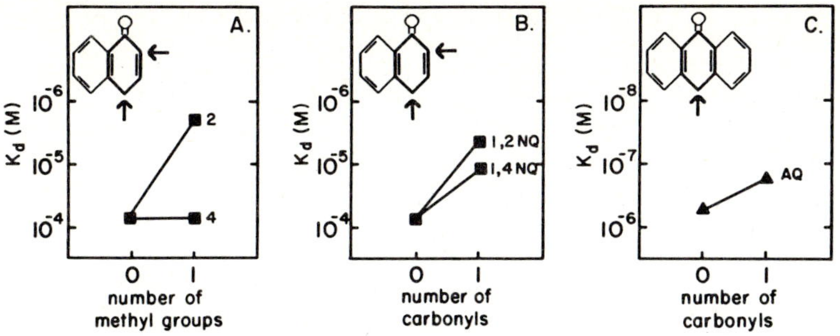

Figure 2: The modification of the affinity of one carbonyl quinone analogues; a) addition of a methyl group to 1-tetralone b) addition of a carbonyl to 1-tetralone; c) addition of a carbonyl to 9-anthrone

contributes a significant fraction of the energy of binding the quinone to the Q_A site, but the second adds little or nothing to the affinity.

The force contributed by one carbonyl in orienting the molecule in the site can be shown by comparing the affinities of quinones and single carbonyl analogues substituted in various positions. Figure 2 shows that the addition of either a methyl or a carbonyl in the para position of tetralone increases the affinity by only a relatively small amount. These changes are no different from those encountered with the first addition to the alpha position of NQ and AQ, which also have little effect on the K_d (see Figure 1a). However, additions made ortho to the lone carbonyl increase the affinity to an extent that is considerably larger than that found with the para substitution: in fact, the effect is comparable to that found with additions made to ortho positions on quinones, suggesting that in each case this position makes the same contact with the protein. Viewed in this manner, the observation that 1,2-NQ actually binds more tightly than 1,4-NQ can be explained simply by regarding the second carbonyl as a substituent which is either ortho or para to the binding carbonyl. This then interacts with the site in a manner that is no different than other small substituents.

Thus, we identify one carbonyl as playing a crucial role in binding the quinone head group to the site. Clearly, it provides not only a major fraction of the free energy of the interaction but is also important in positioning the quinone in the site. It seems reasonable to suggest that the interaction is derived from a hydrogen bond between the quinone carbonyl and some group on the protein; certainly the nature of the carbonyl and the value of the free energy contributed by the group are consistent with this possibility. Indeed, the lack of any hydrogen bond between the second carbonyl and the protein is surprising; surveys of various proteins have shown that there are hydrogen bonds to the large majority of groups that are capable of participating (11). It is possible that the presence of only one hydrogen bond to Q_A may be important in regulating its function, as it might serve to stabilize the semi-quinone while keeping the doubly-reduced Q_A at much higher energy.

4 Acknowledgements

This work was supported by a grant from the National Science Foundation (PCM 82-09292)

5 References

1. Cogdell, R.J., Brune, D.C. and Clayton, R.K. (1974) FEBS Letts. 45: 344.

2. Okamura, M.Y., Isaacson, R.A., and Feher, G. (1975) Proc. Natl. Acad. Sci. U.S.A. 72: 3491-3495.

3. Deisenhofer, J., Michel, H., and Huber, R. (1985) Trends Biochem. Sci. 10: 243-248.

4. Clayton, R.K. and Wang, R.T. (1971) Methods in Enzymology 23, 696-704.

5. Braun, B.S., Benbow, U., Lloyd-Williams, P., Bruce, J.M. and Dutton, P.L. (1985) Methods in Enzymology, (eds. S. Fleischer and B. Fleischer) Proton Motive Force and Secondary Transport. Part B: Electron Transport. Academic Press, New York. In Press.

6. Gunner, M.R., Tiede, D.M., Prince, R.C., and Dutton, P.L. (1982) Quinones in Energy Conserving Systems (ed. B.L. Trumpower) Acad. Press, New York, pp. 265-270.

7. Parson, W.W., and Ke, B. (1982) In Photosynthesis Volume 1: Energy Conversion in Plants and Bacteria (ed. Govindjee) Acad. Press, New York, pp. 331-385.

8. Collins, M.D., Jones, D. (1981) Microbiological Reviews 45: 316-352.

9. Okamura, M.Y., Feher, G. and Nelson, N. (1982) in "Photosynthesis Vol. 1: Energy Conversion by Plants and Bacteria" (ed. Govindjee) Acad. Press, New York, pp. 195-272.

10. Gunner, M.R., Braun, B.S., Bruce, J.M. and Dutton P.L. To be submitted to Biochim. Biophys. Acta.

11. Baker, E.N. and Hubbard, R.E. (1984) Prog. Biophys. molec. Biol. 44: 97-179.

Part V

Model Systems on Structure of Antennas and Reaction Centers

Structure and Energetics in Reaction Centers and Semi-synthetic Chlorophyll Protein Complexes

S.G. Boxer

Department of Chemistry, Stanford University, Stanford, CA 94305, USA

I. Introduction

The initial photochemistry of photosynthesis involves a highly efficient electron-transfer reaction that generates an anion-cation radical pair. This primary charge-separation is followed by a cascade of electron-transfer steps in which the electron and hole become increasingly spatially separated and stabilized with respect to recombination, allowing the formation of stable chemical products. At each step, the rate of the forward electron transfer reaction exceeds the rate of the charge-recombination reaction by several orders of magnitude. In this paper I discuss results related to two aspects of the initial events in photosynthesis: the relationship between the energetics and kinetics of charge-separation and recombination in bacterial photosynthesis, and the distance and orientation dependence of excitation energy-transfer in semi-synthetic chlorophyll-protein complexes.

II. Energetics and Kinetics of Primary Charge Separation and Recombination

The initial electron-transfer steps in bacterial reaction centers (RCs) are illustrated schematically in Fig. 1. The symbols P and I represent the primary electron donor and acceptor, respectively, with the chemical natures of P and I left intentionally vague; they are interacting chlorophyll-type molecules, but sizeable inter-chromophore interactions make more specific chemical identification difficult. The initial electron-transfer is between chlorophyll-type species (not chlorophyll and quinones), and the initial intermediate, $P^{+}I^{-}$, therefore is essentially a pair of aromatic radicals.

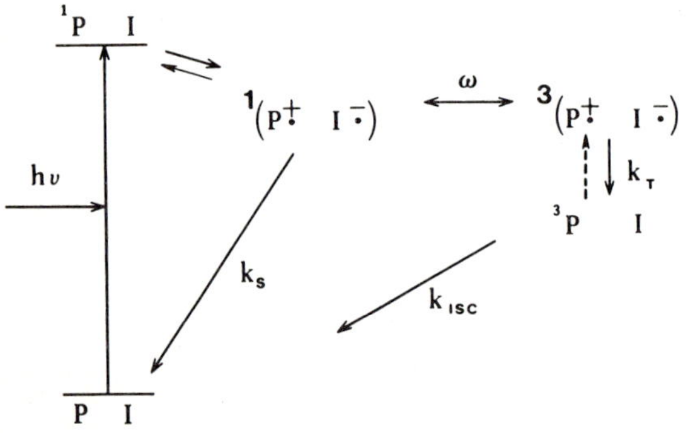

Figure 1 Scheme for the initial events in bacterial photosynthesis when secondary electron-transfer is blocked by removal of ubiquinone

In RCs blocked by removal of quinones, $P^{+}I^{-}$ (formed with unit quantum yield) decays rapidly by recombination (~20ns) to either the ground or excited triplet state of P. Formation of triplets requires that the electron spin angular momentum of $P^{+}I^{-}$ equal one prior to recombination to ^{3}P [1]. Care is taken to distinguish the initially formed singlet radical pair, $^{1}(P^{+}I^{-})$, which can decay via spin-allowed recombination to the ground-state with rate-constant k_S, from the triplet radical pair, $^{3}(P^{+}I^{-})$, which can decay via spin-allowed recombination to the locally excited triplet state, ^{3}P, with rate-constant k_T. Because ^{3}P lives for many µs, radical pair decay by these two pathways can be distinguished. The key point is that the interconversion of $^{1}(P^{+}I^{-})$ and $^{3}(P^{+}I^{-})$ is characterized by a frequency, ω, which depends on the magnetic properties of P^{+}, I^{-}, and $P^{+}I^{-}$ and the magnitude of an externally applied magnetic field. These include the electron-nuclear hyperfine interactions and g-factors for both radicals, which can be measured when the radicals are separately trapped [2], and the electron-electron exchange (J) and dipolar interactions, which can only be measured when both radicals are present. The latter are related to the degree of orbital overlap between P^{+} and I^{-} and the distance between their magnetic centers, respectively. The time-dependent branching to the two products, ^{3}P and ground-state, then depends on the time-evolution of singlet and triplet character in the radical pair and the recombination rate-constants k_S and k_T. The effects of magnetic field on this branching have proven to be very useful in studying the mechanism of photosynthesis, as first shown by Parson [3], Hoff [4], Haberkorn [5], Schulten [6] and their co-workers. A relatively recent review is given in [7].

A measurement of the field-dependence of the quantum yield of ^{3}P and fits to a detailed theory for the effect can provide estimates for the magnetic properties of the radical pair and the recombination rate-constants. Qualitatively, we expect a drop in the ^{3}P quantum yield on application of a

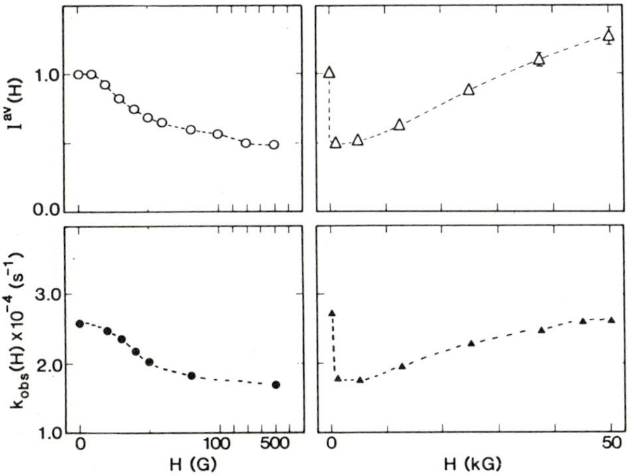

Figure 2 Upper panels (open symbols): magnetic field-effect on the relative quantum yield of formation of ^{3}P; lower panels (filled symbols): magnetic field-effect on the observed decay rate of ^{3}P. The circles on the left panels show the effect of low field on both observables, while the triangles on the right panels show the field-dependence between 0 and 50 kG (Q-depeleted R-26 RCs, room temperature).

small field, and an increase in the ^3P yield at higher field as the g-factor difference between the radicals becomes significant. These effects can be very large, and are illustrated in the upper panels of Fig. 2.

Approximate numerical treatments have been used to analyze the effects at low field [5-7], while an analytical theory has been developed for the high field case [8]. An interesting complication is that all of the magnetic interactions which drive and impede singlet-triplet mixing are tensor quantities, consequently we expect and have shown that the magnetic field effects are strongly anisotropic [9] (it is possible to avoid these effects by judicious choice of experimental conditions, as was done for the data in Fig. 2). Analyses of the field-dependent steady-state and time-resolved quantum yield of ^3P [5,6,8,10], of the related RYDMR spectrum [11,12], and of the radical pair lifetime [13] have led to a consensus among laboratories for the values for k_S ($\sim 3 \cdot 10^7 s^{-1}$), k_T ($\sim 5 \cdot 10^8 s^{-1}$), and J ($\sim 10^{-3} cm^{-1}$). Considerable disagreement remains on the value for the electron-electron dipole-dipole interaction ($0 - 5 \cdot 10^{-3} cm^{-1}$), the sign of J, the importance of the back-reaction $^1(P^+I^-) \rightarrow {}^1PI$, and the role of intermediate states in the forward and recombination reactions.

We were recently surprised to discover that the <u>decay kinetics</u> of ^3P also depend on magnetic field strength, as shown in the lower panels of Fig. 2 [14]. The observed drop in decay rate with application of a small field and increase at higher field, similar to the changes in the ^3P quantum yield (upper panels), is a signature of the involvement of a radical pair in the ^3P decay. We hypothesize that the predominant decay route for ^3P at room temperature exactly parallels its formation route, <u>i.e.</u>, the decay is via the radical pair, P^+I^-, as illustrated with the dotted vertical arrow in Fig. 1. This hypothesis leads to the following predictions: (i) the ^3P decay should be strongly temperature-dependent since the radical pair decay pathway is highly activated, and (ii) the magnetic field-effect on the ^3P decay should vanish as the temperature is lowered (the magnetic field-effect on the ^3P quantum yield persists until much lower temperatures). Both of these predictions are confirmed by experiment; the temperature-dependence of the ^3P decay rate is shown in Fig. 3.

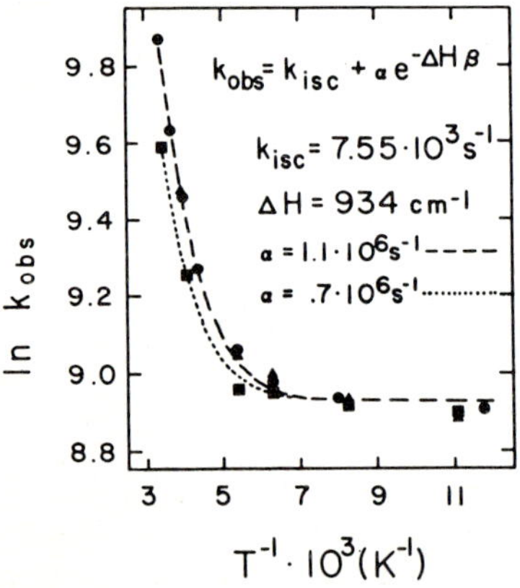

Figure 3 Arrhenius plots of ^3P decay rate as a function of temperature at several magnetic field-strengths (Q-depeleted R-26 RCs). Data taken at H = 0 (•), 5 kG (□) and 45 kG (Δ). $\alpha = (1/3) k_S \Phi_T$.

A quantitative analysis of this new effect requires several assumptions which are discussed in detail in [14]. Taking the scheme in Fig. 1 as complete, we assume that the entropies of ^3P and $^3(P^+I^-)$ are the same, so that the rate of reformation of $^3(P^+I^-)$ from ^3P is $k_T\exp(\Delta H\beta)$, where ΔH is the enthropy difference between $^3(P^+I^-)$ and ^3P and $\beta = (kT)^{-1}$. We further assume that this step is the only temperature-dependent process and that both electron and nuclear spin-lattice relaxation rates in ^3P are faster than the ^3P decay rate. Given these assumptions, it can be shown [14] that the observed decay rate of ^3P is:

$$k_{obs} = k_{isc} + (1/3)k_S\Phi_T\exp(\Delta H\beta),$$

where k_{isc} is the intrinsic decay rate of ^3P due to ordinary intersystem crossing. A fit of the temperature-dependent k_{obs} to this equation is shown in Fig. 3. This gives a value of approximately 0.12eV for ΔH, a reasonable value for the pre-exponential (which can be estimated independently from other data), and k_{isc} ~$7\cdot10^3 s^{-1}$ (consistent with other low-temperature data [2]). The value of 0.12eV for the $^3(P^+I^-)$ - ^3P enthalpy difference can be combined with the measured (somewhat uncertain) value for the ^1P - ^3P enthalpy difference of 0.40eV [15] to obtain an energy-gap of 0.28eV for the primary charge-separation step, ^1P - $^1(P^+I^-)$ (this places the radical pair about 1.1 eV above the ground-state). All previous estimates of the ^1P - $^1(P^+I^-)$ energy-gap have been considerably smaller than the value estimated from the present work [16-18].

A few brief comments on assumptions in this analysis are in order. The assumption that the entropies of ^3P and $^3(P^+I^-)$ are equal is clearly an oversimplification, and buries any opportunity to learn something about important entropy effects in the charge-separation process. Although measurements of the temperature-dependence of k_T and its inverse could provide information, it should be recognized that k_T is not directly measured. This is because the reaction dynamics in RCs is fundamentally tied to the spin dynamics (see Fig. 1), and an estimate for any rate constant requires collective consideration of the entire scheme. A second key assumption concerns the effects of electron and nuclear spin-lattice relaxation in the product state ^{3}PI which becomes the starting material for the reverse reaction. While it is reasonable to suggest that the electron relaxation rate exceeds the triplet decay rate at room temperature (i.e. the three triplet levels in ^3P are at equilibrium prior to charge-separation), it is unlikely that this is also true for the nuclear spins. The problem is further complicated by the fact that the nuclear relaxation rate in ^3P may be quite rapid since ^3P is paramagnetic, but it is certain that nuclear relaxation in I (part of the product state ^{3}PI) is much slower since this is a diamagnetic species. The point is that the nuclear spin states in products PI and ^{3}PI are expected to be strongly polarized by the spin selection process. Thus, it may well be that the distribution of ω for the reaction initiated from ^{3}PI is different from that initiated from ^{1}PI. There has been no previous consideration of the effects of nuclear spin polarization in RCs, however, it is clear that such effects can be very substantial and have very interesting consequences. We have undertaken extensive theoretical and experimental explorations of such effects on several observables in RCs which are presented elsewhere [19].

Given reasonable estimates for the energetics and kinetics of these different electron-transfer reactions in RCs, we can consider their relationship in the context of electron-transfer theories [20,21] and recent experimental work which tests this relationship [22,23]. Assuming that the scheme in Fig. 1 fully describes all relevant states, the energetics and kinetics are as follows: the primary charge-separation step, $^1PI \rightarrow P^+I^-$, has a rate in excess of $2\cdot10^{12}s^{-1}$ [24,25] and an energy-gap of 0.28eV; triplet radical pair recombination, $^3(P^+I^-) \rightarrow {}^3P$ has a rate $k_T\sim5\cdot10^8s^{-1}$ and an energy-gap of 0.12eV; singlet radical pair recombination, $^1(P^+I^-) \rightarrow PI$ has a rate $k_S\sim3\cdot10^7s^{-1}$ and an energy-gap of 1.1eV. Although the ground and excited singlet and triplet states of P are different, we will consider them to be the same for the purpose of a broad argument. It is seen that the fastest reaction occurs when the energy-change is on the order of 0.3eV and that the rate falls for recombination reactions whose energy-gaps are greater and lesser than this value. This is reminiscent of predictions of the Marcus theory [20] of electron-transfer in which the rate increases with increasing free-energy change to a maximum and falls off for a larger free energy change (the inverted region).

The change in free energy corresponding to the rate maximum gives the total reorganization energy. To date, the reorganization energies in model systems tend to cluster around 1eV [22,23], much larger than the value found in photosynthesis. We expect the inner sphere contribution to the reorganization energy to be relatively small, since both P and I are large, aromatic molecules whose bond lengths are likely to change relatively little on loss or addition of one π-electron. The outer sphere contribution is unknown, however, it is again expected to be relatively small as the charge on P^+ or I^- is spread out over large macrocycles. A precise description of outer sphere reorganization when the solvent is a protein matrix is not yet available. Although proteins exhibit complex dynamics in solution, the range of motions available to amino acid residues or dipolar α-helices is limited by the high internal viscosity of the protein and mechanical constraints. (Note: it is possibly relevant that side-chain motion in myoglobin has been shown to be smallest for residues in the immediate vicinity of the heme macrocycle relative to residues closer to the protein surface [26].) It is possible that the protein structure in the vicinity of the reactive macrocycles is predisposed such that the formation of charged intermediates causes minimal reorganization. Ideas related to this have been outlined previously by Warshel and co-workers [27]. A physical situation of this sort will not arise in model systems where the solvent is randomly distributed about the reactants, chemically homogeneous, and, of course, not subject to evolutionary pressures.

III. Energy-Transfer in Semi-Synthetic Chlorophyll-Protein Complexes

Since all of the redox components *in vivo* are tightly associated with proteins, we thought it might be interesting to construct novel environments for chlorophyll-type chromophores using proteins [28]. The heme proteins, especially myoglobin (Mb) and hemoglobin (Hb), are good choices because of the obvious similarity between Chl and heme, the ease of removal and reconstitution of heme, and the large body of crystallographic data for these proteins. Our initial and continuing interest was to prepare crystals of these complexes for analysis by a variety of spectroscopic methods [29]. A second direction is to use these proteins as a molecular tweezer, to physically separate Chl-type chromophores in order to study long-distance energy and electron-transfer in defined arrays [30,31]. A third direction considers these complexes as very highly ordered media which position the chromophore near a protein/water interface for photoinduced electron transfer. Finally, the fact that a protein is the environment for photochemical studies suggests modification of the gene which codes for the protein by site-directed mutagenesis, leading to selective modifications of the environment around the chromophore. The latter two directions are motivated by the discussion at the end of the previous section.

An example of the use of proteins to organize Chls is our recent work on singlet excitation energy-transfer in Chl-substituted Hb [30,31]. Hb consists of four polypeptide chains, two α-chains and two β-chains. The α- and β-chains can be separated, the hemes removed, and each type of chain can be independently reconstituted with other chromophores. We exploited this to prepare the two complementary hybrid proteins in which either α- or β-chains are substituted with Chl-type chromophores while the other chains contain heme, or the tetramer in which all four chains contain Chl. Note that in the hybrid proteins not only the chromophores, but also their environments, are identical for the same chains. Detailed analytical and biochemical studies of these complexes show that they are tetrameric, and exhibit all of the well-known T-state tertiary and quaternary structural features of deoxyHb itself. This structure has been studied by x-ray crystallography and exhibits the following center-to-center separations:

$$\alpha_1\text{-}\alpha_2=34.8Å,\ \alpha_1\text{-}\beta_1=24.8Å,\ \alpha_1\text{-}\beta_2=36.4Å,\ \beta_1\text{-}\beta_2=40.2Å.$$

If a dilute sample of non-rotating hybrid protein is excited with linearly polarized light into the lowest excited state, the polarization of fluorescence will be that of a monomer if energy-transfer is absent, or it will be degraded relative to this value to an extent which depends on the rate of energy-transfer to the other Chl. This rate depends on the known distance between the Chls, the known relationship of the macrocyle planes, and the unknown relative orientations of their transition dipole moments, where the unknown is the orientation of the ^1S transition dipole moment, θ, as defined in Fig. 4. Steady-state measurements of fluorescence anisotropies in the multichromophoric systems demonstrate this energy-transfer:

$$\alpha_2{}^{Chl}\beta_2{}^{heme}=0.289\pm0.005;\ \alpha_2{}^{heme}\beta_2{}^{Chl}=0.287\pm0.003;\ \alpha_2{}^{Chl}\beta_2{}^{Chl}=0.178\pm0.004;$$

Chl-Mb(monomer) = 0.366±0.004 (Chl = Zn-pyrochlorophyllide a; heme = deoxy protoporphyrin IX; maximum theoretical anisotropy = 0.4).

A quantitative analysis of this data has been used to provide ranges for the possible orientations of the Chl singlet transition moment within the macrocycle plane [30]; however, the steady-state data does not suffice to

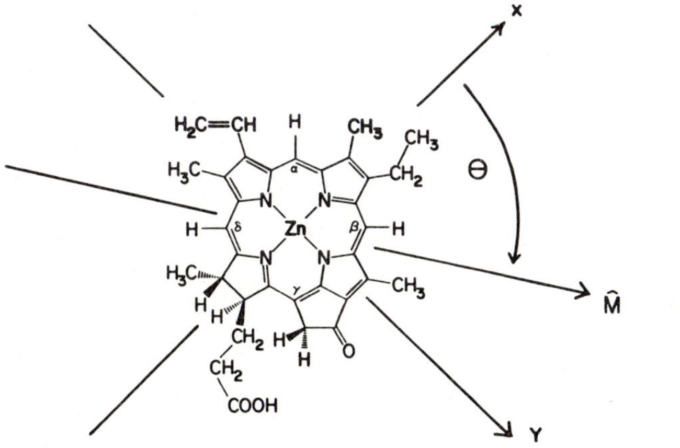

Figure 4 Structure of zinc pyrochlorophyllide a molecule. $\hat{M}$ denotes the unit vector of the transition dipole moment for the lowest excited state

decide among several choices. The orientation of the transition moment can be specified with considerable certainty by supplementing the steady-state data with ps time-resolved depolarization measurements on the same proteins (work done in collaboration with Professor Fayer) [31]. As expected, the form of the anisotropy decay is a single exponential decaying to a non-zero asymtote (the residual anisotropy once the excitation energy has equilibrated between the chromophores). Combined with considerations of the orientation-dependence of energy-transfer predicted by the Forster theory [32], we have obtained the orientation of the transition moment for the ^1S excited state. Defining the X-molecular axis as the line between the nitrogen atoms in rings B and D (Fig. 4), the ^1S transition moment in Zn-pyrochlorophyllide a was shown to be at $\theta = 95\pm2°$ from the X-axis, i.e. nearly along the molecular Y-axis. Given that the transition moment is predicted by theory to lie roughly along the Y-molecular axis (the Q_Y transition), the logic of these experiments can be reversed to provide support for the orientation-dependence predicted for dipole-dipole resonant transfer by Forster. Thus, calculations of energy-transfer rates for structurally defined arrays can be made reliably provided that the chromophores interact weakly. However, it is unlikely that measurements of fluorescence anisotropy decays can ever yield reliable or unique structures for multichromophoric systems, though many would like to believe otherwise.

The influence of the protein environment on the spectral and redox properties of photosynthetic pigments *in vivo* is not known. It is obvious that the primary redox components are held relatively rigidly in the protein matrix. More subtle roles, such as protonation or deprotonation of charged reaction intermediates, modulation of redox or spectral properties by charged amino acid side chains, or mediation of electron-transfer by orbitals on the protein, are tempting possibilities which await crystallographic data. It is conceivable that the RC protein catalyzes the loss of Mg to generate the bacteriopheophytin electron-acceptor. Another possibility for environmental control arises in *R. viridis* RCs, where the C_2 symmetrical relationship of the two potential electron-transfer pathways demonstrated in the crystal structure [33] is likely to be broken only at the level of individual amino acid side chains.

We have recently begun to explore some of these medium effects in a model system, using the structurally defined Chl-Mb complexes with systematic modifications of amino acid residues in the vicinity of the heme pocket. In order to accomplish this, we have recently cloned the cDNA which codes for human Mb [34] and inserted it into an expression system developed by Nagai and Thogersen [35]. In this system, the Mb cDNA is fused to the DNA which codes for the first 31 amino acids of the λ cII gene via a linker which codes for a peptide sequence which can be selectively cleaved in the translated fusion protein. Very high level expression of this fusion protein has been obtained; it has been reconstituted with heme, the extra protein has been cleaved, and the resulting human Mb has been purified to homogeneity. Very large (gram) quantities of synthetic Mb can be produced this way. Since the cDNA is cloned, it is straightforward to modify individual bases leading to site-specific mutations in the primary protein sequence. Mutants in which charged groups are selectively positioned in the vicinity of the chlorophyll, or in which the axial ligands have been varied are presently being constructed.

Acknowledgement: The author gratefully acknowledges the work of his students and collaborators on each of the projects described above. Magnetic field-effects: M. Roelofs, C.E.D. Chidsey, L. Takiff, R. Goldstein, D. Holten and C. Kirmaier; Chl-Hb: R. Moog and A. Kuki; Mb cloning: R.

Varadarajan, A. Szabo, and D. Lambright. This work is supported by NSF, NIH, and GRI grants PCM 8303776/8352149, GM27738 and 82-260-0089, respectively.

1 M.C. Thurnauer, J.J. Katz, J.R. Norris: Proc. Natl. Acad. Sci. USA, 72, 3270 (1975)
2 A.J. Hoff: Phys. Reports, 54, 75 (1979)
3 R.E. Blankenship, T.J. Schaafsma, W.W. Parson: Biochim. Biophys. Acta, 461, 297 (1975)
4 A.J. Hoff, J.H. Rademaker, R. van Grondelle, L.N.M. Duysens: Biochim. Biophys. Acta, 460, 547 (1977)
5 R. Haberkorn, M.E. Michel-Beyerle: Biophys J., 26, 589 (1979)
6 H.-J. Werner, K. Schulten, A. Weller: Biochim. Biophys. Acta, 502, 255 (1978)
7 S.G. Boxer, C.E.D. Chidsey, M.G. Roelofs: Ann. Rev. Phys. Chem., 34, 389 (1983)
8 C.E.D. Chidsey, M.G. Roelofs, S.G. Boxer: Chem. Phys. Lett., 74, 113 (1980)
9 S.G. Boxer, C.E.D. Chidsey, M.G. Roelofs: Proc. Natl. Acad. Sci. USA, 79, 4632 (1982)
10 A. Ogrodnik, H.W. Kruger, H. Orthuber, R.A. Haberkorn, M.E. Michel-Beyerle, H. Scheer: Biophys. J., 39, 91 (1982)
11 M.K. Bowman, D.E. Budil, G.L. Closs, A.G. Kostka, C.A. Wraight, J.R. Norris: Proc. Natl. Acad. Sci. USA, 78, 3305 (1981)
12 J.R. Norris, M.K. Bowman, D.E. Budil, J. Tang, C.A. Wraight, G.L. Closs: Proc. Natl. Acad. Sci. USA, 79, 5532 (1982)
13 C.E.D. Chidsey, C. Kirmaer, D. Holton, S.G. Boxer: Biochim. Biophys. Acta, 766, 424 (1984)
14 C.E.D. Chidsey, L. Takiff, R. Goldstein, S.G. Boxer: Proc. Natl. Acad. Sci. USA, in press (1985)
15 V.A. Shuvalov, W.W. Parson: Proc. Natl. Acad. Sci. USA, 78, 957 (1981)
16 R. vanGrondelle, W.G. Holmes, H. Rademaker, L.N.M. Duysens: Biochim. Biophys. Acta, 503, 10 (1978)
17 H. Rademaker, A.J. Hoff: Biophys. J., 34, 325 (1981)
18 N.W.T. Woodbury. W.W. Parson: Biochim. Biophys. Acta, 767, 345 (1984)
19 R. Goldstein, S.G. Boxer: submitted for publication
20 R.A. Marcus: Annu. Rev. Phys. Chem., 15, 155 (1964)
21 J. Jortner: J. Am. Chem. Soc., 102, 676 (1980)
22 R.K. Huddleston, J.R. Miller: J. Phys. Chem., 86, 200 (1982)
23 L.T. Calcaterra, G.L. Closs, J.R. Miller: J. Am. Chem. Soc., 105, 670 (1983)
24 D. Holten, C. Hoganson, M. Windsor, C.C. Schenck, W.W. Parson, A. Mignus, R.L. Fork, C.V. Shank: Biochim. Biophys. Acta, 592, 461 (1980)
25 V.A. Shuvalov, A.V. Klevanik, A.V. Sharkov, Y.A. Matueetz: FEBS Letts., 91, 135 (1978)
26 H. Frauenfelder, G.A. Petsko, D.Tsernoglou: Nature 280, 558 (1979)
27 A. Warshel, D.W. Schlosser: Proc. Natl. Acad. Sci., USA, 78, 5564 (1981)
28 S.G. Boxer, K.A. Wright: Biochemistry, 20, 7546 (1981)
29 S.G. Boxer, A. Kuki, K.A. Wright, B.A. Katz, N.H. Xuong: Proc. Natl. Acad. Sci., USA, 79, 1121 (1982)
30 A. Kuki, S.G. Boxer: Biochemistry, 34, 389 (1983)
31 R. Moog, A. Kuki, M.D. Fayer, S.G. Boxer: Biochemistry, 23, 1564 (1984)
32 Th. Forster: Ann. Phys. (Leipzig), 2, 55 (1948)
33 J. Deisenhofer, O. Epp, K. Miki, R. Huber, H. Michel: J. Mol. Biol., 180, 385 (1984)
34 R. Varadarajan, A. Szabo, S.G. Boxer: Proc. Natl. Acad. Sci., in press (1985)
35 K. Nagai, M.C. Thogersen: Nature, 309, 810 (1984)

Small Oligomers of Bacteriochlorophylls as *in vitro* Models for the Primary Electron Donors and Light-Harvesting Pigments in Purple Photosynthetic Bacteria

A. Scherz, V. Rosenbach, and S. Malkin*

Biochemistry Department, The Weizmann Institute of Science, Rehovot 76100, Israel

1. Introduction

There are four obvious forms of bacteriochlorophyll a (Bchla) in purple photosynthetic bacteria with Q_y transitions at 800, 850, 860 and 870 nm [1,2] and three spectral forms of Bchlb with Q_y transitions at 830, 960 and 1020 nm [3]. The Bchl molecules participate in two types of protein - chromophore complexes, serving as antennae pigments in the light-harvesting devices and primary electron donors in the photoreaction centers. All *in vivo* forms have their lowermost transition (Q_y) bathochromically shifted, intensified by a factor of 1.5 - 2.0 and optically active with respect to the Q_y transition of the *in vitro* monomers [5].

The generality of the spectral variations exhibited by the *in vivo* Bchls may be accidental, but may also reflect a common effect of the *in vivo* setting. This idea is further supported when chromatophores which lack reaction centers are exposed to ferricyanide; there, a band at 1200 nm appears, due to oxidized antennae pigments like the one formed when P-860 is oxidized [4]. Since the protein moiety for the light-harvesting pigments is different from that for the photoreaction centers, and since the Bchls occur in stochiometry of several molecules per functional unit, the common effect probably originates in similar types of chromophore-chromophore interactions. Here we expect exciton formation due to interactions between transition dipoles that originate in the separated molecules and also, if there is an overlap of p-orbitals, in charge-transfer states (PARSON? SCHERZ AND WARSHEL, this volume).

For studying the effect of chromophore-chromophore interactions, yet to avoid multiplicity of pigments with overlap transitions (e.g. the Q_x bands of B800 and B850 in B800-850) homogenous systems made of Bchl oligomers that resemble specific *in vivo* forms are expected to be of high value. We therefore developed a novel procedure whereby small oligomers of bacteriopheophytins (Bph) and Bchls can be trapped and studied [5,6]. In these preparations the Q_y transition is shifted to 850-880 nm, gains oscillator strength and becomes optically active. The minimum number of monomers required to produce these spectral variations was found to be 3 - 10 for Bpha in micelles of LDAO and 2 for Bchla and Bchlb in micelles of triton X-100.

2. Methods

For the preparation of large oligomers of Bpha, Bchla, Bphb and Bchlb, mixtures of water and formamide (FA) were used, e.g. in FA:H_2O - 3:1 most Bchla was converted into aggregates with a narrow (450 [cm^{-1}]) absorption band at 853-860 nm [5]. By adding small amounts of triton X-100, which forms micelles in this solvent mixture, smaller oligomers were trapped [6].

*Bat-Sheva de Rothschild fellow.

Preparation of large oligomers from Bchla (Bchl-860) and Bchlb (Bchl-882) was accomplished upon sonication of variable pigments concentration in FA:H_2O = 3:1 (v:v) for 4 minutes. Small oligomers of Bchla (Bchl-853), Bchlb (Bchl-878) and Bpha (Bph-847), were isolated by adding 5 [μl] and 100[μl] of 100% triton X-100 (Fluka) to 12 - 80 μl pyridine solution of Bchls or Bpha respectively in FA:H_2O - 3:1 (total volume = 4[ml]).

For measuring the free energy changes involved in the formation of Bchl-853, Bchl-860 and Bph-847 from the corresponding monomers, their solutions were heated to the desired temperature and the resulted spectra were recorded. The concentration of monomers versus the oligomers was calculated from the corresponding spectra using the extinction coefficients for Bchl-853 [5,6], Bchla, Bph-850 and Bpha [5]. After each measurement, the sample was cooled and resonicated at 0°[C] for 4 minutes. Full recovery of the recorded spectra was observed up to 65°[C] at which point partial bleaching occurred.

Calculation of spectral parameters (the oscillator strength-f, the rotational strength-R and the dipole strength-D) was done as previously described [7]. Simulation of the spectral parameters was carried on an Apple II computer using an adaptation of the previously reported algorithm [7].

3. Results

a. Oligomerization of Bchla in FA:H_2O

The spectrum of Bchla in FA (that contains 1.5% pyridine) is shown in Fig. 1A (dashed line). Dropwise addition of water into FA that contained more than 0.1 [μM] Bchla resulted in the partial replacement of the 780 nm transition by a new one at 853 nm.

From above 0.3 [μM] of Bchla, the 853 nm band shifted to 860 nm (Fig. 1A - solid line) and its intensity increased linearly with the overall concentration of the pigment (Fig. 2A). Centrifugation at 48,950 [g] discarded 96% of

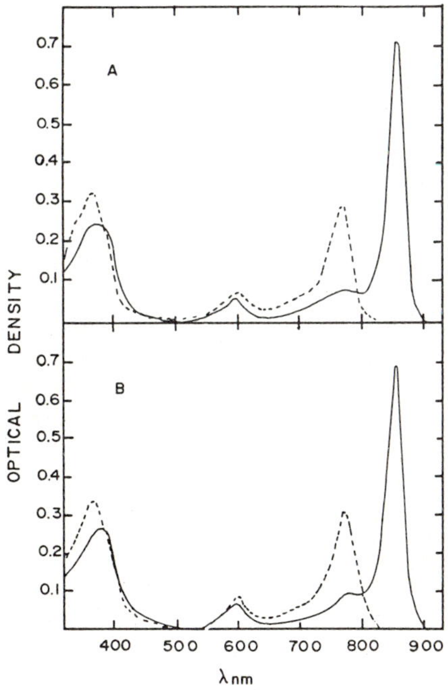

Fig. 1. (A) Spectrum of 5.14 [μM] [Bchla] in FA that contains 1.5% Pyridine (dashed line) and in FA:H_2O - 3:1 (v:v) that contains 1.5% Pyridine (solid line). (B) Spectrum of 5.14 [μM] [Bchla] in FA:H_2O - 3:1 (v:v) that contains 1.5% Pyridine and 1.5% Triton X-100 (dashed line): Spectrum of 5.14 [μM] (Bchla) in FA:H_2O - 3:1 (v:v) that contains 0.125% Triton (solid line).

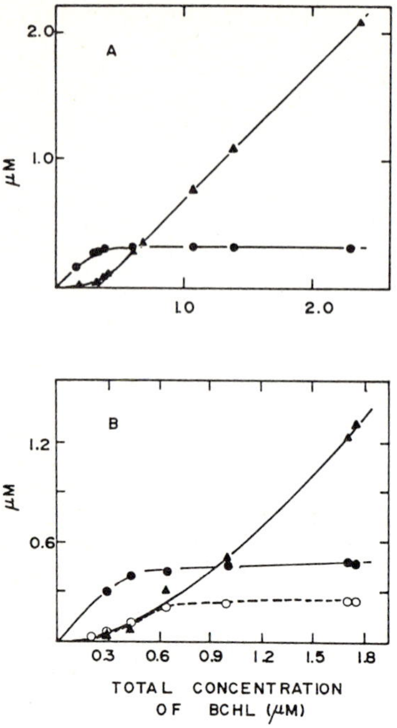

Fig. 2. (A) Aggregation of Bchla in tritonless solutions of FA:H_2O = 3:1. (-●-) The change in the concentration of Bchla monomers absorbing at 780 nm as the total concentration of Bchla is increased. (-▲-) The change in the concentration of monomeric units within oligomers that absorb at 860 nm. (B) Aggregation of Bchla in the presence of 0.125% Triton in FA:H_2O - 3:1 (v:v) that contain 1.5% Pyridine. (--o--) The change in the concentration of Bchla monomers absorbing at 780 nm as the total [Bchla] is increased. (- ▲ -) The change in the concentration of monomeric units in the form of oligomers absorbing at 853 nm and 860 nm as the total concentration of Bchla is raised. (o-o-o-) The calculated contribution of Bchla dimers at various [Bchla] concentration for K_a = 5 x 10^5[M^{-1}] (see text).

the 860 nm species in the form of pale green precipitant,and the supernatant was greatly enriched by the 780 nm component. We therefore concluded that the 860 nm band reflects the Q_y transition of a large oligomer of Bchla (Bchl-860), presumably in equilibrium with monomers that absorb at 780 nm (Bchl-780).

In the presence of 0.125% triton X-100, the formation of Bchl-860 was inhibited and one was left with Q_y transition at 853 nm assigned to Bchl-853.The conversion of Bchl-780 to Bchl-853 under such conditions was not linear (Fig. 2B

Upon 30 minutes of centrifugation at 48,950 [g], no more than 50% of the aggregate was discarded. Concomitantly,there was a drop in the 780 nm transition,and the residual absorption was stable for hours. We therefore deduced that Bchl-853 is an aggregate of Bchla with lower molecular weight than Bchl-860. This conclusion is further confirmed by the fact that up to 0.3 [μM] Bchla, the concentration of the oligomer in the presence of triton could be well fitted by dimers in equilibrium with monomers having equilibrium-constant K_a = 5 x 10^5 [M^{-1}]. When the concentration of Bchla was further increased,the experimental curve that describes the fraction of Bchla in the form of Bchl-853, could be fitted by adding trimers, tetramers, etc. with the same equilibrium-constant (5 x 10^5 [M^{-1}]) for their formation. At concentrations of Bchla that exceed 5 [μM], the 853 nm transition shifted to 860 nm.

The Q_y transition of Bchla in the aggregated form is optically active with net positive rotation (Fig. 3) and has higher intensity (ε_{853} = 1.4 x 10^5 [M^{-1}][cm^{-1}]) and oscillator strength (f = 0.38) per each monomeric unit than the respective isolated monomers. It is also narrowed by 50% (460 [cm^{-1}] versus 950 [cm^{-1}] respectively). The absorption increase and non-conservative optical activity (hyperchromism) of the oligomer in the Q_y region are compensated by absorption decrease and net negative rotation (hypochromism) in

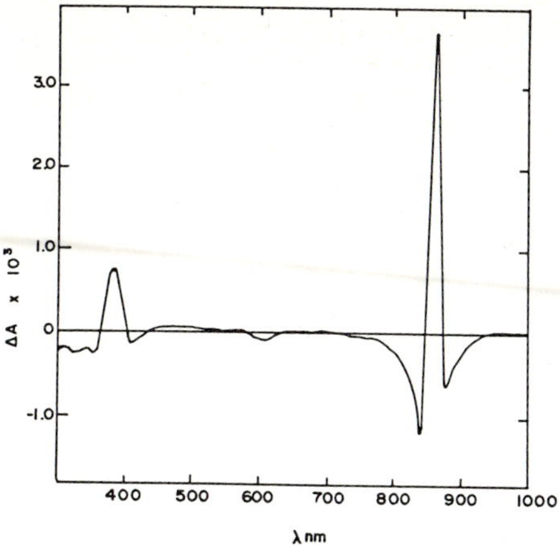

Fig. 3. C.D. Spectrum of 2 μM Bchl-853 in FA:H_2O = 3.1 in the presence of 0.125% Triton X-100.

the Q_x and B_{xy} transitions, the second of which is also red-shifted. The overall oscillator and rotational strengths seem to be conserved, as observed previously for Bph-850 [5]. The Q_x transition (596 nm) slightly shifts to the blue.

Absorption increases in 780 nm at the expense of the 853 nm or the 860 nm bands, upon heating mixtures of Bchl-853 and Bchl-780, was observed. We have shown [6] that under such conditions

$$\ell n(K_a) = \ell n([\text{Bchl-853}]+[\text{Bchl-860}])-2\ell n([\text{Bchl-780}])-\ell n2 \qquad (1)$$

A plot of $\ell n(K_a)$ versus 1/T is given in Fig. 4. The calculated ΔH and ΔS are -47 [Kcal]/mole and 0.12 [Kcal]/([mole][deg]) respectively.

The concentration of Bchl-853 or Bchl-860 critically depends on the pyridine content and the FA:H_2O ratio. An increase of the pyridine content to above 5% in FA:H_2O - 3:1 induces dissociation of the aggregates of Bchla as re-

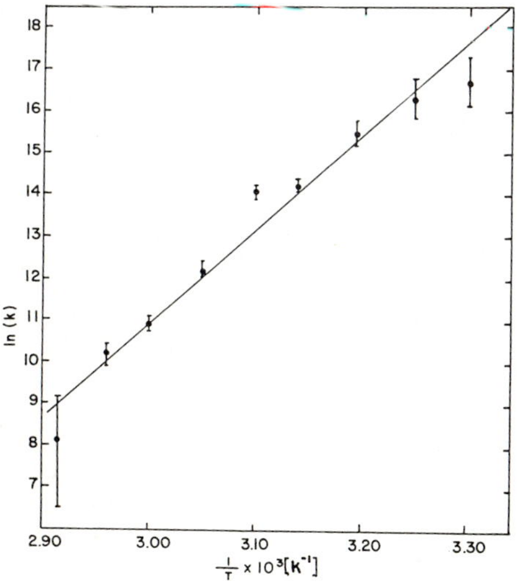

Fig. 4. ℓn K_a versus (1/T) $[K^{-1}]$. The concentration of Bchla in the form of Bchla 853 was calculated from the absorption at 853 nm (ε_{853} = 1.4 x 10^5 $[M^{-1}][cm^{-1}]$. The concentration of Bchla in the form of Bchl-780 was calculated from the net absorption at 780 nm after substructing the contribution of Bchl-853.

flected by an absorption decrease at 853-860 nm and a concomitant absorption increase at 780 nm. The resulted monomers are probably biligated with pyridine since their Q_x transition shifts to 610 nm.

b. Oligomerization of Bph

In pure FA, in pure H_2O or in mixtures of both (regardless of the exact ratio), Bpha forms oligomers with threshold concentration of 0.01 [μM]. When triton is added, the threshold concentration is greatly increased (e.g. 0.4 [μm] at 0.125% triton X-100). Neither high percentage of water, nor the presence of up to 30% pyridine, can prevent the oligomerization of Bpha, which has ΔH and Δs equal to -50 [Kcal]/[mole] and 0.13 [Kcal]/([mole][deg]) respectively.

c. Oligomerization of Bchlb in FA:H_2O

Bchlb forms large oligomers with Q_y transition at 882 nm in 3:1 FA:H_2O and small oligomers (probably dimers or trimers) with Q_y transition at 878 nm in the presence of triton X-100. The Q_y transition is hyperchromic with respect to that of the isolated monomer. This is well compensated by the large hypochromism of the Q_x and B_{xy} transitions (Fig. 5). There is no indication for a charge-transfer band at 630-880 nm with an appreciable oscillator strength.

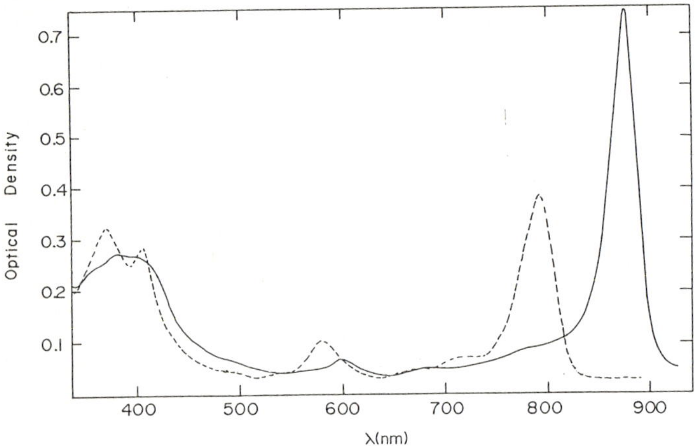

Fig. 5. Optical absorption of Bchlb monomers in ether (dashed line) and 3:1 FA:H_2O that contains 1.5% pyridine with 0.125% triton X-100 (solid line). $\varepsilon_{795} = 1.05 \times 10^5$ mol^{-1} cm^{-1} (for the monomers).

In general, the oligomerization of Bchls or Bphs in H_2O: Amide solutions is prevented if the FA is substituted by n-methyl amide, Acetamide and dimethyl formamide [6].

4. Discussion

The aggregation of Bpha, Bchla and probably that of Bchlb in aqueous formamide, is a biphasic process. It is non-linear in low concentrations of the Bchls and linear at higher concentrations, when the Q_y transition shifts to a longer wavelength (Fig. 2). The phenomena probably reflects cooperative oligomerization of the Bphs or the Bchls [8 a-b].

A scheme for cooperative aggregation is given in ref. 8ab. This scheme applies for, e.g. the aggregation of Bchla in tritonless solutions (Fig. 2A). At Bchla concentrations lower than 0.3 [μM] the concentration of Bchl-780 grows, with the total concentration of Bchla in a linear fashion having no contribution at longer wavelengths. At 0.3 [μM] it approaches saturation, and the concentration of monomeric units within aggregates (Bchl-853 and Bchl-860) becomes linear with [Bchla].

Consequently, the overall concentration of Bchla in all forms is given by [6, 8 ab].

$$n = m + \sum_{i=2}^{j} iK_a^{i-1}m^i + K_a^{-1}\sigma\{(K_b m)^j[\frac{j}{(1-K_b m)} + \frac{K_b m}{(1-K_b m)^2}] - \frac{m[1-(mK_b)^{j-1}]}{(1-K_b m)}\} \quad (2)$$

where: $\sigma = (\frac{K_a}{K_b})^{j-1}$. S and $S = \frac{[\text{ j-mer }]'}{[\text{ j-mer }]}$.

The third term on the r.h.s. (right-hand side) equals the concentration of Bchla in the form of Bchl-860, the second term has to be identified with Bchl-853, which is found to be smaller than Bchl-860 and the first term stands for the concentration of monomers. Since [Bchl-780] approaches a saturation value of 0.3 [μM] in Fig. 2A, it follows that $K_b = 3.3 \times 10^6$ $[M^{-1}]$.

The addition of triton micelles reduces the effective concentration of the [j-mer]', with respect to [Bchla]. The first stage of oligomerization is therefore accomplished at higher total concentrations of Bchla, and may be easily followed for the derivation of K_a. This is demonstrated in Fig.2B where the concentration of Bchl-853 can be well fitted by K_a[Bchl-780] (having $K_a = 5 \times 10^5$ $[M^{-1}]$, up to total concentration of Bchla equal to 0.5 [μM]. We therefore deduced that $(Bchla)_2$ is the smallest unit of Bchl-853. Since the spectral properties of Bchl-853 and Bchl-860 are almost identical, we further assumed that Bchl-860 is actually made of many pairs of Bchla that are spectrally isolated one from the other.

By following the procedure of SCHERZ and PARSON [6,7] we calculated a C_2 configuration for the two molecules of Bpha Bchla and Bchlb that make the basic units of Bph-847, Bph-850, Bchl-853, Bchl-860 Bchl 878 and Bchl-882 with an overlap of either pyrrole rings I or III (Fig. 6 and Table 1). In having overlap of pyrole rings III one can think of either mutual hydrogen bonding between ring V carbonyls of the solvated molecules (structure a) or between the C-9 keto of one molecule and the central Mg of another one via water bridge as previously suggested for chla and Bchla structure [9] . However, for both structures ΔH should be no more than 10 - 15 [Kcal]/[mole] as it involves two hydrogen bonds at most. More than that, structure a may be applied to Bph-847 but it is not obvious how, in the case of Bchla, it is disturbed by ligation of pyridine to the central magnesium and a particular solvent structure. On the other hand, structure b, in which the Mg is essential, accounts for the ligand effect but is incompatible with the formation of Bph-847 if it has the same geometry as Bchl-853. We were therefore left with structure c, in which pyrrole rings I overlap and the C-2 acetyl of one molecule lies above the macrocycle plane of the other. This structure is consistent with disturbing the formation of Bchl-853 by adding pyridine, since it prevents proper alignment of the molecules and still doesn't require the participation of the central metal in forming the dimers. However, the enthalpy change for that structure, is expected to be mainly contributed by π-π interactions and possibly, for Bchl-853 and Bchl-878, a ligation of the

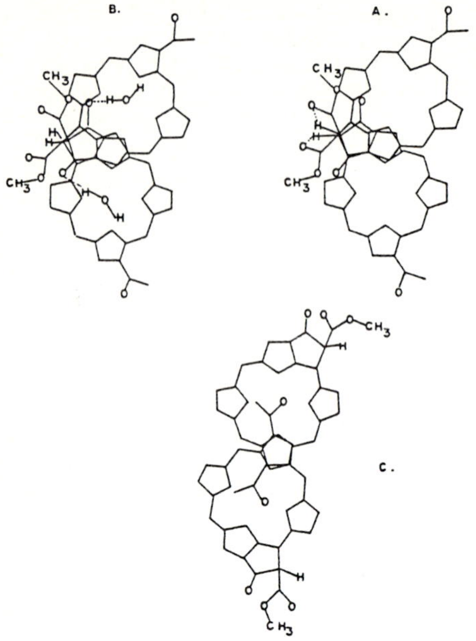

Fig. 6. Structures of Bchla dimer for which the parameters of Table 1 were derived. The centers of the two molecules are separated by 3.5 Å in the Z direction, - 5.5 Å in the y direction and - 1.7 Å in the X direction. The y polarized transitions are mutually rotated by 160° and the x polarized transitions are mutually rotated by 14°. The vector of one molecule is 75° of the Z vector of the other molecule. In structure A the β-oxygen of C-10 is bonded to the α- proton of C-10 on adjacent molecule. In structure B the C-9 keto is hydrated and the water molecule serves as ligand to the central Mg of the other molecule. In structure C the pyrrole rings I overlap and the C-2 acetyls may interact with the central Mg in the case of Bchla dimers.

C-2 acetyle of one molecule to the central Mg of the other. The contribution of π-π interaction cannot exceed a few Kcal/[mole][10]. The ligation of the C-2 acetyl may add a few more Kcal. However, since ΔH for the formation of Bchl-853 and Bph-847 are similar 45-50 [Kcal]/[mole]), the vast part of it cannot originate in the suggested ligation.

One possible way to account for the observed high enthalpy change, the solvent effect in the case of Bchl-853 formation, the cooperativity and the spectral properties of Bph-847, Bph-850, Bchl-853, Bchl-860 Bchl-878 and Bchl-882 is by allowing active participation of the solvent molecules (FA or H_2O) in the process of oligomerization. We may assume that each molecule of Bph or Bchl is solvated, so that the carbonyl groups of ring V and possibly, the C-2 carbonyls are hydrogen-bonded with several molecules of FA (Fig. 7a). As the concentration of Bpha is increased, chains of FA [11] or three-dimensional structures of H_2O with more than one chromophore are formed (Fig. 7b). Dimerization of the pigments driven by hydrophobic interaction of pyrole rings I brings FA polymers or water structures together, assisting hydrogen-bonding of solvent molecules (Fig. 7c). Thus, the free energy change for the association is mainly contributed by self-organization of the solvent molecules, and the enthalpy for their being hydrogen bonded -(30-40[Kcal]/[mole]) and some of it by the π-π interactions of pyrrole rings I. The process of oligomerization is apparently a synergetic one, and therefore should be cooperative in both the concentrations of pigments and solvents.

Bchla-853 and Bchl-860 are quite attractive as models for the primary electron donors in the reaction centers of the purple photosynthetic bacteria that contain Bchla and their light-harvesting complexes. It is possible that a mechanism similar to the cooperative oligomerization of Bchls in FA:H_2O may apply to the self-organization of chromophor-protein matter in the light-harvesting and reaction centers complexes. There, the Bchls are attached to the protein moiety in their hydrophylic array (see OLSON et al. and ZUBER et

Table 1: Spectral parameters derived for Bchls monomers, Bchl-853, Bchl-878 hypothetical dimers of structure c and B850. The parameters for B850 were derived from ref.[17,18]. [a]The formula used to derive the dipole strength and rotational strength from experimental values are given in ref. [5,7]. The parameters are given for one unit of Bchl in each form.

Compound	$\lambda^{Q_y}_{max}$ (nm)	Dipole strength (Debye^2)	Rotational strehgth (Debye-BM)	$\lambda^{Q_x}_{max}$ (nm)	[a]Dipole strength (Debye^2)	Rotational strength Debye-BM)	$\lambda^{B_{xy}}_{max}$ (nm)	Dipole strength (Debye^2)	Rotational strength (Debye-BM)
Bchla in FA:H_2O-3:1 contains 1.25% Triton	776	39.5	+0.05	600		-0.05	357	125	+0.2
Bchla in FA:H_2O-3:1 contains 0.125% Triton	853	60.0	+0.55	596	8.0	-0.15	377	107	-0.5
Calculated spectrum for structure c (of Bchla)							331	3.5	-3.5
	856	57.0	+1.0	596	7.0	+0.41	374	44.0	+2.4
	732	0.725	-0.4	604	0.0	-0.55	384	65.0	+3.8
							393	1.0	-3.3
Bchla in B850[a]	850	60.0	+0.65	585	7.0	0	376	100.0	
	865	0	-0.13	612	0	-0.2			
Bchlb in ether	795	50.0		580	7.7		370	74.0	
							408	41.5	
Bchlb in FA:H_2O - 3:1 contains 0.125%. Triton X-100	878	∿70		600	5.0		415	∿112	
Calculated spectrum for structure C (of Bchlb)							344	4.3	-1.9
	890	69.0	+1.17	584	5.5	-0.21	391	62	0.08
	733	2.2	-0.24	576	0.63	0.10	399	1.6	-1.0
							419	44.0	2.8

al. this volume) and may partly overlap in their hydrophobic region. The calculated configuration of the coupled chromophores in all *in vitro* preparations, including Bchl-878 (Bchlb dimer) is strikingly similar to the structure of P-965 (the primary electron donor in R. viridis) as recently analyzed by X-rays defraction [12] and the structure suggested for P-860 (the primary electron donor in purple bacteria that contain Bchla) by ENDOR analysis [13]. These structures rule out the necessity of Mg bridging,but point to the importance of a structural medium which allows the molecules a proper orientation.

It should be noted however, that the Q_y shift of the Bchl-878 with respect to Bchlb monomers *in vitro* is lower by 500 [cm^{-1}] than the inspected shift for P-965 (where one may assume a monomeric Q_y at 830 nm (see PARSON, SCHERZ and WARSHEL, this volume). Yet the calculated geometry is strikingly similar. This may indicate (a) specific charge-transfer effects *in vivo* (PARSON, SCHERZ and WARSHEL. This volume). (b) Proximity of charged groups in the protein moiety which should differently effect Bchlb and Bchla if the C-4 vinyl is close to the protein walls,as pointed out by ACCLLES and HONIG [14]. The possible effect of charged amino residues on the optical absorption of P-965 gains some support from the fact that C-4 vinyl in the two Bchls of the R. virid primary donor are close to the protein moiety [12]. It is clear,however,that their affect is minor to the exciton coupling.

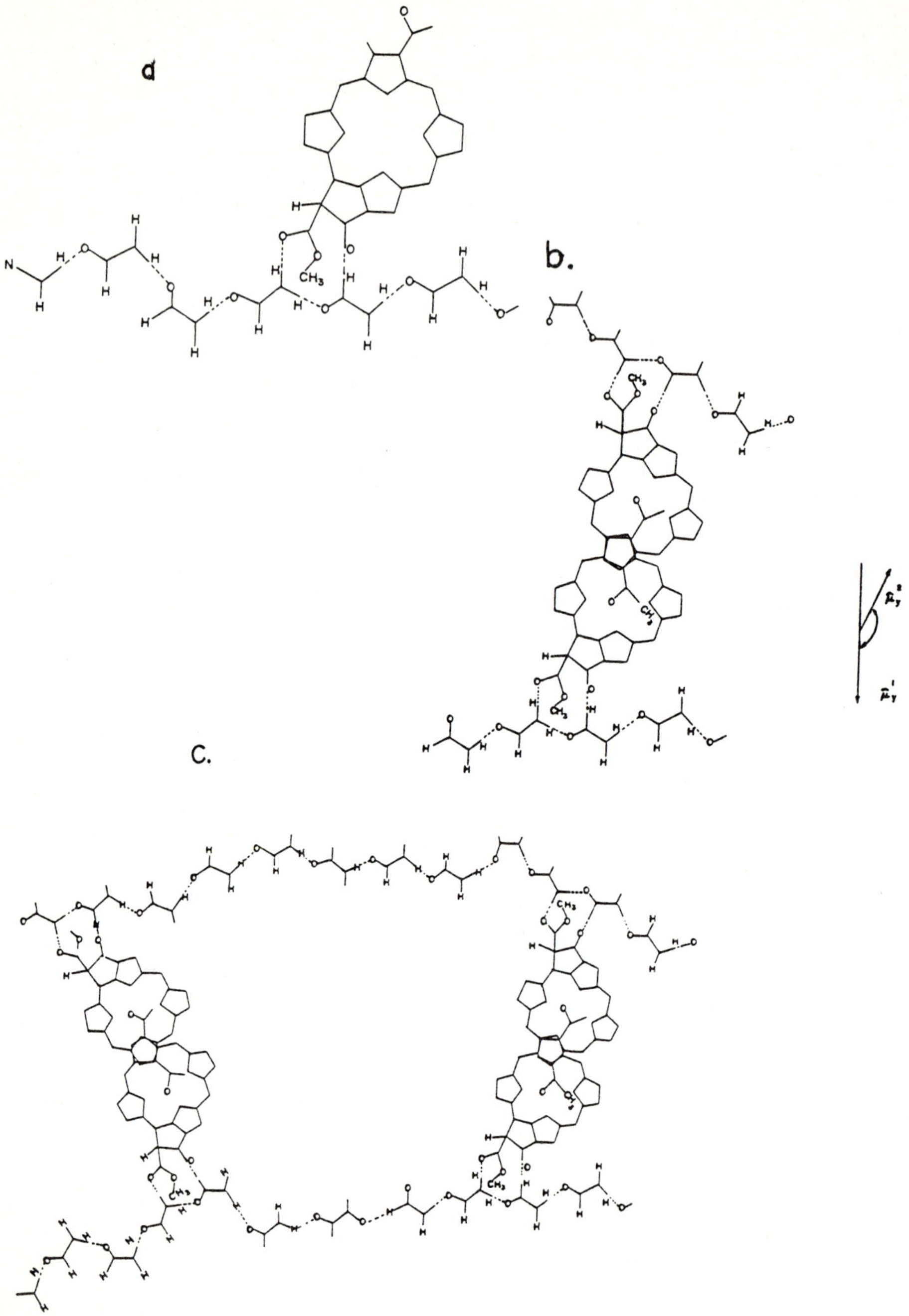

Fig. 7. Suggested oligomerization of Bpha in FA. (a) Solvation of Bpha, (b) Formation of FA-Bpha dimer, (c) Cooperative aggregation via bridging of two FA polymers.

The identification of P-860 as a dimer of Bchls, although supported by various techniques, remained controversial: very recently, monomers of Bchla or Chla attached to protein as Schiff bases or near to opposite charges were suggested to be primary donors of the photosynthetic bacteria or higher organisms respectively [14,15]. Monomers with Q_y transitions at 850 nm were also suggested for the light-harvesting complexes [16]. The assignment of the long wavelength transition *in vitro* with Bchla dimer is therefore of great importance. The similarity in spectral properties of the *in vitro* preparations and the *in vivo* primary donors and light-harvesting pigments shows that they may be used for the study of energy and electron-transfer by artificial compounds analogous to the *in vivo* systems.

5. References

1. J.P. Thornber, J. Barber in "Topics in Photosynthesis in relation to model systems" ed. J. Barber, Elsevier (1979).
2. M.Y. Okamura, G. Feher, N. Nelson in "Photosynthesis: Energy-Conversion by Plants and Bacteria" ed. Govindjee, Academic Press, N.Y. (1982).

3a. D. Holten, M.W. Windsor, W.W. Parson, J.P. Thornber, Biochem. Biophys. Acta 501, 112 (1978).

3b. A. Jensen, O. Aasmundred, K.E. Eimhjellen, Biochem. Biophys. Acta 88, 466 (1964).

4. I. Gomez, R. Picorel, J.M. Remirez, R. Perez, F.F. del Campo, Photochem. Photobiol. 35, 399 (1982).
5. A. Scherz, W.W. Parson, Biochim. Biophys. Acta 766, 653 (1984).
6. A. Scherz, V. Rosenbach, S. Malkin, submitted for publication (1985).
7. A. Scherz, W.W. Parson, Biochim. Biophys. Acta 766, 665 (1984).

8a. C.R. Cantor, P.R. Schimmel, "Biophysical Chemistry" Freeman, San Francisco, Vol. I. p. 145 (1980).

8b. F. Oosawa, M. Kasai, J. Mol. Biol. 4, 10 (1962).

9. J.J. Katz, L.L. Shipman, T.M. Cotton, T.R. Janson in "The Porphyrins", Vol. V. ed. D. Dolphin, p. 401, McGraw Hill, New York, (1977).
10. R.J. Abram, P.A. Barbridge, A.H. Jackson, D.B. Macdonald, J. Chem. Soc. (B), 620 (1966).
11. J.F. Hinton, R.D. Harpool, J. Am. Chem. Soc. 99, 349 (1977).
12. J. Deisenhofer, O. Epp, K. Miki, R. Huber, H. Michel, J. Mol. Biol. 180, 385 (1985).
13. W. Lubitz, F. Lendzian, H. Scheer, J. Gottstein, M. Plato, K. Mobius, Proc. Nat. Acad. Sci. U.S.A. 81, 1401 (1984).
14. J. Ecceles, B. Honig, Proc. Nat. Acad. Sci. U.S.A. 80, 4959 (1983).
15. L.L. Magiora, M.M. Magiora, Photochem. Photobiol. 39, 847 (1984).
16. C.N. Rafferty, J. Bolt, K. Sauer, R.K. Clayton, Proc. Natl. Acad. Sci. U.S.A. 76, 4429 (1979).
17. K. Sauer, L.A. Austin, Biochem. 17, 2012 (1978).
18. R.J. Clayton, B.J. Clayton, Proc. Nat. Acad. Sci. U.S.A. 78, 5583 (1981).

Experimental, Structural and Theoretical Models of Bacteriochlorophylls a, d and g

J. Fajer[1], K.M. Barkigia[1], E. Fujita[1], D.A. Goff[2], L.K. Hanson[1], J.D. Head[3], T. Horning[1], K.M. Smith[2], and M.C. Zerner[3]

[1]Department of Applied Science, Brookhaven National Laboratory, Upton, NY 11973, USA
[2]Department of Chemistry, Univ. of California, Davis, CA 95616, USA
[3]Quantum Theory Project, Univ. of Florida, Gainesville, FL 32611, USA

The model studies presented here seek to probe possible effects that the protein environment may induce or impose on the conformation and electronic configuration of photosynthetic chromophores. Results are presented for the effect (or lack of it) of axial ligation and hydrogen bonding on the electron spin resonance (ESR) properties of bacteriopheophytin (BPheo) and bacteriochlorophyll (BChl) a cation radicals. X-ray diffraction results on single crystals illustrate (1) the significant conformational changes that different crystal habits enforce on the same skeleton of bacteriopheophytins d and (2) the structural consequences of oxidation on a magnesium porphyrin. Finally, theoretical calculations are described that predict spectral and redox properties of bacteriochlorophyll g, the recently discovered chromophore of the anoxygenic bacterium, Heliobacterium chlorum.

1. Ligand and Hydrogen Bonding Effects in Bacteriochlorophyll and Bacteriopheophytin a Cation Radicals (with Etsuko Fujita and Tex Horning)

Assignments of the radicals observed on oxidation of bacterial reaction centers to dimeric bacteriochlorophylls a (P870$^+$) or b (P960$^+$) are based on comparisons of ESR and ENDOR characteristics in vivo with those observed for BChl$^+$ a and b in vitro [1-5]. Recent X-ray results of the R. viridis reaction center (BChl b) [6] support the dimer formulation for its primary donor, P960, and provide evidence of possible interactions between the magnesium and the oxygen of a nearby 2-acetyl group, as well as of ligation by neighboring protein residues. Chlorophylls further interact with their environment by hydrogen bonding of their peripheral carbonyl groups, as evidenced by resonance Raman data [7]. To separate ligation and/or hydrogen bonding effects from the "intrinsic" unpaired spin distributions of BChl radicals, the ENDOR parameters of the cation radical of BPheo a have been measured as a function of solvent and temperature.

1) In CH_3OH/CH_2Cl_2, in which the Mg is complexed and the oxygen functions of the macrocycles are hydrogen-bonded by the methanol [8], ENDOR results are comparable for BChl a^+ and BPheo a^+ (Table 1). The differences that do exist in the β proton coupling constants are also observed in synthetic bacteriochlorins, and are predicted theoretically [9].

Table 1. Major ENDOR coupling constants (Gauss)

T°K	1-CH_3	5-CH_3	Hβ	Hβ	Hβ	Hβ
	BChl a^+ in 1:6 CH_3OH/CH_2Cl_2					
129	1.77	3.41		Not resolved		
153	1.78	3.38		NR	4.74	6.02
163	1.79	3.40	4.23	NR	4.79	5.78
173	1.76	3.45	4.26	NR	4.74	5.79
183	1.76	3.44	4.24	NR	4.71	5.80
213	1.75	3.43	4.19	4.66	4.79	5.83
280	1.70	3.37	4.11	4.65	4.89	5.86
	BPheo a^+ in 1:6 CH_3OH/CH_2Cl_2					
129	1.81	3.20		Not resolved		
153	1.88	3.18	Not Resolved		5.86	6.58
163	1.86	3.22	4.57	4.94	5.76	6.55(8.29)
183	1.80	3.20	4.52	4.95	5.76	6.73
213	1.79	3.17	4.59	5.08	5.75	6.72
253	1.75	3.17	4.60	5.09	5.71	6.69
	BPheo a^+ in CH_2Cl_2					
163	1.79	3.20	4.51	4.91	5.74	6.70
173	1.79	3.19	4.52	4.94	5.75	6.69
183	1.78	3.20	4.52	4.97	5.76	6.73
213	1.76	3.16	4.60	5.09	5.80	6.72
253	1.74	3.10	4.61	5.11	5.71	6.69
Comparison of BPheo a^+ at 183°K (well-resolved spectra)						
CH_3OH/CH_2Cl_2	1.80	3.20	4.52	4.95	5.76	6.73
CH_2Cl_2	1.78	3.20	4.52	4.97	5.76	6.73

2) In "non-bonding" CH_2Cl_2 solutions, the coupling constants of BPheo a^+ agree within 2% with those observed in the alcoholic solvent, indicating that hydrogen bonding has little effect on the spin distributions of the radicals (Table 1). (Comparison of BChl a^+ in CH_2Cl_2 and CH_3OH is complicated by the formation of aggregates in CH_2Cl_2.)

The combination of 1) and 2) suggests therefore that oxygen ligation of the Mg in BChls and/or hydrogen bonding of the oxygen peripheral groups do not significantly alter the spin profiles of BChl a^+ in vitro and, by extrapolation, the profiles of P870+ and P960+ in vivo.

3) Computer signal averaging yields [3] a partially resolved ESR spectrum of BPheo a^+ with 40 or more lines. This spectrum provides a sensitive test for the assignments of the ENDOR coupling constants. A satisfactory facsimile of the experimental BPheo a^+ spectrum is obtained with a computer simulation that assumes two methyl groups with a_H=1.74 G (1-CH_3) and 3.10 G (5-CH_3), four protons with a_H=4.6, 5.1, 5.7 and 6.7 G (β protons on rings II and IV) and four nitrogens with a_N=1.1 G (approximated from the sole N splitting that is resolved). These results thus provide additional support for the ENDOR assignments and for the similar spin profiles predicted by M.O. calculations for the BPheo and BChl cations [3,4,9].

4) Changes in the coupling constants of BPheo a^+ can be induced near the freezing point of CH_3OH/CH_2Cl_2 mixtures (Table 1). However, only the β protons of rings II and IV are significantly affected. The effect observed is attributable to conformational changes (twisting) of the flexible saturated rings [10] induced by the glassy matrix. Similar effects could be induced by packing and protein interactions in vivo, and may explain some of the ESR variations observed in different photosynthetic bacteria.

2. Structural Studies: Hydrogen Bonding and Conformational Changes Induced by Crystal Packing and Oxidation (with Kathleen M. Barkigia, Dane A. Goff and Kevin M. Smith)

X-ray diffraction studies of single crystals of *Chlorobium* pheophytins provide striking examples of the variety of hydrogen bonding that is observed in vitro and, by extra-

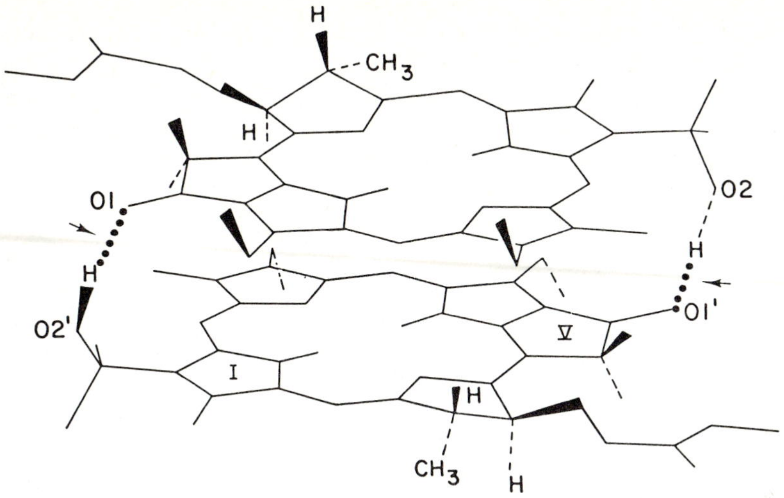

Figure 1. Dimer of methyl 4,5-diethylbacteriopheophorbide d. The arrows point to the hydrogen bonds between the protons of the hydroxy groups and the oxygen of the 9-keto groups. A similar dimeric arrangement is found with methyl 4-isobutyl,5-ethylbacteriopheophorbide d when it is crystallized from the same solvent mixture of dichloromethane/hexane.

polation, may be expected to occur in vivo. Chlorobium chlorophylls are pyrochlorophyll derivatives found in photosynthetic green bacteria. They are distinguished from other chlorophylls by the presence of a 2-(1-hydroxyethyl) group on ring I (as opposed to the 2-vinyl group of Chl a) and by a series of homologous substituents at positions 4 and 5 on rings II and III respectively.

Figures 1, 2 and 3 illustrate three types of hydrogen-bonded aggregates observed in single crystals. 1) A cofacial dimer formed by hydrogen bonding between the hydroxy groups and the 9-keto groups. These structures are found when either methyl 4,5-diethyl or 4-isobutyl,5-ethyl bacteriopheophorbides d are crystallized from dichloromethane/hexane. (Triclinic space group P1, Z=2) [11] (Fig. 1).

2) An arrangement in which the hydroxy group also hydrogen bonds to the 9-keto group of a neighbor but now in an extended linear array. The compound, methyl 4-neopentyl,

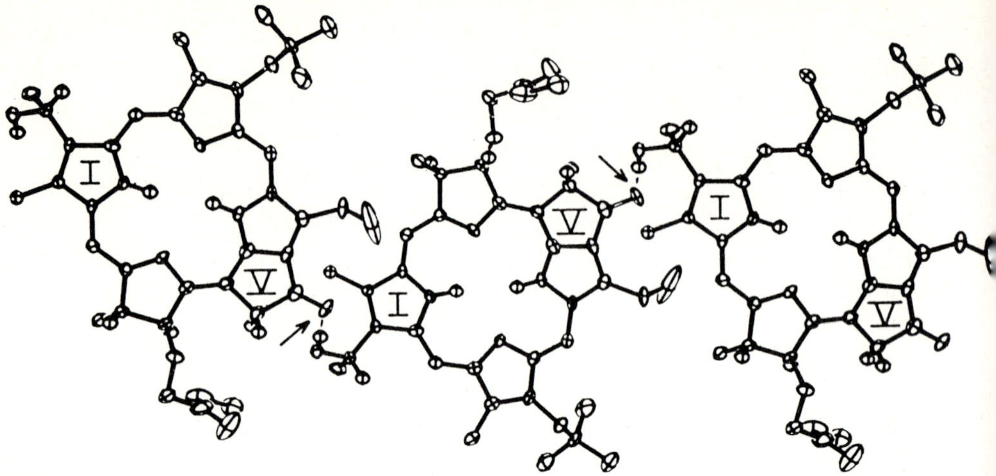

Figure 2. "Linear" aggregation of methyl 4-neopentyl, 5-ethylbacteriopheophorbide d. The arrows indicate the hydrogen bonds between the hydroxy groups and the oxygens of the 9-keto groups.

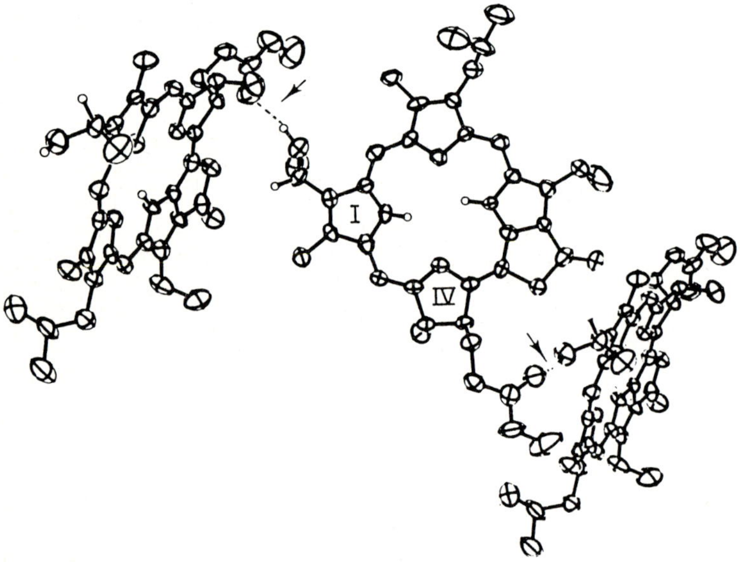

Figure 3. Aggregation of methyl 4-isobutyl, 5-ethylbacteriopheophorbide d, when crystallized from acetone/ethanol. The arrows show the hydrogen bonds between the protons of hydroxy groups and the carbonyl oxygens of the methyl propionate.

5-ethylbacteriopheophorbide d, crystallizes from methanol with Z=4, in the orthorhombic space group $P2_12_12_1$ [10] (Fig. 2).

3) An "orthogonal" stacking of adjacent molecules induced by hydrogen bonding of the hydroxy group to the carbonyl oxygen of the propionic ester chain on ring IV. The molecule is again methyl 4-isobutyl, 5-ethylbacteriopheophorbide d, and it crystallizes from ethanol/acetone in the space group $P2_12_12_1$ with Z=4 (Fig. 3).

Particularly interesting, besides the three different types of aggregation observed, are the different conformations assumed by the chlorophyll derivatives. Ring IV of the neopentyl derivative (Fig. 2) is effectively planar, i.e., the dihedral angle defined by the α, β, β, α carbons of the ring is ~2°. In the other crystal forms, however, rings IV of the isobutyl or diethyl derivatives exhibit dihedral angles that range from 23 to 27° in the dimers (Fig. 1) to 13° in the "orthogonal" isobutyl crystal (Fig. 3).

As a whole, the molecules are not planar. Indeed, an increasing body of structural data on chlorophylls, chlorins, bacteriochlorins and isobacteriochlorins points to the remarkable flexibility of the skeletons of these molecules [12]. If such significant variations in conformations can be induced by crystal forces then, by extrapolation, similar effects may presumably also be caused by protein environments. The flexibility of these macrocycles may become even more significant when the effects of oxidation and reduction following electron-transfer are taken into account.

Recent structural studies of single crystals of meso-tetraaryl porphyrin cation radicals reveal that their Zn(II), Cu(II), Fe(III)Cl and Mg(II) complexes are all saddle-shaped [13-16]. Since the unoxidized species are either planar or slightly domed, oxidation to the radicals results in a major conformational change. An example of this effect is shown in Fig. 4 which presents the displacements of the 24 atoms that comprise the skeleton of the cation radical of Mg tetraphenyl porphyrin (perchlorato 5,10,15,20-tetraphenyl porphinato magnesium (II)) [16].

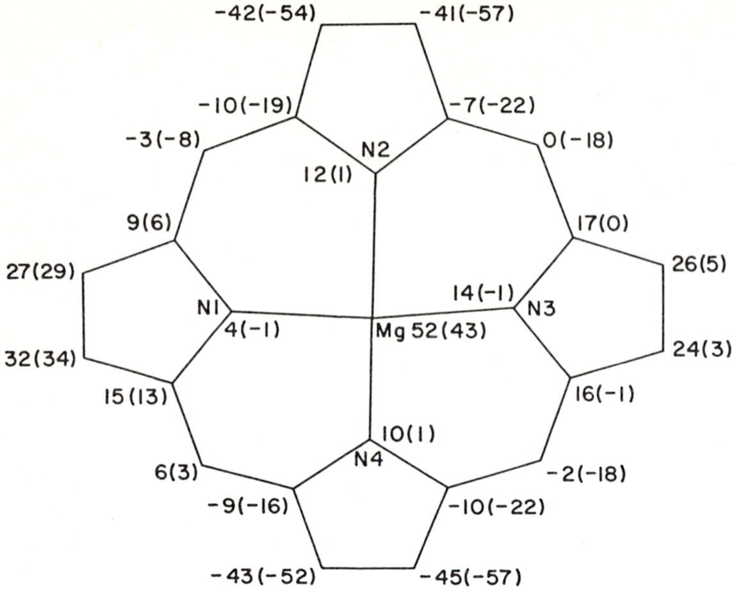

Figure 4. Deviations (A x 10^2) from the least squares plane of the macrocycle and, in parentheses, from the plane of the four nitrogens. The average uncertainty in a nitrogen displacement is 0.005 A (from ref. 16).

The point to be raised, in light of the above results, is that the chromophores within the reaction center are flexible and may well undergo rearrangements following electron-transfer that alter both the distances between donor and acceptor and, perhaps, their relative orientations as well. Such rearrangements need not be slow: resonance Raman and other data for photodissociated heme complexes indicate that significant movement of the iron occurs on a subpicosecond time-scale [17].

3. Theoretical Predictions for Bacteriochlorophyll g, the Chromophore of Heliobacterium chlorum (with Louise K. Hanson, John D. Head and Michael C. Zerner)

The anoxygenic photosynthetic bacterium Heliobacterium chlorum contains [18] a new bacteriochlorophyll labelled "g" with a proposed structure [19] similar to that of BChl b except for the substitution of a vinyl group for the acetyl function found on ring I in BChls b and a (Fig. 5).

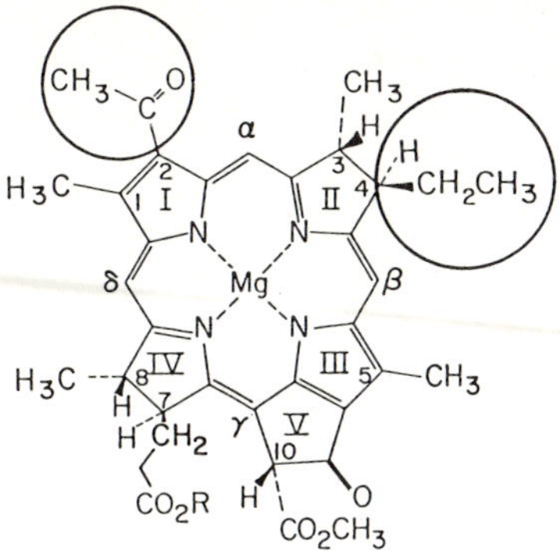

a

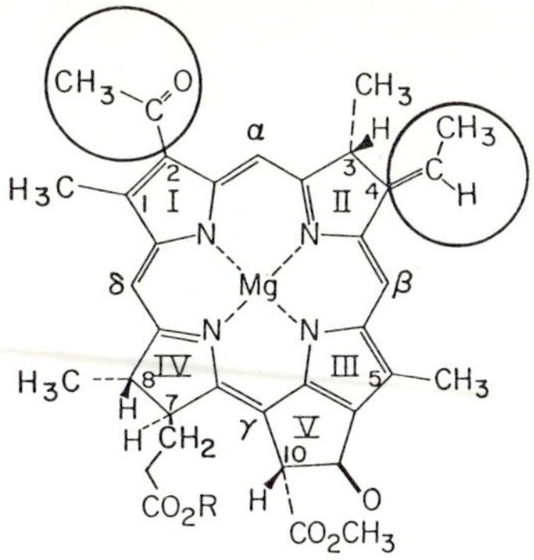

b

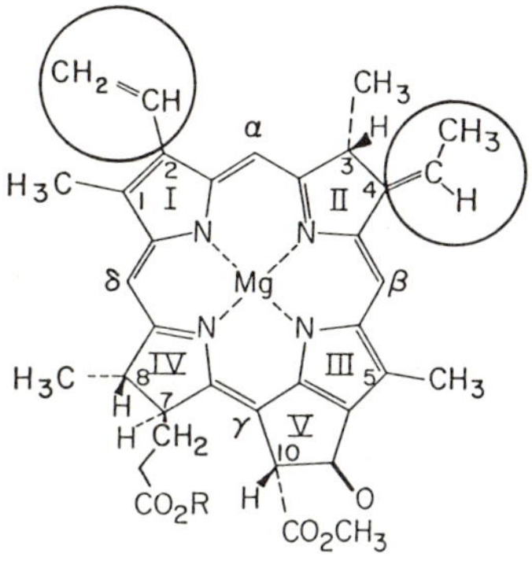

g

Figure 5. Structural formulas of BChls a, b and g.

Optical spectra of BChls b and g do not differ significantly in vitro. In vivo, however, reaction centers containing BChl b exhibit [4] red absorption maxima at 960 nm (P960) whereas the first low-energy band of H. chlorum is found at 798 nm [18,20]. The small spectral differences observed for BChl g in vitro (λ_{max}=763 nm in dioxane) and in vivo suggest, at first glance, that H. chlorum contains a monomeric BChl in its reaction center. (Green plant chlorophylls, which are also postulated to exist as monomers in P680 [21] and P700 [22], similarly possess vinyl instead of acetyl groups on ring I. Acetyl groups could thus help stabilize the "special pairs" in bacteria that contain BChls a and b [6]).

Molecular orbital calculations are presented that address the following questions: a) Are significant differences expected in the optical and redox properties of BChl g, b and a? b) What are the predicted spin distributions of BChl g radicals? c) Can a protein modification, in the form of neighboring charges, explain the unusual optical [20] and redox [23] properties of H. chlorum, and preliminary EPR results for its triplet [23] and oxidized donor [23,24]?

A. Neutral Species

Optical transitions were obtained by INDO-CI calculations for neutral BChls g, b and a. Small variations in the transitions are predicted (and could be further modulated by axial ligand and/or different orientations of the 2-acetyl or 2-vinyl groups) but, in accord with the general trend observed experimentally, no major differences in optical features are predicted for BChls g, b or a.

Reduction potentials deduced from IEH calculations are predicted to be effectively the same, with BChl g sligh+ly harder to reduce (0.01 V). Experimentally, $E_{1/2}$'s for bacteriopheophytins a and b agree to within 20 mV ($E_{1/2}$= -0.74 V vs SCE in DMF). $E_{1/2}$ for BChl a in DMF=-1.05 V) [4].

Oxidation potentials suggest that BChl g should be slightly easier to oxidize than BChl b, which should itself be easier to oxidize than BChl a by ~0.1 V. (These again could be altered by different ligands or substituent orientation.) No experimental $E_{1/2}$ (oxid.) data are available for BChls g and b. $E_{1/2}$ for BChl a=0.40 V in CH_2Cl_2 (vs. SCE) [9].

B. Radicals

Unpaired spin density distributions (IEH calculations) show the g, b and a anions to be very similar. (Variations at the 1-CH_3 position can be induced by rolling the 2-substituents in or out of plane.) Experimentally, a and b are analogous. The ESR linewidth, ΔH, of BChl g^- is therefore predicted to be ~13 G at low temperatures. Resolved ESR and ENDOR spectra should be nearly identical to those of BChl b^- [4]. BPheo g^-, if a transient acceptor, should also be similar to BPheo b^-.

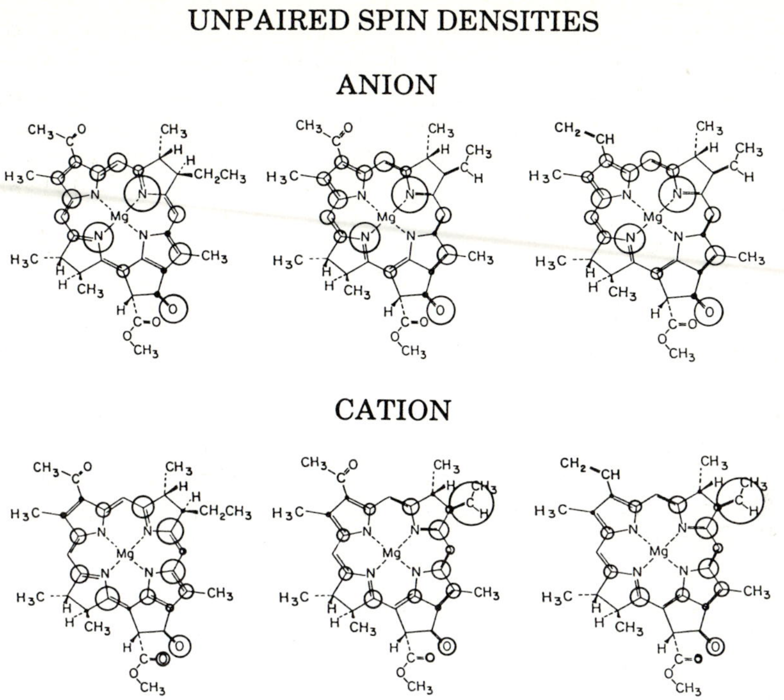

Figure 6. Unpaired spin densities of BChl a, b and g anion (top) and cation (bottom) radicals from charge-iterative extended Hückel calculations (IEH). Shown are results for models with the C2 substituent oriented perpendicular to the macrocycle plane. Rolling the C2 substituent into the plane alters the spin density at C1. The diameters of the circles are proportional to the spin densities.

Spin distributions for the cation radicals (Fig. 6) also predict similar profiles for BChl b and g. By analogy with BChl b^{+} [4], ΔH for BChl g^{+} is predicted at ~14 G.

The M.O. calculations do not reveal any major modulations of the properties of BChl g relative to BChl a or b that would obviously favor its biological selection. The most salient experimental differences are the significantly blue-shifted first absorption maximum in vivo and the similarity between the in vivo and in vitro optical spectra. These would suggest that the chromophore is monomeric in the reaction center, perhaps because it lacks the 2-acetyl group of BChl b that helps form the special pair in R.

viridis [6]. If the donor of H. chlorum is indeed monomeric, then the optical and ESR properties of the BChl g radicals generated by the primary charge separation should be quite similar to those of the corresponding monomeric BChl b derivatives [4]. The spin-polarized triplet that results from the recombination of P^+ and I^- would also be expected to exhibit D and E parameters of a monomeric species.

Preliminary results by Blankenship et al. [23] and Norris et al. [24] yield reversible, light-induced g=2 ESR signals with ΔH=9.5 to 11 G at 5-10° K for H. chlorum. If these signals are due to P^+ and not to decomposition products [4], they suggest a dimer for the oxidized donor. Triplet data for H. chlorum [24] are also similar to those found in organisms that contain BChl a, i.e., dimeric donors.

An intriguing additional protein effect may reconcile the observed optical and ESR properties of H. chlorum. Charged protein residues, placed near the periphery of bacteriochlorophylls have been invoked theoretically [25] (with some experimental support) to explain optical red shifts observed for (B)Chls in vivo. As shown in Table 2, the present INDO calculations predict not only red shifts but blue shifts as well, depending on the sign of the charge and on its location relative to the BChl skeleton.

The unpaired spin densities of BChl radicals also differ with the position and/or sign of a nearby charge (Table 3) but the charges do not appear to be sufficient to explain ESR and ENDOR features associated with dimers, i.e., a neighboring charge will not make a monomer look like a dimer.

Negative and positive charges would tend to stabilize or destabilize a cation radical and thereby render a chromophore respectively easier or harder to oxidize. The low oxidation potential found for H. chlorum [23] would thus suggest a nearby negative charge. Such a charge near ring I is predicted to induce a blue shift of 2900 cm^{-1} in addition to lowering the oxidation potential.

Table 2. Effect of point charges on the Q_y (red) band of BChl g (INDO calcs)

Point Charge	Position	Freq. shift of Q_y relative to no charge	
+0.6	2.4A below C2b	$-1407\ cm^{-1}$	(red shift)
+1.0	3.5A above C1	-1754	red
-1.0	3.5A above C1	+2929	blue
+1.0	3.5A above C4a	+1154	blue
-1.0	3.5A above C4a	-1541	red
+1.0	3.5A above O1	+1937	blue
-1.0	3.5A above O1	-1110	red

(Charges placed at analogous positions of BChls a, b produce similar shifts.)

The structure shown is the model used in the calculations.

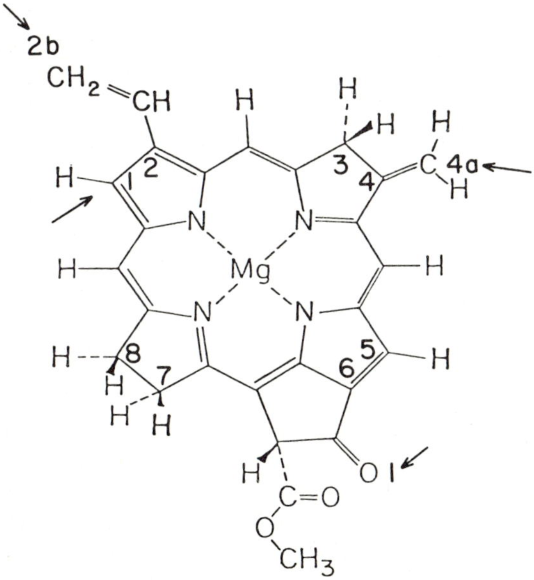

It is possible therefore that <u>H. chlorum</u> may be a dimer whose optical spectrum is blue-shifted by nearby protein charges.

Table 3. Effect of point charges on BChl g^+ unpaired spin densities (INDO calcs)

Pt. charge, position (see Table 2)	%change relative to no charge					
	C1*	C5	saturated Ring Cα's			
+0.6, C2b	188%	-21%	28,	- 7,	- 4,	41%
+1.0, C1	175%	-33%	48,	-14,	-11,	56%
-1.0, C1	-50%	21%	-31,	- 1,	- 4,	-35%
+1.0, C4a	-12%	-35%	- 3,	2,	12,	- 7%
-1.0, C4a	38%	24%	2,	- 8,	- 8,	6%
+1.0, 01	-50%	56%	-30,	- 2,	-19,	-39%
-1.0, 01	138%	-50%	43,	- 8,	3,	54%

*The spin density calculated for C1 is very small; this may exaggerate the % change.

If peripheral charge effects can be substantiated experimentally (Schiff bases of chlorins and bacteriochlorins already provide examples of the influence of positive charges [25]), they add to the modes available to control redox, optical and ESR properties of (B)Chls in vivo: i.e., aggregation, axial ligation, conformation and electrostatics. Combinations of all of these may be required to explain the range of properties exhibited by the purple and green photosynthetic bacteria that contain BChls a, b or g, and the apparently disparate characteristics of the chlorophylls of photosystems I and II of green plants.

Acknowledgments

This work was supported by the U.S. Department of Energy, Divison of Chemical Sciences, Washington, D.C., under Contract No. DE-AC02-76CH00016 at BNL, by the National Science Foundation (CHE-81-20891) at UCD, and by the Division of Sponsored Research of the University of Florida at UF.

References

1. J.F. Norris, H. Scheer, and J.J. Katz: In The Porphyrins, edited by D. Dolphin (Academic Press, New York) Vol. 4, pg. 159.
2. G. Feher, A.J. Hoff, R.A. Isaacson, and L.C. Ackerson: Ann. N.Y. Acad. Sci. 244, 239 (1975).
3. D.C. Borg, A. Forman, and J. Fajer: J. Amer. Chem.Soc. 98, 6889 (1976).
4. M.S. Davis, A. Forman, L.K. Hanson, J.P. Thornber, and J. Fajer: J. Phys. Chem. 83, 3325 (1979).
5. W. Lubitz, F. Lendzian, H. Scheer, J. Gottstein, M. Plato, and K. Mobius: Proc. Natl. Acad. Sci. USA 81, 1401 (1984).
6. J. Deisenhofer, O. Epp, K. Miki, R. Huber, and H. Michel: J. Mol. Biol. 180, 385 (1984).
7. M. Lutz: Biochim. Biophys. Acta 460, 408 (1984). In Advances in Infrared and Raman Spectroscopy, edited by R.J.H. Clark and R.E. Hester, (Wiley-Heyden), Vol. II, pg. 211.
8. T.M. Cotton, K.D. Parks, and R.D. VanDuyne: J. Am. Chem. Soc. 102, 6399 (1980). R.G. Brereton and J.K.M. Sanders J. Chem .Soc. Perkin Trans. I. 423 (1983).
9. J. Fajer, D.C. Borg, A. Forman, R.H. Felton, D. Dolphin, and L. Vegh: Proc. Nat'l. Acad. Sci. USA 71, 994 (1974).
10. K.M. Smith, D.A. Goff, J. Fajer, and K.M. Barkigia: J. Am. Chem. Soc. 105, 1674 (1983).
11. K.M. Smith, D.A. Goff, J. Fajer, and K.M. Barkigia: J. Am. Chem. Soc. 104, 3747 (1982).
12. K.M. Barkigia, J. Fajer, C.K. Chang, and R. Young: J. Am. Chem. Soc. 106, 6457 (1984) and references therein.
13. L.D. Spaulding, P.G. Eller, J.A. Bertrand, and R.H. Felton: J. Am. Chem. Soc. 96, 982 (1974).
14. W.F. Scholz, C.A. Reed, Y.J. Lee, W.R. Scheidt, and G. Lang: J. Am. Chem. Soc. 104, 6791 (1982).
15. G. Buisson, A. Deronzier, E. Duee, P. Gans, J.C. Marchon, and J.R. Regnard: J. Am. Chem. Soc. 104, 6793 (1982).
16. K.M. Barkigia, L.D. Spaulding, and J. Fajer: Inorg. Chem. 22, 349 (1983).
17. J.M. Friedman: Science 228, 1273 (1985).

18. H. Gest and J.L. Favinger: Arch. Microbiol. 136, 11 (1983).
19. H. Brockmann and A. Lipinski: Arch Microbiol. 136, 17 (1983).
20. R.C. Fuller, S.G. Sprague, H. Gest, and R.E. Blankenship: FEBS Lett. 182, 345 (1985).
21. M.S. Davis, A. Forman, and J. Fajer: Proc. Nat'l. Acad. Sci. USA 76, 4170 (1979).
22. P.J. O'Malley and G.T. Babcock: Proc. Nat'l. Acad. Sci. USA 81, 1098 (1984).
23. R.E. Blankenship, H. Gest, and R.C. Prince: private communication.
24. J.R. Norris, S. Kolaczkowski, and M. Bowman: private communication.
25. J. Eccles and B. Honig: Proc. Nat'l. Acad. Sci. USA 80, 4959 (1983). L.L. Maggiora and G.M. Maggiora: Photochem. Photobiol. 39, 847 (1984). L.K. Hanson, C.K. Chang, B. Ward, P.M. Callahan, G.T. Babcock, and J.D. Head: J. Am. Chem. Soc. 106, 3950 (1984).

ENDOR Characterization of Hydrogen-Bonding to Immobilized Quinone Anion Radicals

P.J. O'Malley[1], *T.K. Chandrashekar, and G.T. Babcock*

Department of Chemistry, Michigan State University, East Lansing, MI 48824, USA

ENDOR spectra have been recorded for the anion radicals of p-benzoquinone, 2-methyl-p-benzoquinone and ubiquinone in frozen isopropanol solution. Hyperfine interaction between solvent -OH protons hydrogen-bonded to the semiquinone oxygens and the unpaired electron of the radical occurs and is detected by selective deuteration of the isopropanol hydroxyl group. The hyperfine tensor of the hydrogen-bonded proton is axial with $A_{\parallel}$ positive and equal to approximately twice the negative of $A_{\perp}$, which indicates that the interaction is essentially dipolar in nature. For $UQ^{\overline{\cdot}}$, $A_{\parallel} = 6.1$ MHz and $A_{\perp} = -2.8$ MHz. These results demonstrate the utility of ENDOR in studying interactions between quinone radicals and their local environments, and should be useful in characterizing quinone binding sites in reaction centers and other quinone binding proteins.

1. Introduction

The local environment of quinone binding sites in reaction centers and other electron-transfer proteins plays a major role in determining the properties of the bound quinone. In bacterial reaction centers, as well as in the reaction center of PSII, the one-electron-reduced acceptor, $Q_A^{\overline{\cdot}}$, is stabilized as the negatively charged anion [1,2]. This site is sensitive to pH changes, but these effects appear to be coupled through the protein matrix to a redox-linked acid/base group, presumably an amino acid side chain; the semiquinone itself does not appear to be susceptible to protonation under normal conditions [3]. On the donor side of PSII, on the other hand, the oxidized donor, $Z^{\dot{+}}$, is apparently stabilized as the cation radical [4].

A mechanism which is likely to contribute to protein control of quinone properties involves hydrogen-bonding to the 1,4 quinone oxygens by amino acid residues in the binding site. Detection of such hydrogen-bond interactions has been addressed by Hales in EPR studies of immobilizied semiquinone radicals [5]. He noted increases in EPR linewidths when model quinones were prepared in hydrogen-bonding solvents, and attributed this observation to hyperfine coupling between the unpaired electron spin of the radical and the nuclear spin of the hydrogen-bonded proton. In the course of our ENDOR characterization of cation and anion quinone radicals [6,7], we have found that this spectroscopy [8] provides an excellent method by which to characterize these weak, hydrogen-bond interactions in more detail. We report here our observations for several model quinone radicals, including the anion radical of ubiquinone.

---

[1] Present address: Dept. of Applied Science, Brookhaven National Lab, Upton, Long Island, NY 11973, U.S.A.

2. Materials and Methods

Quinones were obtained from standard commercial suppliers and reduced to the corresponding hydroquinone by using sodium borohydride. Perdeuterated benzoquinone was prepared from the protonated species as described in [9]. X-band EPR measurements were carried out with a Bruker ER200D; magnetic field and microwave frequencies were measured by using a Bruker NMR gaussmeter and a Hewlett Packard 5245L frequency counter (5255A plug-in frequency converter), respectively. ENDOR measurements were carried out as in [6,7].

3. Results and Discussion

Figure 1a shows the proton ENDOR spectrum of benzoquinone (BQ) anion radical, immobilized at -150°C, in isopropanol. In addition to interactions

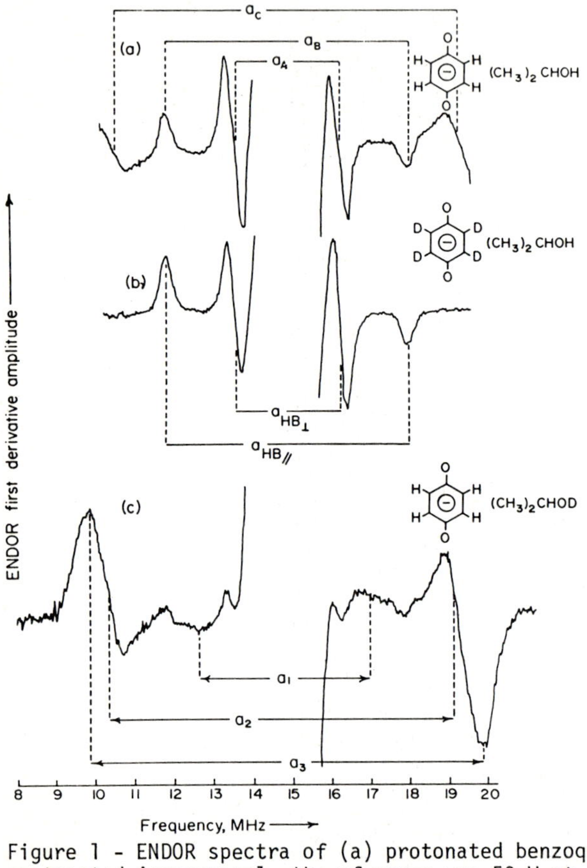

Figure 1 - ENDOR spectra of (a) protonated benzoquinone anion radical in protonated isopropanol, the rf power was 50 W at 10 MHz; (b) perdeuterated benzosemiquinone anion in protonated isopropanol, the rf power was 45 W at 10 MHz; and (c) protonated benzoquinone anion radical in isopropanol-d_1, the rf power was 60 W at 12 MHz. For all three spectra the temperature was 123 K, the microwave power was 1 mW and the fm deviation was ±150 KHz. The magnetic field was set to the zero crossing of the first derivative EPR spectrum.

due to the weakly coupled protons in the matrix region, at least three distinct proton hyperfine splittings, a_A, a_B and a_C, are apparent[2]. The spectrum can be resolved into couplings which arise from solvent -OH protons and those due to ring protons by selective deuteration. In Fig. 1b, the ring hydrogens have been replaced by deuterium and coupling a_C is no longer apparent. Couplings a_A and a_B remain (relabelled as $a_{HB\perp}$ and $a_{HB\parallel}$, see below) and are attributed to solvent -OH protons hydrogen-bonded to the 1,4 quinone oxygens. That this assignment is correct is confirmed by the fact that both lines disappear when protonated BQ anion radical is dissolved in isopropanol deuterated at the hydroxyl group (Fig. 1c). In this spectrum coupling a_C is apparent, which indicates that it arises from the α protons on the quinone ring. A more detailed study of this anion/solvent system shows that coupling a_C is actually the composite of two hyperfine interactions, labelled a_2 and a_3 in Fig. 1c; moreover, a third coupling, a_1, is also apparent once the strong hydrogen-bond interactions have been removed. All three couplings in Fig. 1c, a_1, a_2 and a_3, arise from the ring protons. The signs, magnitudes and orientations of these tensor components have been determined, and are reported in detail elsewhere [10].

Returning to Fig. 1b, we note several aspects of the hyperfine couplings from the hydrogen-bonded protons. First, the shape of the spectrum is typical of that expected for an axial hyperfine tensor, i.e., $a_\perp = a_y = a_z \neq a_\parallel = a_x$. Although ENDOR signals from hydrogen-bonded protons have not been detected and interpreted previously for powder samples, some single crystal work has been done on hydrogen-bonded systems and axial tensors are generally observed [11,12]. Second, the magnitude of $a_\perp$ (2.8 MHz) is essentially half that of $a_\parallel$ (5.9 MHz). This is consistent with the expectation that the hydrogen bond hyperfine interaction should be purely dipolar and hence average to zero under isotropic conditions. Thus, the hydrogen bonded proton hyperfine tensor should be traceless, which requires that the sign of $a_\perp$ or of $a_\parallel$, but not both, be negative. Third, an axial tensor is expected under conditions in which the point dipole approximation holds. In this approximation, the <u>ith</u> component of the hyperfine tensor (A_i, i = x, y or z) is given by

$$A_i = 78.4\ \rho \left(\frac{3\cos^2\Theta_i - 1}{r^3} \right) \qquad (1)$$

where A_i is expressed in MHz, ρ is the unpaired electron spin density on the hydrogen-bonded atom (the 1,4 oxygens in this case), Θ_i is the angle between the applied field (i = x y or z) and the line joining the proton and its hydrogen-bonded partner, and r is the distance in Å between the two atoms [13]. This equation predicts that the major coupling ($a_\parallel$) is along the O···H direction and that its sign is positive; the minor coupling ($a_\perp$) is two-fold degenerate and negative. Thus we assign $a_\parallel$ = 5.9 MHz and $a_\perp$ = -2.8 MHz for the BQ anion radical of Fig. 1. A knowledge of ρ allows one to calculate the O···H hydrogen-bond distance. By using the g-value anisotropy of BQ^- orientation selection is possible and the directions of $a_\parallel$ and $a_\perp$ with respect to the g-tensor axes may be assigned. For $BQ^{\overline{\cdot}}$ we find that the O···H distance is 1.55 Å, the O···H

[2] In our earlier work on BQ anion radical [6], we attributed all three couplings in Fig. 1a to ring protons; the results we present here and elsewhere [9] indicate that the situation is more complex, owing to hydrogen-bonding effects.

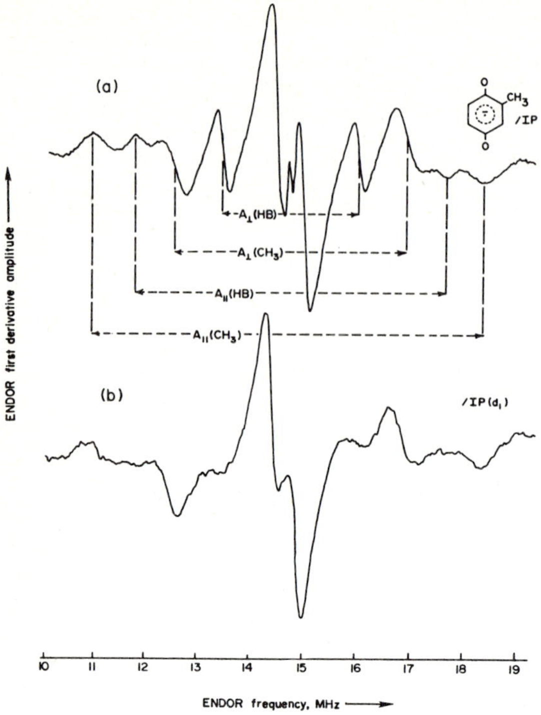

Figure 2 - ENDOR spectra of 2-methyl benzoquinone anion radical in (a) isopropanol and (b) isopropanol-d_1. For both spectra the temperature was 123 K, the microwave power was 1 mW, the rf power was 100 W at 10 MHz and the fm deviation was ±150 KHz. The magnetic field was set to the zero crossing of the first derivative EPR spectrum.

bond direction is in the plane of the quinone ring and is essentially colinear with the 1,4 oxygen direction [10].

With the hydrogen-bond interactions characterized by ENDOR in the relatively simple BQ system, extension of the technique to more complex quinones is informative. Figure 2 shows the ENDOR spectrum of 2-methyl benzoquinone anion radical immobilized in isopropanol (Fig. 2a) and in isopropanol-d_1 (Fig. 2b). The $A_{\parallel}$ (5.9 MHz) and $A_{\perp}$ (2.6 MHz) components of the hydrogen-bonded proton hyperfine tensor are easily identified by their disappearance upon -OH deuteration of the solvent. The characteristic axial signal associated with the rotating methyl group at the 2-position is also easily identified [7]. An isotropic coupling for the $-CH_3$ of 5.41 MHz, which agrees well with the solution value of 5.7 MHz, can be calculated from the parallel and perpendicular values of the hyperfine tensor.

Figure 3 shows the ENDOR spectra of ubiquinone anion radical ($UQ^{\cdot-}$) immobilized in isopropanol (Fig. 3a) and in isopropanol-d_1 (Fig. 3b). In this case, the assignment of the proton hyperfine tensor for the hydrogen-bonded proton is more complex owing to the more complicated nature of the ring substituents which adds more lines to the spectra. Nonetheless, a

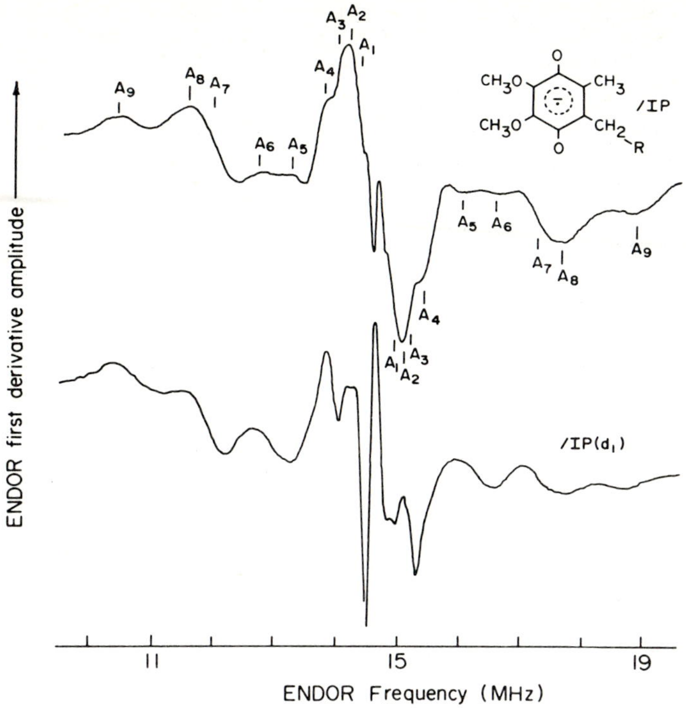

Figure 3 - ENDOR spectra of ubiquinone anion radical in (a) isopropanol and (b) isopropanol-d_1. The temperature was 123 K; microwave power was 6.3 mW; rf power was 110 W at 9.5 MHz, fm deviation ±150 KHz. Magnetic field was set to the zero crossing of the first derivative EPR spectrum.

comparison of Figs. 3a and b shows that the lines marked A_8 and A_5 in the isopropanol spectrum disappear in the isopropanol-d_1 spectrum and thus we assign $A_8 = A_{HB_{\parallel}} = 6.1$ MHz and $A_5 = A_{HB_{\perp}} = 2.8$ MHz for the $UQ^{\overline{\cdot}}$ species. The structure of the spectrum near the matrix region also changes upon deuteration of the solvent -OH, which suggests that a weaker class of hydrogen bonds may also occur in $UQ^{\overline{\cdot}}$. We have noted similar behavior for the anion radical of the sterically crowded duroquinone species [10]. The methyl at the -2 position is evident in the spectrum as couplings A_7 and A_9. From these we calculate $A_{iso} = 6.33$ MHz = 2.26 G which is in reasonable agreement with the value of 2.04 G reported by DAS et al. for $UQ^{\overline{\cdot}}$ in ethanol solution [14]. These authors also reported a coupling of ~1 G for the $-CH_2$-protons of the isoprenoid chain. The A_6 coupling is in this range but its assignment is unclear; in α-tocophenol quinone anion radical we observe no coupling in this region even though the rest of the spectrum resembles that of $UQ^{\overline{\cdot}}$ (not shown). Efforts to resolve and assign lines which arise from both the methylene protons and from the methoxy protons are currently underway.

4. Conclusions

The experiments reported here for hydrogen-bonded proton hyperfine interactions in powder samples and those reported earlier for α and β proton couplings [6,7] show that ENDOR spectroscopy is a powerful tool for unraveling hyperfine interactions in immobilized quinone radicals. Recent work by KURRECK et al. [15,16] on flavin radicals indicates that this utility is likely to extend to a broad range of organic radicals of biological interest. The fact that the anisotropic components of the hyperfine tensors can now be measured directly in powders facilitates the analysis considerably, in that the directional properties of the interaction can be determined and interpreted. The resolution of the hydrogen-bond interaction indicates that interactions which occur between organic cofactors and their protein binding sites can be explored in detail. These interactions, as noted above, may be key in controlling the function of the organic species. The model compound results and analysis should be useful to the interpretation of quinone spectra in bacteria [17], in PSII [18] and in cytochrome bc-type complexes from bacteria, mitochondria and chloroplasts [19].

Acknowledgements This research was supported by the Photosynthesis Program of the Competitive Research Grants Office of USDA. TKC is a McKnight Foundation Postdoctoral Fellow. We thank Drs. G. Feher and W. Lubitz for useful discussions and for communicating results in advance of publication.

References

1. P.A. Loach and R.L. Hall: Proc. Nat'l. Acad. Sci. USA 69, 786 (1972).
2. H.J. van Gorkom: Biochim. Biophys. Acta 347, 439 (1974).
3. C. Wraight: Functions of Quinones in Energy-Conserving Systems (Trumpower, B., ed., Academic Press, New York 1982) pp. 181-197.
4. P.J. O'Malley and G.T. Babcock: Biochim. Biophys. Acta 765, 370 (1984).
5. B.J. Hales: J. Am. Chem. Soc. 98, 7350 (1976).
6. P.J. O'Malley and G.T. Babcock: J. Am. Chem. Soc. 106, 817 (1984).
7. P.J. O'Malley and G.T. Babcock: J. Chem. Phys. 80, 3912 (1984).
8. K. Möbius, W. Fröhling, F. Lendzian, W. Lubitz, M. Plato and C.J. Winscom: J. Phys. Chem. 86, 4491 (1982).
9. E. Charney and E. Becker: J. Chem. Phys. 42, 3314 (1972).
10. P.J. O'Malley and G.T. Babcock, in preparation.
11. H. Muto and M. Iwasaki: J. Chem. Phys. 59, 4821 (1973).
12. M.V.V.S. Reddy, K.V. Lingam and T.K. Gundu Rao, J. Chem. Phys: 76, 4398 (1982).
13. D.J.E. Ingram: Free Radicals (Butterworths, London, 1958) pp. 102-134.
14. M.R. Das, H.D. Connor, D.S. Leniart, and J.H. Freed: J. Am. Chem. Soc. 92, 2258 (1970).
15. H. Kurreck, M. Bock, M. Elsner, H. Kraus, W. Lubitz, F. Müller, J. Geissler and P.M.H. Kroneck: J. Am. Chem. Soc. 106, 737 (1984).
16. H. Kurreck, B. Kirste and W. Lubitz: Angew. Chem. Int. Ed. Eng. 23, 173 (1984).
17. W. Lubitz, R.J. Debus, M.Y. Okamura and G. Feher: Biophys. J. 47, 3a (1985).
18. V.V. Klimov, E. Dolan, E.R. Shaw and B. Ke: Proc. Nat'l. Acad. Sci. USA 77, 7227 (1980).
19. A.R. Crofts and C.A. Wraight: Biochim. Biophys. Acta 726, 149 (1983).

Concluding Remarks

Some Aspects of Energy Transfer in Antennas and Electron Transfer in Reaction Centers of Photosynthetic Bacteria

J. Jortner

Department of Chemistry, Tel Aviv University, 69978 Tel Aviv, Israel

M.E. Michel-Beyerle

Institut für Physikalische und Theoretische Chemie,
Technische Universität München, D-8046 Garching, F. R. G.

Introduction

We shall focus on the interrelationship between the structure, interactions and dynamics in photosynthetic bacteria. The relative simplicity of the functional units, antennas and reaction centers , renders these bacteria to play the role of the "hydrogen atom of photosynthesis", providing basic systems for the critical scrutiny of novel concepts and theories. The structural, spectroscopic and dynamic data which have been accumulated recently provide a wealth of information on the two basic functional units:

(A) Antennas
- (A1) Structure and organization
- (A2) Electronic spectroscopy
- (A3) Electronic energy transfer (EET)

(B) Reaction Centers
- (B1) Interrelationship between structure and dynamics
- (B2) Spectroscopic methods
- (B3) Mechanisms of electron transfer (ET)

The two central questions addressed are:

(1) How does the structure determine and control the acquisition, storage and useful disposal of energy, both in antennas and in reaction centers?

(2) What is the general conceptual framework and which are the principles for the description of the primary dynamic processes of EET and ET?

Our discussion attempts to be selective rather than comprehensive. Rather than providing an overview of the papers and discussions at the Feldafing conference, we shall focus on several central issues, endeavouring to examine what is understood , what is not understood and what novel experimental data and theoretical concepts are required for the understanding of antenna und reaction center function.

(A1) Structure and organization of antennas

Bacterial antennas fall into two classes:

1) The phycobiliproteins, in which bile pigments, e.g. the phycocyanin, are covalently bound to the protein. These antennas are in most cases highly aggregated into the phycobilisomes /1,2/ and located at the surface of the photosynthetic membrane. Pigment-pigment interaction is only moderately strong due to the large average inter-pigment distances of 20 to 50 Å.

2) Bacteriochlorophyll containing antennas of purple (e.g. Rp.sphaeroides, Rp.viridis, Rp.capsulata, Rs.rubrum) and green bacteria (e.g.Chloroflexus aurantiacus, Chlorobium limicola) /3-8/. These antennas are also highly aggregated but in contrast to class 1) located either within the photosynthetic membrane as in purple bacteria or within extra-membrane vesicles, the chlorosomes, as in green bacteria. In both cases, the pigment-pigment interactions are much stronger than in the antennas of class 1).

Many of the various chromoproteins comprising bacterial antennas can now be isolated in spectrally and functionally intact states /1,2,6,9,10/. In addition model systems such as lipid films containing bacteriochlorophyll-proteins /11/, bacteriochlorophyll aggregated in micellar solution /12-14/, covalently and hydrogen bound tetrapyrrol oligomers /15,16/ and complexes with heme-apoproteins /17,18/ provide relevant information regarding spatial organization and EET processes.

Of central interest are the basic mechanisms of energy acquisition, transport and transfer in the antennas of classes 1) and 2), in particular questions as

(a) what are the organizing factors for the interactions between proteins, between pigments and between proteins and pigments?

(b) How does EET relate to these structures? Exciton transfer, role of orientation.

(c) How does EET depend on the topology of pigment molecules? Regularity, dimensionality, homogeneity of spatial distribution of pigments and limiting size of cooperative units in the antennas.

(d) Coupling of antenna pigments to the reaction center. Structural organization of the antennas with regard to the reaction center.

First answers to these questions emerged from the interplay of recent topographic, structural and spectroscopic data.

Information on the primary structure of the apoproteins is now available for light-harvesting pigment-protein complexes of several purple bacteria /3,19/, of one green bacterium /19/ and of a series of biliproteins /19/. The high resolution X-ray structures of several water soluble antennas, as one bacteriochlorophyll-containing complex /20/ and some phycobiliproteins /21/ have put many of the earlier spectroscopic results into a new perspective, and opened the road to a much more detailed correlation of structure and spectroscopy. The recent breakthrough in crystallization /22-24/ of membrane proteins is expected to provide a similar basis also for the hydrophobic antennas /25-27/. The structural data have to be complemented by the following spectroscopic and photophysical information.

(A2) Electronic spectroscopy

The electronic energy levels of the isolated pigments are fairly well known. However, there are remarkable changes induced by the interactions with the proteins and the neighbouring pigments. A detailed knowledge of the level ordering is necessary to relate the dynamic studies with the assignment to EET among individual pigments. The nature and order of the lowest-lying spin-allowed electronically excited state exerts a major effect on the excited-state dynamics,i.e. the radiative decay,intramolecular radiationless transitions and EET.

Special attention has to be focussed on the remarkable level ordering in the carotenoids, which - when present in bacteriochlorophyll complexes - perform functions in both, EET and degradation of excess energy /28/. Regarding these functions, the small S_0- T_1 energy gap as well as the ordering of the two close-lying spin-allowed electronic excitations in the singlet manifold are important. The latter are the symmetry-forbidden $1^1A_1 \rightarrow 2^1A_1$ transition and the symmetry-allowed $1^1A_1 \rightarrow 1^1B$ transition /29/. Extensive spectroscopic information on related compounds, e.g. polyenes and diphenyl-polyenes, point towards the proximity of these two excited states whose energetic ordering is strongly solvent dependent /29/. This ordering, however, decides on the mechanism and efficiency of excitation transfer from the carotenoid to the porphyrin.

Some interesting medium effects on intramolecular electronic-vibrational energy degradation within the phycocyanine antennas have been exhibited. While the flexible, cyclic-helical structure of bile-pigments in solution results in very efficient radiationless decay, presumably internal conversion with a fluorescence quantum yield of $\Phi \cong 10^{-3}$, the extended rigid structure of phycocyanins in the protein makes them immune towards internal conversions as reflected in $\Phi \cong 1$ /8/. The dynamics of intramolecular radiationless transitions, i.e. internal conversion and intersystem crossing, in large rigid molecules is practically invariant with respect to an "inert" solvent, which does not modify the intramolecular coupling and/or the energy levels. On the other hand, radiationless transitions in "floppy" large molecules sometime exhibit a strong solvent effect, which shows a marked dependence of the the rate of internal conversion on the solvent viscosity. A preliminary theoretical description of viscosity effects on radiationless transitions in non-rigid large molecules is provided in terms of Kramer's theory /30/. The marked effect of the antenna protein on the retardation of the radiationless process in phycocyanins can tentatively be attributed to the effects of high local viscosity of the protein and to the specific binding of the pigment to the protein.

(A3) Electronic energy transfer in antennas

The intermolecular pair interaction between electronically excited and ground state pigment molecules provides the electronic coupling mechanism which induces intermolecular electronic energy transfer. The conventional coupling mechanism involves dipolar interactions with the transition probability for EET between a pair of molecules separated by a distance R, $W(R) = \tau^{-1}(R_0/R)^6$, with τ being the excited state lifetime and R_0 the "critical radius" for EET /31/. EET in bacteriochlorophyll containing antennas proceeds via this dipole-dipole coupling mechanism. The dipole-dipole transition probability vanishes for EET involving a change in the electronic spin and it becomes very small for EET from a weak, spin-allowed electronically excited state. Under these circumstances, intermolecular electronic energy transfer is induced by a two-centre, two-electron exchange interaction /32,33/. On the basis of quantum-mechanical calculations /33/ one can assert that these short-range intermolecular coupling terms exhibit an exponential dependence on the intermolecular separation and are characterized by a marked dependence on the relative orientation of the two molecules. Provided that the 2^1A_1 configuration of the carotenoid constitutes the lowest excited singlet state in the carotenoid-chlorophyll system, EET from the carotenoid to the bacteriochlorophyll will be induced by the exchange mechanism.

The above examination of the intermolecular excited state pair interactions brings us to consider some of the gross features of EET in the

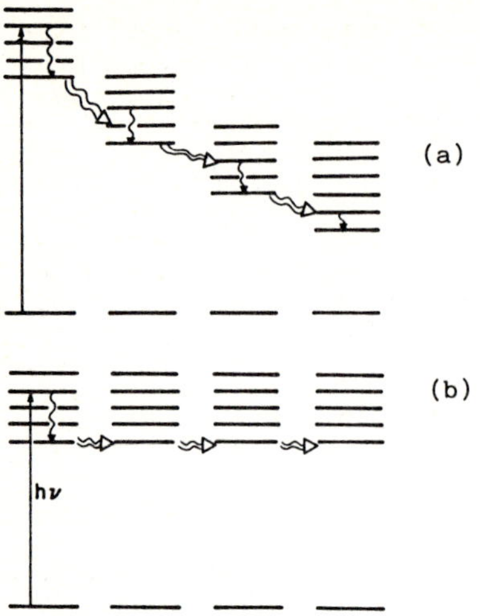

Fig.1
Schemes of ETT in antennas

vibrational relaxation

EET

(a) Energy funneling

(b) Near-resonant EET

antennas (for recent reviews see /34,35/).Two major mechanisms of EET prevail in antennas (Fig.1), being determined by the electronic level structure of the pigments.

(A) Energy funnelling in phycobiliproteins and carotenoid-bacteriochlorophyll complexes

These antennas contain pigments with distinctly different electronic energies. The electronic energy is partially degraded during the EET process.

(B) Near-resonant EET in bacteriochloropnyll antennas
This situation involves EET between nearly isoenergetic excited states of the pigments. It should however be noted that in the one-pigment bacteriochlorophyll antennas diagonal and off-diagonal disorder effects which result in a shift of the electronic energy levels from resonance may be induced as follows. Firstly, the local environment of the proteins may shift the energy levels. Second, dispersive and excitonic interactions between the chlorophyll pigments will result in splitting of the electronic energy levels of a dimer and of an aggregate within the antenna, giving rise to a low-lying electronic state which may in turn be instrumental for the optimization of the EET from the antenna to the reaction center.

At present very little quantitative information is available concerning EET in chlorophyllous antennas, in contrast to phycobiliproteins where the pigment-protein cyclic unit structures invite for a detailed treatment /36/.
Information is needed on diverse aspects of the energy transport process:

(1) Spectral diffusion reflecting the effects of diagonal disorder, i.e. inhomogeneous broadening, on the electronic excitations of the pigments.

(2) Spatial diffusion of the electronic energy .

(3) The diffusion coefficient of the electronic energy , which characterizes the spatial diffusion during the excited state lifetime.

(4) Localization and trapping of the electronic energy, which can be quantitatively characterized by the trapping rates.

(5) Dynamics of energy transfer from the antenna to the reaction center.

The following experimental techniques have already been adopted to interrogate EET in antennas:

(a) Singlet-singlet annihilation. Collisions involving pairs of electronically excited states can be realized only at high intensities of the exciting light as shown for chlorophyllous antennas /37/ and phycobiliproteins /38/. The annihilation rate constant k_{SS} for incoherent, diffusive exciton motion can be related to the diffusion coefficient, D, of the electronic excitation /39/ by $k_{SS} = 4 \pi D \langle R \rangle$, where $\langle R \rangle$ is the encounter radius. These experiments provide some indirect information on spatial energy transfer in antennas.

(b) Picosecond time-resolved spectroscopy /40-42/. The dynamics of the funnelling of the EET process in phycobiliproteins was traced by monitoring the time evolution of the fluorescence decay and/or the ground state recovery. The decay of the fluorescence from the initially excited state of the high-energy pigment is now generally agreed to be fit best by a sum of exponentials /40-42/ rather than by a function of the form $\exp(-\alpha \sqrt{t})$, as implied by the classical Förster mechanism for dipole-dipole EET to a randomly distributed assembly of acceptors /31/. Inhomogeneous broadening effects may also be responsible for the deviation from the Förster law.

(c) Time-resolved polarized fluorescence. This well-established technique, which dates back to Förster`s pioneering work /31/ , provides direct information on the initial-site population probability of the electronic excitation . The data analysis has been critically discussed /41 /.

(d) Hole burning experiments providing information on spectral diffusion of resonant EET and on the rate of EET in energy cascading antennas have been performed /43/.

During the last twenty years, several sophisticated experimental techniques were advanced to probe EET in molecular crystals, in doped solids and in solutions. The application of this arsenal of modern techniques, which are based on laser spectroscopy, for the exploration of EET in antennas will be extremely pertinent. Some of these new methods are:

(i) Spatial intermittency methods advanced in 1965 /44,45/ culminating in the transient grating method /46/, which provides direct information on the spatial diffusion coefficient of the electronic excitation , on the initial localization process and on the nature of transport (i.e. incoherent diffusive or coherent motion of electronic energy).

(ii) Spectral diffusion studies /47,48/ which rest on simultaneous time-resolved and energy-resolved interrogation, are expected to probe inhomogeneous broadening effects on EET.

(iii) Coherent optical effects, e.g. free induction decay and photon echoes /49/ are expected to yield interesting information on the retention of phase memory in EET.

Some of the outstanding open questions in the area of EET in antennas are:

(a) The nature of transport of electronic energy. Two limiting situations can be distinguished in this context involving coherent motion, with the mean free path of the electronic excitation considerably exceeding the interpigment spacing, and the incoherent (i.e. strong scattering diffusive) motion, with the mean free path being equal to the interpigment spacing. Extensive theoretical and experimental information on EET in impurity bands of mixed molecular crystals /50/ indicates the existence of an effective coupling of the electronic excitations with intramolecular vibrations and phonons, which result in the erosion of coherence of exciton transport. EET at room temperature and at moderately low temperatures proceeds via an incoherent transport mechanism. Only at exceedingly low temperatures, i.e. below 4 K, coherent EET between pigments may prevail. Accordingly, in antennas at room temperature the exciton motion will be incoherent provided that the analogy to molecular crystals holds.

(b) The role of diagonal and off-diagonal disorder on EET between pigments in antenna and on the antenna-reaction center transfer.

(c) The transfer of electronic energy to the reaction center. The available information concerning this important process in chlorophyllous antennas is rather meager in chlorophyll-containing antennas where the electronic energy is similar to or even lower than that of the dimer in the reaction center. There is a distinct possibility that this direct EET process is sufficiently fast in some cases to occur from a vibrationally non-relaxed state. A notable example is the ultrafast primary charge separation process in the reaction center /51,52/, to which we shall allude in section (B3). The EET process from the antenna pigments to the reaction center may constitute another example for an electronic process occurring from a vibrationally unrelaxed manifold.

(d) Dimensionality effects on EET in antennas. The geometrical dimensionality exerts a dominating effect on the dynamics of an unistep EET event, say from one initially excited pigment to another, or from an antenna pigment to the reaction center. Similarly, spectral and spatial energy diffusion strongly depend on the dimensionality.

A digression on EET on fractals

The last issue in the previous section regarding the control of EET by the dimensionality of the system, brings up the very interesting matter of EET on fractals /53/. The fractals introduced by MANDELBROT /53 / constitute objects of non-integer effective dimensionality, which are characterized by self-similar structure and by dilatation symmetry, while they are devoid of translational symmetry. In Fig.2, we have reproduced some typical two-dimensional and three-dimensional popular fractal structures, which are referred to as Sarpinskii gasket in two-dimensional Euclidean space and the Sarpinskii pyramid in three-dimensional Euclidean space.

Fractals were invoked for the description and characterization of a multitude of self-similar structures from galaxies /54/ to proteins /55/. In the area of condensed matter systems, fractals were used for the description of percolation clusters, porous glasses, epoxy resins, polymers and proteins. The fractal constitutes one of the very few well-defined precise concepts for the quantitative characterization of strangely-shaped objects. Fractals are characterized by at least three dimensionalities. These are the Euclidean dimensionality d, the fractal dimensionality $\bar{d}$ which determined the size (R) dependence of the site density N(R), i.e. $N(R) \propto R^{\bar{d}}$ /54/ and the

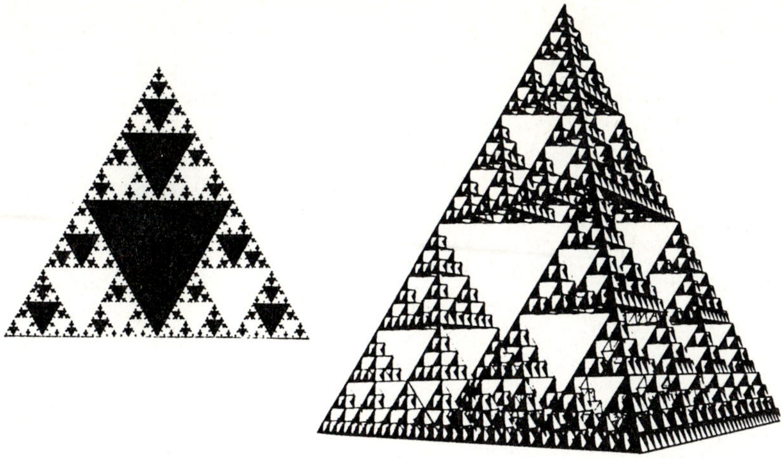

Fig. 2 Two- and three-dimensional fractal structures. Two-dimensional Sarpinskii gasket (left) and three-dimensional Sarpinskii pyramid (right)

spectral dimensionality $\tilde{d}$, which determines the density of phonon states and the characteristics of (long time) random walk on fractals /55/.

Concerning the interesting issue of EET on fractals, it has recently been shown /56/ and experimentally demonstrated /57/ that for EET induced by multipolar interactions of the order s, i.e. $W(R) \propto R^{-s}$, the initial-site population probability is $P(t) \propto \exp\left[-\gamma\, t^{(\tilde{d}/s)}\right]$, with the temporal behaviour being determined by the fractal dimensionality divided by the order of the multipolar interaction. For conventional three-dimensional space $\tilde{d}=d=3$, the celebrated Forster result, the dependence of $P(t) \propto \exp(-\gamma t^{1/2})$ is recovered. For $d<3$, the decay of P(t) is slower than the three-dimensional Förster limit, reflecting the confinement of the EET process on the fractal. It is a very interesting open question whether one can define fractal structures for antennas and related model systems, which will exhibit the confinement of the EET process.

(B1) Reaction centers - Interrelationship between structure and dynamics

The fundamental function of reaction centers (RCs) in photosynthesis is the conversion of a light quantum into an oxidized porphyrin and a reduced quinone species. That this process is vectorial in the thylakoid membrane of chloroplasts was demonstrated /58/ as early as 1968 through the measurement of electrochromic shifts reflecting the transmembrane electric potential gradient. Moreover, only limited by the instrumental time resolution of 20 ns, it was shown /59/ that the process of charge separation had to be very fast. Soon thereafter analogous evidences were reported also for bacteria /60,61/.

Another "quantum step" in the history of photosynthesis was the successful isolation of RCs of Rp. sphaeroides, R-26 /62/. On the basis of spectroscopic and electrochemical data /63/ the first heuristic model (Fig. 3) for the RC was advanced in 1974 /64/.

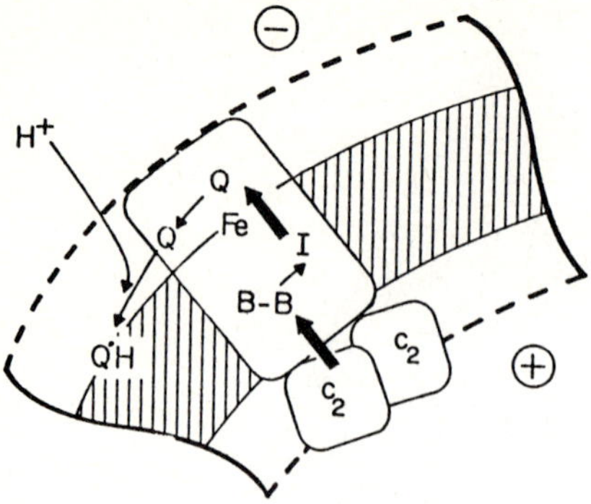

Fig. 3

Schematic representation of the reaction center and immediate reactants (c_2 cytochrome c) in the chromatophore membrane/64/. Since there is no evidence for their position in the RC, the two accessory bacteriochlorophyll (BChl) monomers and the second bacteriopheophytin (BPh) are not depicted. B-B, I and Q refer to the dimer $(BChl)_2$, BPh and quinone, respectively.

This model was confirmed and superseded by the 1984 structure of the RC of Rp.viridis (Fig.4) which rests on the recent crystallographic data /65,66/.

Let us at this stage address the question:

(1) What do we learn from the new crystallographic data that we did not know before? In this context , the following information emerged:

(a) A quasi-symmetric (C2) arrangement of pigments and quinones on the L and M protein which together with the heavy subunit H constitute the reaction center.

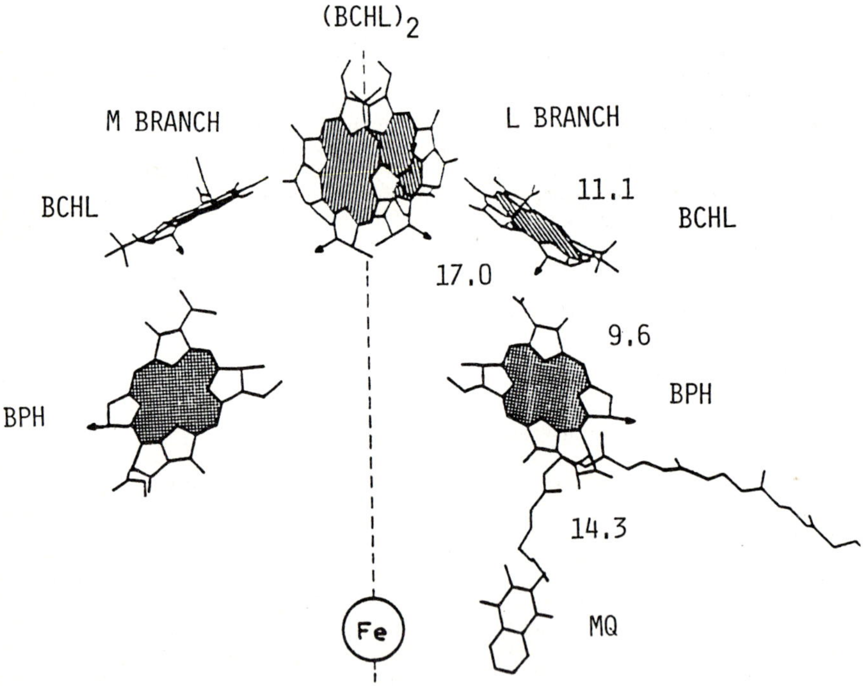

Fig. 4 The structure of the prosthetic groups in the RC of Rp.viridis. Center-to-center distances in Å.

(b) Detailed configuration of the dimer $(BChl)_2$. The dimer "special pair" model for the primary donor was originally proposed on the basis of ESR spectroscopy /67/ and later confirmed in ENDOR measurements /68/.

(c) The spatial ordering of the pigments; one of the BChl monomers mediates structurally between the dimer and the bacteriopheophytin. A crude spatial sequence of the dimer $(BChl)_2$, BPh and the quinone was already evident from psec-time-resolved spectroscopy /69-71/.

(d) The location of the tightly bound menaquinone and the non-heme iron.

(e) The spatial order of the cytochromes and their coupling site to the RC.

This impressive wealth of detailed structural information raised the two fundamental questions:

(2) What are the structural implications of the gross unidirectional charge separation in the RC? As the ordering of the redox components in the two branches is practically identical (Fig.4), the protein environment seems to determine the dominating route of the charge separation on the L branch. The control of charge separation may be accomplished by polar and charged groups in the protein which energetically stabilize the ion pair, i.e. charge transfer states along one branch.

(3) What is the reason for the redundancy in the structure of the RC, which effectively utilizes for forward electron transfer only one branch?

Proceeding to mechanistic issues, which pertain to the correlation between the structure and the ET dynamics, the following further questions can be raised:

(4) What is the role of the dimer? Two implications of the unique structure of the dimer may be relevant for its photophysical function. Firstly, the low excitation energy of the dimer, which originates from exciton and dispersive interactions between its components, as well as the protein environmental shift, may ensure efficient energy harvesting via EET from the antenna. It is not clear at present whether this function of the dimer is significant, as in Rp.viridis the peak excitation energy of the dimer exceeds the lowest peak excitation energy of the antenna pigments. Secondly, the dimer plays a crucial role in the process of charge separation.

(5) What is the route of electron transfer within the RC? Role of the BChl monomer (see section B3); role of the non-heme bound iron.

(6) What are the effects of the protein environment and nuclear motion on the energetics and dynamics of ET?

In the context of the latter question, the following characteristics of the structure, vibrational excitations and nuclear motions of proteins are pertinent:

(a) Distribution of local configurations, originating from the coexistence of a large number of local conformations. Each of the distinct conformations performs the same function but with a different rate. Accordingly, inhomogeneous broadening effects on dynamics processes, e.g. the recombination of diatomic ligands with haemoproteins /72/, are substantial /73/. Such inhomogeneous broadening effects may also prevail in ET processes, being manifested by differences in local energy levels, by differences in medium reorganization energies and by a spread in the distribution of the donor-acceptor distances.

(b) The occurrence of low-frequency nuclear modes in proteins /74/ which may couple to the electronic states.

(c) The possible existence of tunnelling states in proteins /75,76/ which are analogous to those postulated for glasses /77/. These tunnelling states may exhibit a profound effect on the dynamics within a protein medium at low temperatures.

Characteristics (a) - (c) are important in controlling the ET dynamics in the RC only provided that coupling of the donor and the acceptor molecules to the medium modes is important. We shall now proceed to inquire about:

(d) The role of charges and polar groups and their motion with respect to the components of the RC. The existence of static charges and polar groups in the vicinity of the pigments of the RC will result in a considerable energetic stabilization of some ion-pair states, exerting a pronounced effect on the energetics and dynamics of ET.

(B2) Spectroscopic methods

The recent X-ray structure /65,66/ of the RC of Rp. viridis confirmed essentially the crude picture which has evolved from the fragmentary structural information (as e.g. the spatial sequence and mutual orientation of RC components and magnetic interactions of paramagnetic states) contained in spectroscopic data. With the knowledge of the structure, we expect that future spectroscopic studies will focus more on pigment-pigment and pigment- protein interactions and dynamics than on the extraction of structural information. Especially with respect to the resolution of the early ET steps in transient absorption measurements, the identification of individual pigments including their electronic and medium induced interactions is still important. Among the spectroscopic techniques represented at the Feldafing meeting are steady state and time-resolving absorption methods as linear and circular dichroism /78-81/ as well as vibrational spectroscopy, including resonance Raman /82/ and more recently IR difference spectroscopy /83-85/, both becoming a promising tool for the probing of changes in the protein environment. Optically Detected Magnetic Resonance (ODMR) allows the determination of zero field splitting parameters of the triplet state, $^3(BChl)^*$ /86,87/. Moreover, the detection of the ODMR signal in absorption as a function of wavelength gives access to highly resolved optical spectra /86/. This ADMR-spectroscopy is of special advantage when applied to a multi-pigment system with strongly overlapping absorption bands. Magnetic resonance techniques /88-92/ (in particular ENDOR) provide the spin density distributions on radical ions as $(BChl)_2^+$ /89/ and Q^- /90/ and open the road to an estimate of charge distributions. These will be extremely useful once the nature and spatial distribution of the aminoacid side groups in the vicinity of the prosthetic groups will be known from X-ray data. Magnetic resonance techniques with optical detection allow the investigation of the ns- recombination dynamics in short-lived radical pairs as $(BChl)_2^+ BPh^-$ /18,92-94/. In the following we discuss two experimental techniques which have so far been only marginally exploited and which might contribute to a deeper understanding of the details of early charge separation.

Inhomogeneous broadening effects constitute a primary characteristic of the protein medium. The broad spectral lines of the optical excitations of the components of the RC and, in particular, the large width of the dimer band may originate from these effects. The elucidation of inhomogeneous broadening effects on the constituents of the RC is of importance because of two reasons. Firstly, the EET process from the antenna to the RC will be affected by these diagonal disorder effects. Secondly, the local energies of the neutral and charged pigments and their energetic spread will have a pronounced influence on the ET processes.

A powerful spectroscopic method for probing the energetic spread of the excitation energies of the components of the RC (i.e. the (static) inhomogeneous broadening) is based on hole burning. The hole burning technique /95,96/ rests on the interrogation of permanent or transient absorption changes that are induced by irradiation of a broad absorption spectrum with a narrow spectral band laser at low temperatures. The occurrence of distinct sites for any pigment in the RC will result in selective excitation of those sites within the laser bandwidth, which can be interrogated by the observation of either photochemical or transient hole burning. Apart from the qualitative information regarding the occurrence of inhomogeneous broadening, this technique provides quantitative information regarding the homogeneous (lifetime and dephasing) width of the impurity excitation and its temperature dependence. There are two versions of the hole burning technique, i.e., transient hole burning and photochemical hole burning. The photochemical hole burning technique applied to the BPh rests on the isomerization of the BPh due to the translocation of its two porphyrin protons. It will be possible to identify the differences in the excitation energies of the two BPh molecules within the two strands of the RC. This energy difference may be responsible for the uni-directionality of the ET across a single strand within the RC. Moreover, the photochemical hole burning technique is applicable to the dimer band, where absorption changes due to the formation of the dimer cation via ET are expected. Important information concerning the spread of local protein configurations around the $(BChl)_2$ dimer will emerge.

The hole burning technique pertains to the interrogation of "conventional" intravalence excitations of the constituents of the RC in the protein medium. These intravalence excitations have their parentage in the electronic excitations of a single "isolated" constituent of the RC. Thus, for example, the dimer absorption band, the BChl absorption band and the BPh absorption band can be attributed to electronic excitations of the individual BChl and BPh components, with the intermolecular interactions resulting in the delocalization of the electronic excitation over the constituents of the RC. Of considerable interest are electronic excitations, which do not have their parentage in the excitation of a single constituent. Such exotic condensed-matter excitations involve ion-pair, charge-transfer states with an electron and hole residing on different molecules. The search for the existence of such ion-pair states in neat crystals of aromatic molecules provided evidence that these excitations are located above the first spin-allowed singlet excitation /97/ and the location of these charge-transfer states remained elusive. Stark spectroscopy provides a powerful method for the identification of such charge-transfer states. The intensity change of absorption bands in the presence of an external electric field is related to the change of the permanent dipole moment μ between the ground and excited state. The measurement of the Stark effect on the absorption spectrum of RCs of Rp. shaeroides, R26 /98/ showed an enhancement of the dimer band, which considerably exceeded the effect observed on the monomer BChl and BPh bands. Accordingly, the electronic excitation of the dimer is characterized by a high value of $|\Delta\mu|$, so that one can infer that the permanent dipole moment of the electronically excited state of the dimer is high. This interesting spectroscopic observation can be explained with the heavy mixing of charge-transfer states with the neutral electronic excitation of the dimer. Two types of charge-transfer states can be considered in this context.

(A) Dimer charge-transfer states,

$$\alpha\, |(BChl^+\ BChl^-)> \quad + \quad \beta |(BChl^-\ BChl^+)> \qquad (B2.1)$$

where (BChl BChl) corresponds to the two components of the dimer. Provided that the local environment of the two BChl components of the dimer is identical the state (B2.1) is characterized by a nearly zero permanent dipole moment.

(B) Dimer-monomer charge-transfer states. These states of the type

$$\psi_{CT} = \alpha\, |(BChl^+\ BChl)\ (BChl)^-> \ + \ \beta\, |(BChl\ \ BChl^+)\ (BChl)^-> \qquad (B2.2)$$

are characterized by a permanent large dipole moment. The mixing of these charge-transfer states with the lowest singlet neutral dimer excitation,

$$\psi_N = |(BChl^*\ BChl)> \ - \ |(BChl\ \ BChl^*)> \qquad (B2.3)$$

will result in the excited state

$$\psi = a\,\psi_N + b\,\psi_{CT} \qquad (B2.4)$$

The transition moment from the electronic ground state ψ_G to the charge-transfer component ψ_{CT} is very small, so that the mixing of the charge transfer state does not contribute to the integrated intensity of the dimer band but rather appreciably to the Stark enhancement and to the redistribution of intensity in the dimer band. In the absence of quantitative intensity data, one cannot provide an estimate of the charge-transfer mixing coefficient $|b|$. Furthermore, the location of the zero-order charge-transfer states cannot be determined. A rough order of magnitude estimate indicates that the mixing coefficient of the charge-transfer states is

$$|b|^2 \cong |< \psi_N\, |H|\, \psi_{CT} > \ / \Delta E|^2 \qquad (B2.5)$$

where the matrix element is the two-electron and one-electron coupling, while ΔE corresponds to the energy gap between the zero-order states. Model calculations for charge-transfer interactions in organic molecular crystals, i.e., naphthalene and anthracene /99/ resulted in

$$|< \psi_N |H|\, \psi_{CT} >| \cong 100\ cm^{-1}$$

so that reasonable mixing of charge-transfer states with $|b|^2 \cong 0.1$ implies that $|\Delta E| \cong$ 300-500 cm-1 for the energy gap between the zero-order neutral excitation and the dimer-monomer charge-transfer excitation in the RC. Two limitations of this crude estimate should be noted. First, one cannot decide whether the zero-order charge-transfer state is located above or below the neutral excitation. Second, as $|\Delta E|$ is of the order of intramolecular vibrational frequencies, the scrambling of individual vibronic states of the neutral and charge-transfer excitations have to be incorporated in a detailed treatment of the problem.

(B3) Mechanisms of electron transfer in the reaction center

The sequence of the primary electron transfer processes in the RC can conveniently be sub-divided into two sets of processes:

(I) Later stages of ET, corresponding to the time scale exceeding 10 psec.
(II) Early ET within the first 10 psec.
These two distinct time domains will be discussed separately.

(I) Later Electron Transfer

The three ET processes corresponding to this class are

ET from BPh to the quinone /100/

$$(BChl)_2^+ \; BChl \; BPh^- Q_A \; Q_B \xrightarrow{10^{10}\,s^{-1}} (BChl)_2^+ \; BChl \; BPh \; Q_A^- \; Q_B \qquad (B3.1)$$

ET between the two quinones /101/

$$(BChl)_2^+ \; BChl \; BPh \; Q_A^- \; Q_B \xrightarrow{10^{4}\,s^{-1}} (BChl)_2 \; BChl \; BPh \; Q_A \; Q_B^- \qquad (B3.2)$$

and cytochrome oxidation /102/:

$$\text{ferro-c}(BChl)_2^+ \; BChl \; BPh \; Q_A \; Q_B \xrightarrow{10^{6}\,s^{-1}} \text{ferri-c}(BChl)_2 \; BChl \; BPh \; Q_A \; Q_B \qquad (B3.3)$$

We believe that the conventional non-adiabatic ET theory /52,103,104/ is applicable to describe these ET processes occuring in a protein medium. The major difference between the mechanism of ET in solutions as well as in glasses on the one hand, and in a protein medium on the other hand, is a matter of details rather than that of principles. As is well known, the non-adiabatic ET rate constant can be expressed in the general form /52,103,104 /.

$$k = \frac{2\pi}{h} |V|^2 \langle F_I \rangle \qquad (B3.4)$$

where $|V|$ is the donor-acceptor one-electron two-centre integral, which is determined by the "intermolecular engineering", i.e. separation and relative orientation of the donor and the acceptor. $\langle F_I \rangle$ is the thermally averaged Franck-Condon vibrational overlap factor, which is determined by the changes in equilibrium nuclear configuration between the donor-acceptor and donor$^+$-acceptor$^-$ states of the "supermolecule", consisting of the donor-acceptor pair and the surrounding medium. Two types of equilibrium configurational changes contribute to the nuclear rearrangement energy, which fall into two categories:

(i) Intermolecular vibrational modes of the exterior medium, which involve the protein polar modes and which are characterized by the intermolecular reorganization energy E_s.
(ii) Intramolecular vibrational modes of the donor and acceptor molecules, which are characterized by the intramolucular reorganization energy E_c.

The available experimental data for the temperature dependence of the ET reactions (B3.1) and (B3.3) are portrayed in Fig. 5. The data indicate that

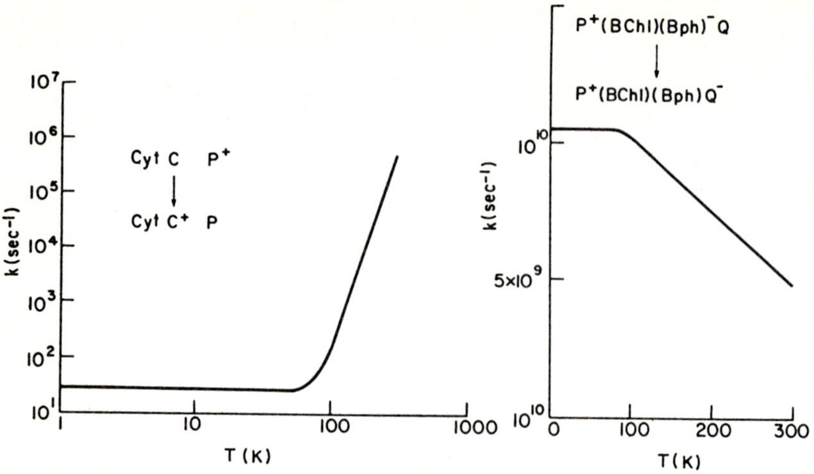

Fig. 5 Temperature dependence of the reaction rates for cytochrome c oxidation /105/ and ET from BPh^- to Q /100/.

reaction (B3.1) corresponds to an activationless process, where the initial and final potential surfaces cross at the minimum of the initial state.

This activationless process is characterized by a frequency of ≅ 150 cm^{-1} which may correspond to intermolecular modes. Reaction (B3.2) has been studied in great detail /90,101/ with focus on the role of the iron and the protein environment. With respect to both ET reactions, (B3.1) and (B3.2), the reconstitution of quinone /106/ and iron /107/ offer specific access to mechanistic studies of ET. The mechanism of the cytochrome c oxidation (B3.3) is not well understood /108/. In particular, it is not clear whether the temperature dependence (Fig. 5) indeed reveals the transition from nuclear tunneling at low temperatures to an activated process at high temperatures.

In contrast to ET in RCs where the relative importance of intermolecular and intramolecular vibrational modes is not yet understood, it is well known that for ET in rigidly linked donor-acceptor systems in solution, $E_c/E_s < 1$. There the major contribution to the nuclear reorganization energy originates from exterior medium modes. The dependence of the effective interaction on the distance in donor-acceptor systems with rigid or semi-rigid bridges has been studied /109-111/. The different theoretical aspects related to experimental tests of ET theory have been extensively treated and reviewed /112/.

In the context of the applicability of the conventional non-adiabatic multiphonon theory to later ET events in the RC, it is interesting to inquire what is the role of the relaxation dynamics of the protein on the later ET processes within the RC. Very recently, considerable theoretical activity has been directed /113-115/ towards the elucidation of the role of dielectric and configurational relaxation times of the solvent on the ET process. For a system where ET is dominated by coupling to polar solvent modes, the following novel features emerge:

(a) The characteristic time scale for medium relaxation is the dielectric relaxation time in the field of a constant charge $\tau = \tau_D \varepsilon_\infty / \varepsilon_s$, where

τ_D is the constant field dielectric relaxation time, while ε_s and ε_∞ correspond to the low frequency and high frequency dielectric constants, respectively.

(b) As long as the ET process is non-adiabatic, its rate is unaffected by τ.

(c) When τ becomes very large the modified Landau-Zener factor, which is proportional to τ, will exceed unity, and the ET process will switch from the non-adiabatic limit to the adiabatic limit.

(d) In the adiabatic limit for large τ the preexponential factor in the adiabatic rate expression will become proportional to τ^{-1}. These interesting novel features, prevailing for the special case of ET dominated by the contribution of solvent modes, may be irrelevant for the later ET processes in the RC, which are overwhelmed by the coupling to the intramolecular modes.

(II) Early Electron Transfer

The events within the first 10 psec involved in the overall ET process

$$^1(\mathrm{BChl})_2^* \ \mathrm{BChl\ BPh} \rightarrow (\mathrm{BChl})_2^+ \ \mathrm{BChl\ BPh}^- \qquad \text{(B3.5)}$$

are not well understood.

Femto- and picosecond time-resolved measurements /116/ on RCs of Rp. viridis (with menaquinone reduced) result in a time constant of $2 \cdot 10^{11}\ s^{-1}$ for the reduction of BPh and give evidence for a series of kinetic events within the time window 150 fsec - 5 psec. A kinetic intermediate involving BChl would be at least compatible with the observations. Experiments with lower time resolution /100/ on Rp. sphaeroides, R-26, yield for the reduction of BPh the rate $2\ 10^{11}\ s^{-1}$ and favour the conclusion that the electron is directly passed from $^1(\mathrm{BChl})^*$ to the BPh. Both experiments were performed under different conditions with respect to the excitation and probing wavelengths and the relative number of RCs excited. The RCs were excited in the second electronic state Q_x high above the lowest electronic excitation of the dimer. Before a definite assignment of the electronic states involved in the primary ET can be made, two additional decay processes of the high energy spectroscopic state in the fsec time domain should be considered. Firstly, EET between RC components, resulting in lower electronic excitations, may be pertinent. Secondly, excitation of the Q_x state may result in the $Q_x \rightarrow Q_y$ internal conversion, which in the "isolated" free-base porphine molecule occurs on the time scale of 500 fsec /117/.

The mechanism of the primary ET events in the RC is determined by the energetics of the dimer-monomer charge transfer state $(\mathrm{BChl})_2^+\ \mathrm{BChl}^-$. Provided that the energy of this equilibrium charge-transfer state is located below the electronic excitation, the $(\mathrm{BChl})_2^+\ \mathrm{BChl}^-$ state constitutes a genuine intermediate state in the ET sequence. On the other hand, when the $^1(\mathrm{BChl})_2^*$ BChl excitation is located somewhat below the $(\mathrm{BChl})_2^+\ \mathrm{BChl}^-$ state but mixes heavily with it (see section B2), then the scrambled initial excited state transfers an electron to the BPh. In such a mechanism a distinct charge transfer state is not involved as a genuine intermediate but rather mediates the primary ET process.

For the microscopic treatment of the ET dynamics in linked donor-spacer-acceptor systems a formalism has been introduced /118/ which is based on a self-consistent description of the initial and final states. It is proposed that the initial state contains already an admixture of the donor-spacer- and the donor-acceptor-states. As applied to ET in RCs, this means that the initial state consists of a superposition of the excited dimer, the intermediate and the final charge transfer states.

This is schematically represented in terms of the relevant wave functions

$$a|{}^1(BChl)^*\ BChl\ BPh> + b\,|(BChl)_2^+\ BChl^-\ BPh> + c\,|(BChl)_2^+\ BChl\ BPh^-> \rightarrow |(BChl)_2^+\ BChl\ BPh^-> \qquad (B3.6)$$

The electronic coupling wich induces the ET is essentially determined by the transfer integral between the states on the left and right hand side of (B3.6). A transfer time of 2 psec from $^1(BChl)^*$ to BPh is achieved with $c \cong 0.01$ /118/ representing in this formalism the most important admixture for the fast ET process (B3.5). Following the discussion of the Stark effect in section (B2) the amplitude $b \cong 0.2 - 0.3$ can be estimated. This is slightly larger than the value calculated /118/.

In spite of the current mechanistic uncertainty regarding the primary charge separation events in the RC, it is evident that the formation of the (BChl) BChl BPh separated ion pair occurs on the time scale of 4-5 psec, and that some additional ET processes may be even faster ($\simeq$ 1 psec). These rates are comparable to or even faster than medium-induced vibrational relaxation rates of the electron donor and acceptor molecules. This state of affairs raises the question of whether the conventional non-adiabatic multiphonon theory is applicable for the description of the primary events of charge separation /51,52/. It has previously been emphasized /51,52/ that a basic necessary validity condition for the applicability of the conventional ET theory is that medium-induced vibrational relaxation and excitation is fast on the time scale of the microscopic rates, W_I, for ET.

$$W_I < \gamma_{VR} \qquad (B3.7)$$

where γ_{VR} represents a typical medium-induced vibrational relaxation rate. Vibrational relaxation of donor and acceptor molecules in proteins presumably occurs with rates $\gamma_{VR} \cong 10^{11} - 10^{12}\ s^{-1}$. The validity of condition (B3.7) allows for the separation of the time scales for (fast) vibrational relaxation and for the (slower) ET, making the conventional ET theory applicable for the later ET processes. For the primary ultrafast ET processes in the RC, the assumption of fast vibrational relaxation cannot be taken for granted and ET can compete with vibrational relaxation.

A treatment of ET during vibrational relaxation has already been provided /51,52/. The major physical aspect of this process (Fig.6) is that, when the donor-acceptor system is excited at the energy E_e, it will start to lose energy via vibrational relaxation. If the initially excited configuration E_e is located below the crossing point E_x of the potential surfaces, the relevant microscopic ET rates are sufficiently low to allow the system to relax vibrationally to a thermally equilibrated configuration from which "conventional" ET will occur. However, if the initially excited state E_e is located above the crossing point E_x, the system will pass on its way downwards through E_x. In the vicinity of the crossing point, the nuclear contribution to the microscopic ET rates is large, the characteristic Franck-Condon factor, being $\cong 10^{-1} - 10^{-2}$ /116/. Thus, provided that the electronic coupling is large, due to spatial proximity of the donor-acceptor pair and/or effective mediating of the electron coupling, efficient ET will occur at the crossing point. Thus, the necessary condition for ET during vibrational relaxation is

$$E_e \geqq E_x \qquad (B3.8)$$

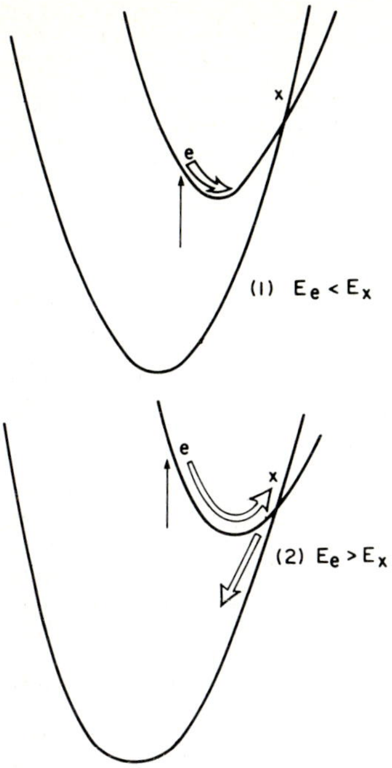

Fig. 6

A schematic descriotion of potential surfaces (DA and D^+A^-) for ET from the initially excited configuration followed by vibrational relaxation

which ensures efficient ET. There is a distinct possibility that the primary ET events in the RC compete with medium-induced vibrational relaxation occuring in a system which does not reach thermal equilibrium. A crucial experimental test of this conjecture will involve an interrogation of the wavelength dependence of the quantum yield for charge separation in the RC following photoselective excitation in the dimer absorption band. At low energies condition (B3.8) may be violated and a drop of the quantum yield will then be exhibited.

The primary ultrafast electron transfer in the RC, which effectively competes with vibrational relaxation /51,52/ may constitute a general rather than an isoteric example for the competition between electronic processes and vibrational relaxation in biophysical systems. Another relevant example in this category of electronic processes occuring from a vibrationally non-equilibrated molecular manifold may involve EET from the antenna to the RC /118/. Our proposal that the ultrafast ET /51,52/ and EET /118/ electronic processes in the RC eclipse vibrational relaxation is extremely appealing, as it provides a general mechanism for the efficient utilization of energy in primary photobiological processes.

Acknowledgements

We are highly indebted to Prof. Hugo Scheer for extensive and stimulating discussions. One of us (M.E.MB) thanks Dipl. Phys. W.Lersch and Dr.A. Ogrodnik for critical reading of the manuscript. Financial support from the Deutsche Forschungsgemeinschaft (SFB 143) is gratefully acknowledged.

References

/1/ E.Gantt (1979), Ann.Rev.Plant.Physiol. 32, 327
/2/ W.Wehrmeyer (1983) in PROTEINS AND NUCLEIC ACIDS IN PLANT SYSTEMATICS (U.Jensen and D.E.Fairbrothers, eds.), Springer-Verlag Berlin
/3/ G.Drews (1985), Microbiol.Rev. 49, 59
/4/ R.J.Cogdell, J.Valentine, J.G.Lindsay, and K.Schmidt (1982), Biochem.Soc.Transact. 10, 334
/5/ J.M.Olson (1978) in THE PHOTOSYNTHETIC BACTERIA (R.K.Clayton and W.R.Sistrom, eds.) p.161, Plenum Press, New York
/6/ J.M.Olson, P.D.Gerola, G.H. van Brakel, R.F.Meiburg and H.Vasmel (1985), this book
/7/ J.P.Thornber, R.J.Cogdell, R.E.B.Seftor, B.K.Pierson and E.M.Tobin (1984) in ADV.PHOTOSYNTH.RES. (C.Sybesma, ed.), Vol. III, p.25, Martinus Nijhoff, Dr.W.Junk Publ., The Hague
/8/ R.Feick and C.Fuller (1984),Biochem. 23, 3693
/9/ H.Scheer (1982) in LIGHT REACTION PATH OF PHOTOSYNTHESIS (F.Fong, ed.) p.7, Springer-Verlag Berlin
W.John, R.Fischer, S.Siebzehnrübl, and H.Scheer (1985), this book
/10/ A.N.Glazer (1983), Ann.Rec.Biochem. 52, 125
/11/ J.Riegler, W.M.Heckl, J.Peschke, M.Lösche, and H.Möhwald (1985), this book
/12/ H.Scheer, B.Paulke, and J.Gottstein (1985) in OPTICAL PROPERTIES AND STRUCTURE OF TETRAPYRROLES (G.Blauer, H.Sund, eds.). p.507, Walter de Gruyter & Co, Berlin
/13/ A.Scherz and W.W.Parson (1984), BBA 766,653,665
/14/ A.Scherz, V.Rosenbach, and S.Malkin (1985), this book
/15/ J.Fajer, K.M.Barkigia, E.Fujita, D.A.Goff, L.K.Hanson, J.D.Head, T.Horning, K.M.Smith, and M.C.Zerner (1985),this book
/16/ M.R.Wasielewski (1982) in LIGHT REACTION PATH OF PHOTOSYNTHESIS (F.Fong, ed.), p.234, Springer-Verlag Berlin
/17/ A.Kuki and S.G.Boxer (1983), Biochem. 34, 389
/18/ S.G.Boxer (1985), this book
/19/ H.Zuber (1985), this book
/20/ W.B.Matthews, R.E.Fenna, M.C.Bolognesi, M.F.Schmidt and J.M.Olson (1979), J.Mol.Biol. 131, 259
/21/ R.Huber (1985), this book
/22/ H.Michel and D.Oesterhelt (1980), PNAS 79, 1283
/23/ H.Michel (1982), J.Mol.Biol. 158, 567
/24/ R.M.Garavito and J.P.Rosenbusch (1980), J.Cell.Biol. 86, 327
/25/ J.P.Allen, R.Theiler, G.Feher (1985), this book
/26/ R.J.Cogdell, K.Woolley, R.C.Mackenzie, J.G.Lindsay, H.Michel, J.Dobler and W.Zinth (1985), this book
/27/ W.Mäntele, K.Steck, T.Wacker, W.Welte, B.Levoir, and J.Breton (1985), this book
/28/ R.J.Cogdell (1985), this book
/29/ B.S.Hudson, B.E.Kohler, and K.Schulten (1982) in EXCITED STATES, VOL. 1, p.1 , Academic Press New York
/30/ H.A.Kramers (1940), Physica (Utrecht), 7, 284
/31/ T.Förster (1959), Disc.Farad.Soc.(London) 27, 7
/32/ D.L.Dexter (1953), J.Chem.Phys. 21, 636
/33/ J.Jortner, J.L.Katz, S.I.Choi, and S.A.Rice (1965), J.Chem.Phys. 42, 309
/34/ R.M.Pearlstein (1982) in PHOTOSYNTHESIS, ENERGY CONVERSION BY PLANTS AND BACTERIA (Govindjee, ed.), Vol.1, p.293, Academic Press New York
/35/ R. van Grondelle (1985), BBA 811, 147
/36/ R.M.Pearlstein and H.Zuber (1985), this book
/37/ A.J.Campillo and S.L.Shapiro (1978), Photochem.Photobiol. 28, 975

/38/ R.R.Alfano (1982) in BIOLIGICAL EVENTS PROBED BY ULTRAFAST LASER SPECTRSOCOPY (R.R.Alfano, ed.), chapter 6,17,18
/39/ A.Bergman, M.Levine, and J.Jortner (1967), Phys.Rev.Lett. 18, 15
/40/ A.Holzwarth (1985), this book
/41/ S.Schneider, P.Geiselhart, T.Mindl, F.Dörr, W.John, R.Fischer, and H.Scheer (1985), this book
/42/ S.Schneider, P.Geiselhart, C.Scharnagl, T.Schirmer, W.Bode, W.Sidler, and H.Zuber (1985), this book
/43/ J.Friedrich, H.Scheer, B.Zickendraht-Wendelstaedt, and D.Haarer (1981), JACS 103, 1030, J.Chem.Phys. 74, 2260
/44/ P.Avakian and R.E.Merrifield (1964), Phys.Rev.Lett. 13, 541
/45/ M.Levine, J.Jortner and A.Szoke (1966), J.Chem.,Phys. 45, 1591
/46/ J.R.Salcedo, A.E.Siegman, D.D.Dlott and M.D.Fayer (1978), Phys.Rev.Lett. 41, 131
/47/ R.Flach, D.S.Hamilton, P.M.Selzer and W.M.Yen (1977), Phys.Rev. B15, 1248
/48/ D.S.Hamilton, P.M.Selzer and W.M.Yen (1978), Phys.Rev. B16, 1858
/49/ D.Wiersma in Adv.Chem.Phys,Series, PHOTOSELECTIVE CHEMISTRY, Part 2, p.421, Academic Press New York
/50/ R.Kopelman (1976) in TOPICS IN APPLIED RESEARCH, (K.F.Fong, ed.), Vol. 15, 298
/51/ J.Jortner (1980), JACS 102, 6676
/52/ J.Jortner (1980), BBA 594, 193
/53/ B.B.Mandelbrot (1982): THE FRACTAL GEOMETRY OF NATURE, Freeman, San Francisco
/54/ H.J.Stapelton, J.P.Allen, C.P.Flynn, D.G.Stimson, and S.R.Kurz, (1980), Phys.Rev.Lett. 45, 1456
/55/ S.Alexander and P.Orbach (1982), J.Phys.Lett. 43, 65
/56/ J.Klafter and A.Blumen (1984), J.Chem.Phys. 80, 875
/57/ U.Even, K.Rademan, J.Jortner, N.Manor, and R.Reisfeld (1984), Phys.Rev.Lett. 52, 2164
/58/ W.Junge and H.T.Witt (1968), Z.Naturf. 23b, 244
/59/ C.Wolff, H.F.Buchwald, M.Rüpel, K.Witt, and H.T.Witt (1969), Z.Naturf. 24b, 1038
/60/ J.W.Jackson and A.R. Crofts (1969), FEBS Lett. 4, 185
/61/ W.Junge and J.B.Jackson (1982), in PHOTOSYNTHESIS OF GREEN PLANTS AND BACTERIA (Govindjee, ed.), Vol. 1, p.589, Acad.Press New York
/62/ D.W. Reed and R.K.Clayton (1968), Biochem.Biophys.Res.Commun 30, 471
/63/ Brookhaven Symposia in Biology, Vol. 28 (1976), on CHLOROPHYLL-PROTEINS, REACTION CENTERS, AND PHOTOSYNTHETIC MEMBRANES (J.M. Olson and G.Hind, eds.), Brookaven National Laboratory, Upton, New York
/64/ P.L.Dutton, K.M.Petty, H.S.Bonner, and S.D.Morse (1974), BBA 347, 126
/65/ J.Deisenhofer, O.Epp, K.Miki, R.Huber, and H.Michel (1984), J.Mol.Biol. 180, 385
/66/ J.Deisenhofer and H.Michel (1985), this book
/67/ J.R.Norris, R.A.Uphaus, H.L.Crespi, and J.J.Katz (1971), PNAS 68, 625
/68/ G.Feher, A.J.Hoff, R.A.Isaacson, and L.C.Ackerson (1975), Ann.N.Y. Acad.Sci. 244, 260
/69/ M.G.Rockley, M.W.Windsor, R.J.Cogdell, and W.W.Parson (1975) PNAS 72, 2251
/70/ K.J.Kaufmann, P.L.Dutton, T.A.Netzel, J.S.Leigh, and P.L.Rentzepis (1975), Science 188, 1301
/71/ W.W.Parson (1982), Annu.Rev.Biophys.Bioenerg. 11, 57-80
/72/ R.H.Austin, K.W.Beeson, L.Eisenstein, H.Frauenfelder, and I.C. Gunsalus (1975), Biochemistry 14, 5355

/73/ H.Frauenfelder (1983) in STRUCTURE AND DYNAMICS: NUCLEIC ACIDS AND PROTEINS (E.Clementi and R.H.Sarma, eds.), p. 369, Academic Press New York
/74/ J.Jortner and J.Ulstrup (1977), J.Am.Chem.Soc. 101, 3744
/75/ V.I.Goldanskii, Yu.F.Krupyanskii, and V.N.Fleurov (1983), Doklady A.N. USSR 272, 978
/76/ G.A.Saigh, H.J.Shink, H.J.Lohneysen, F.Parak, and S.Hunklinger (1984), Z.Phys. B55, 23
/77/ P.W.Anderson (1974)
/78/ W.Zinth, M.Sander, J.Dobler, W.Kaiser, and H.Michel (1985), this book
/79/ E.W.Knapp and S.F.Fischer (1985), this book
/80/ W.W.Parson, A.Scherz, and A.Warshel (1985), this book
/81/ J.Breton (1985), this book
/82/ M.Lutz and B.Robert (1985), this book
/83/ F.Siebert, W.Mäntele, and U.Gerwert (1983), Eur.J.Biochem. 136, 119
/84/ F.Siebert (1984) in SPECTROSCOPY OF BIOLOGICAL MOLECULES (C.Sandorfy, T.Theophanides, eds.), p.341
/85/ M.Lutz (1985), private communication
/86/ A.J.Hoff (1985), this book
/87/ A.Angerhofer, J.U.von Schütz, and H.C.Wolf (1985), this book
/88/ F.G.H.van Wijk, P.Gast, and T.J.Schaafsma (1985), this book
/89/ W.Lubitz, F.Lendzian, M.Plato, K.Möbius, and E.Tränkle (1985), this book
/90/ G.Feher, R.A.Isaacson, M.Y.Okamura, and W.Lubitz (1985), this book
/91/ P.J.O.Malley, T.K.Chandrashekar, N.G.T.Babcock (1985), this book
/92/ J.R.Norris, D.E.Budil, S.V.Kolaczkowski, J.H.Tang, and M.K.Bowman (1985), this book
J.R.Norris,D.E.Budil,H.L.Crespi,M.K.Bowman,P.Gast,C.P.Lin,C.H.Chang, and M.Schiffer (1985), this book
/93/ A.Ogrodnik, W. Lersch, M.E.Michel-Beyerle, J.Deisenhofer, and H.Michel (1985), this book
/94/ R.Bittl,H.Treutlein and K.Schulten (1985),this book
/95/ H.P.H.Thijssen, A.I.M.Dicker, and S.Völker (1982), Chem.Phys.Lett. 92, 7
/96/ H.P.H.Thijssen, S.Völker, M.Schmidt, and H.Port (1983), Chem.Phys.Lett 94, 537
/97/ R.S.Berry, J.Jortner, J.C.Mackie, E.S.Pysh, and S.A.Rice (1965) J.Chem.Phys. 42, 1535
/98/ D.de Leeuv, M.Malley, G.Butterman, M.Y.Okamura, and G.Feher (1982), Biophys.J. 37, 111a
/99/ R.Silbey, J.Jortner, S.A.Rice, and M.Vala (1965), J.Chem.Phys. 42 733
/100/W.W.Parson,N.W.T.Woodbury, M.Becker, C.Kirmaier,and D.Holten(1985). this book
/101/ M.Y.Okamura, R.J.Debus, D.Kleinfeld, and G.Feher (1982) in FUNCTION OF QUINONES IN ENERGY CONSERVING SYSTEMS (B.L.Trumpower, ed.), p.299, Academic Press New York
/102/ P.L.Dutton and R.C.Prince (1978) in THE PHOTOSYNTHETIC BACTERIA (R.K.Clayton and W.R.Sistrom, eds.), p.525, Plenum Press New York
/103/ J.J.Hopfield (1974), PNAS 71, 3640
/104/ J.Jortner (1976), J.Chem.Phys.64,4860
/105/ D.DeVault and B.Chance (1966), Biophys.J. 6, 825
/106/ M.R.Gunner, B.S.Braun, J.M.Bruce, and P.L.Dutton (1985), this book
/107/ R.J.Debus, G.Feher, and M.Y.Okamura (1985), submitted
/108/ P.L.Dutton (1985), private communication
/109/ J.R.Miller (1985), this book

/110/ N.R.Wasielewski, M.P.Niemczyk, W.A.Svec, and E.P.Pewitt (1985), this book
/111/ H.Heitele and M.E.Michel-Beyerle (1985), this book; JACS, in press
/112/ R.A.Marcus and N.Sutin (1985), this book; BBA 811, 265
/113/ M.Ya.Ovchinnikova (1981), Theoret.Exper.Chem. 17, 507
/114/ D.F.Calif and P.G.Wolynes (1983), J.Phys.Chem. 87, 3787
/115/ J.L.Friedman (1963), J.Chem.Soc.Farad.Trans. 2, 79
/116/ W.Zinth, M.C.Nuss, M.A.Franz, W.Kaiser, and H.Michel (1985), this book
/117/ U.Even, J.Jortner (1982), J.Chem.Phys. 77, 4391
/118/ S.F. Fischer,I.Nussbaum, and P.O.J.Scherer (1985), this book
/119/ J.Jortner (1979), Phil.Mag. B40, 317
/120/ M.E.Michel-Beyerle and J.Jortner, to be published

Index of Contributors

Allen, J.P. 82
Angerhofer, A. 78

Babcock, G.T. 339
Barkigia, K.M. 324
Becker, M. 278
Bittl, R. 264
Bode, W. 36
Bowman, M.K. 147,190
Boxer, S.G. 306
Braun, B.S. 298
Breton, J. 88,109
Bruce, J.M. 298
Budil, D.E. 147,190

Chandrashekar, T.K. 339
Chang, C.H. 147
Cogdell, R.J. 62,85
Crespi, H.L. 147

Dörr, F. 26
Deisenhofer, J. 94,198
Depka, B. 216
Dijkema, C. 81
Dobler, J. 85,97
Dutton, P.L. 298

Fajer, J. 324
Feher, G. 82,174
Feick, R. 74
Fischer, R. 17,26
Fischer, S.F. 103,131,256
Franz, M.A. 286
Fujita, E. 324

Gast, P. 146,147
Geiselhart, P. 26,36
Gerola, P.D. 67
Göbel, W. 292
Goff, D.A. 324
Gunner, M.R. 298

Hanson, L.K. 324
Head, J.D. 324
Heckl, W.M. 207
Heitele, H. 250
Hörber, J.K.H. 131,292
Hoff, A.J. 150
Holten, D. 278
Holzwarth, A.R. 45
Horning, T. 324
Huber, R. 15

Isaacson, R.A. 174

John, W. 17,26
Jortner, J. 345

Kaiser, W. 97,286
Kirmaier, C. 278
Knapp, E.W. 103,292
Kolaczkowski, S.V. 190

Lösche, M. 207
Lendzian, F. 164
Lersch, W. 198
Levoir, B. 88
Lin, C.P. 147
Lindsay, J.G. 85
Lubitz, W. 164,174
Lutz, M. 138

Mackenzie, R.C. 85
Mäntele, W. 88
Malkin, S. 314
Marcus, R.A. 226
Meiburg, R.F. 67
Michel, H. 85,94,97,131,
198,286
Michel-Beyerle, M.E.
131,198,250,292,345
Miller, J.R. 234
Mindl, T. 26
Möbius, K. 164
Möhwald, H. 207

Niemczyk, M.P. 242
Norris, J.R. 147,190
Nuss, M.C. 286
Nussbaum, I. 256

O'Malley, P.J. 339
Ogrodnik, A. 198,292
Okamura, M.Y. 174
Olson, J.M. 67

Parson, W.W. 122,278
Pearlstein, R.M. 53
Peschke, J. 207
Pewitt, E.B. 242
Plato, M. 164

Riegler, J. 207
Robert, B. 138
Rosenbach, V. 314

Sander, M. 97
Schaafsma, T.J. 81,146
Scharnagl, C. 36
Scheer, H. 17,26
Scherer, P.O.J. 131,256
Scherz, A. 122,314
Schiffer, M. 147
Schirmer, T. 36
Schneider, S. 26,36
von Schütz, J.U. 78
Schulten, K. 264
Searle, G.F.W. 81
Sidler, W. 36
Siebzehnrübl, S. 17
Smith, K.M. 324
Steck, K. 88
Sutin, N. 226
Svec, W.A. 242

Tang, J.H. 190
Theiler, R. 82
Tränkle, E. 164
Trebst, A. 216
Treutlein, H. 264

van Brakel, G.H. 67
van Wijk, F.G.H. 146
Vasmel, H. 67

Wacker, T. 88
Warshel, A. 122
Wasielewski, M.R. 242
Welte, W. 88
Wolf, H.C. 78
Woodbury, N.W.T. 278
Woolley, K. 85

Zerner, M.C. 324
Zinth, W. 85,97,286
Zuber, H. 2,36,53

pp58-59 exciton coherence + storage